The Development Process

Akin L. Mabogunje, MA, Ph.D (LOND.). has been Professor of Geography and Director of the Planning Studies Programme at the University of Ibadan, Nigeria. He has been concerned for many years with the spatial aspects of development planning in developing countries. He has served as consultant and adviser to governments in Nigeria on urban and regional planning, especially on the plans to establish a new federal capital. He has also held a number of national and international offices, such as Chairman of the Western State Forestry Commission, Chairman of the Nigerian Council on Management Development, President of the Governing Council of the Pan-African Institute for Development, First Vice-President of the International Geographical Union and Member of the Board of Trustees of the Population Council in New York. Professor Mabogunje has taught in many universities in Britain, the United States, Canada, Sweden and Brazil. He is author of a number of books, including *Urbanization in Nigeria, Regional Mobility and Resource Development in West Africa*, and *Reginal Planning and National Development in Africa.* He has also published many academic papers on problems of urbanization, migration, regional planning and development in developing countries, and has been awarded the Anders Retzius gold medal for his services to geogaphical science.

Other titles of interest

Africa: Geography and Development
Alan B. Mountjoy and David Hilling

The African City
Anthony O'Connor

African Philosophy: Myth and Reality
Paulin J. Hountondji

African Women: Their Struggle for Economic Independence
(in Association with Zed Press)
Christine Obbo

Agricultural Development and Nutrition (by arrangement with FAO and UNICEF)
Edited by Arnold Pacey and Philip Payne

Climatology of West Africa
Derek Hayward and Julius Oguntoyinbo

The Development of African Drama
Michael Etherton

Forced Migration: The Impact of Export Slave Trade on African Societies
Edited by J. E. Inikori

A History of Africa
J. D. Fage

Indigenous Agricultural Revolution
Paul Richards

Peasants and Proletarians: The Struggles of Third World Workers
Edited by Robin Cohen, Peter C. W. Gutkind and Phyllis Brazier

Rural Development: Theories of Peasant Economy and Agrarian Change
Edited by John Harriss

Rural Settlement and Land Use
Michael Chisholm

Twelve African Writers
Gerald Moore

West Africa Under Colonial Rule
Michael Crowder

The Development Process

A spatial perspective

Akin L. Mabogunje

London
Unwin Hyman
Boston Sydney Wellington

Published by the Academic Division of

Unwin Hyman Ltd
15/17 Broadwick Street, London W1V 1FP, UK

Unwin Hyman Inc.,
8 Winchester Place, Winchester, Mass. 01890, USA

Allen & Unwin (Australia) Ltd,
8 Napier Street, North Sydney, NSW 2060, Australia

Allen & Unwin (New Zealand) Ltd in association with the
Port Nicholson Press Ltd,
60 Cambridge Terrace, Wellington, New Zealand

First published in 1980
Reprinted 1981, 1982, 1984
Second edition 1989

British Library Cataloguing in Publication Data

Mabogunje, Akin Ladipo
The development process.
1. Africa – Economic conditions – 1945 –
2. Underdeveloped areas – Economic conditions – Case studies
3. Geography, Economics – Case studies
I. Title
330.9′6 HC502
ISBN 0 04 445302 7

Typeset in Times and printed in Great Britain by
Billings and Sons Ltd, London and Worcester

Contents

Tables

Figures

Preface

All over the Third World, the development process is the cause of great concern. In country after country, development 'planning' has created conditions for a substantial proportion of the population which are far from satisfactory or dignifying. Evidence of acute deprivation and almost inhuman degradation is sharply presented everywhere, both in urban and rural areas. Their conditions stand out most sharply because of the presence of a minority of really wealthy individuals concentrated particularly in the metropolitan centres. The depressing economic conditions of the masses have been compounded by physical disasters such as droughts, floods and earthquakes.

Although this gloomy picture is relieved to some degree in certain countries, especially in Latin America and South-East Asia, there can be no doubt that it has provoked a serious reappraisal of ideas and thoughts about the development process. Incisive analysis of the victims and beneficiaries of the current strategies of development are becoming common. New and alternative approaches are being presented in the literature. One of the reassuring aspects of this recent intellectual activity in the development field is an increasingly vigorous participation of scholars of the Third World who in their daily lives experience the realities of some of these harsh conditions.

In my case, for instance, various isolated events have conjoined to direct my attention to the viability of an approach deriving from a spatial perspective. For instance, in 1960, during a field survey in the northern parts of the Republic of Benin, I was confronted by a distraught French agricultural officer smarting at the refusal of local farmers to continue to produce for the export market. The farmers had argued that for them political independence meant an end to the strain and stresses of export agricultural production. In later years, I was to reflect on this situation, in which labour and resources are available, but in which farmers have lacked motivation and have not been mobilized to strive for their own development. In 1970, on a visit to Sweden, I was struck in a discussion with a colleague by the argument that the manipulation of spatial forms has always been a means used by governments to secure popular co-operation. I was reminded of simple road blocks to ensure traffic diversion or to change the status of a road.

This particular discussion brought to my mind an incident in my undergraduate days at the University College, Ibadan, when a group of us challenged a friendly Catholic priest to explain why the Catholic religion encouraged its congregations to pray to statuaries of the Madonna and other saints, when the Bible clearly enjoins that 'God is a spirit and they who worship him must worship him in spirit and in truth.' His retort was memorable. 'Can you tell me', he said, 'how better you can concentrate the attention and spiritual energy of the masses in prayers? Certainly not by asking them to gaze at thin air.'

The thrust of this volume is thus on how to use spatial forms, structures and organizations to concentrate the energies of people in underdeveloped countries to engage in their own development. In articulating my thoughts and ideas on this theme, I have profited immensely from discussions with a large number of scholars and policymakers, to all of whom I acknowledge my indebtedness. In particular, I wish to express sincere appreciation to Professor O. Aboyade, my colleague and former co-director of the Planning Studies Programme, University of Ibadan, for the continuous exchange of ideas on the direction of the development process in Nigeria; to Professor Michael Chisholm of the Department of Geography, University of Cambridge, for offering me both the facilities of his department and the benefit of his critical comments on the first draft of this volume; and to Dr Paul Richards, my former colleague at Ibadan and now of the Department of Anthropology, University College, London, whose perceptions of rural conditions in Nigeria I have found most insightful and refreshing and whose comments on the first draft were also invaluable. All of them must take credit for any merit found in this volume. They are, however, absolved from its shortcomings for which I alone must be held responsible.

I would like also to thank very sincerely my secretary, Mrs Joke Adedokun, and Mr Kassim who have worked tirelessly to produce the manuscript in record time; also Mrs Aderogba for producing the maps and diagrams.

1 Introduction

The second half of the twentieth century has been characterized in its opening years by the retreat of colonialism all over the world. Within a space of twenty years between 1945 and 1965, some fifty-seven countries, mainly in Africa and Asia, became politically independent. During the same period, the convention was established that as each country became politically independent it sought and was accorded membership of the United Nations. Thus, the size of the membership of the United Nations came to represent roughly the number of independent political entities in the world. Between 1945 when the organization was established by charter and 1965, membership rose from 51 to 118 nations. By 1971, the number had further risen to 132. Today, membership of the United Nations stands at 142 which is just sixteen short of the total number of territorial entities which presently cover the whole earth surface, apart from the sixty-five dependencies of various European powers which mostly comprise small islands, continental enclaves or portions of the desolate Antarctic continent.

With the issue of overt colonialism largely resolved, the most important global preoccupation, particularly of the last three decades, has been the question of the relatively poor living standards and the degrading quality of life of the vast majority of the population of the world. A close relationship was observed between relative poverty and the status of being an erstwhile colony. This relationship was noticed not only in respect of the newly independent countries of Africa and Asia, but also those countries of Central and South America which became politically independent as far back as the first quarter of the nineteenth century. All of these countries came to be regarded as underdeveloped and their problems of absolute and relative poverty became an overriding international concern. Indeed, at no time in human history has so much intellectual energy and humanitarian concern been directed towards understanding the basis for such poverty and evolving strategies for its eradication.

Not unexpectedly, therefore, the academic community, particularly in the rich and more developed countries of the world, accepted the challenge of this unprecedented situation. Its members sought, on the basis of their understanding of past conditions in their own countries and the prevailing circumstances in the underdeveloped territories, to erect theoretical frameworks and operational models of how the poverty situation could be transformed. The discipline of economics came to dominate the intellectual effort in this direction and tried heroically to adapt the tools of analysis which were developed for the management of the complex economies of the rich countries, to the needs of the underdeveloped ones.

The United Nations took the leadership in focusing world attention on the extent of global poverty and in monitoring changes in the situation through continuous and comprehensive data collection. To give visibility to its effort and to rally all countries of the world to co-operate in tackling this problem, the United Nations declared the 1960s its First Development Decade. The 1970s were in turn declared the Second Development Decade and it was hoped that the unimpressive performance of most of the underdeveloped countries of the world during the first decade would be improved upon in the second. In the second half of the Third Development Decade the story remains far from inspiring. In many countries, although a minority of the population seems to have prospered almost beyond belief, the miserable conditions in which the majority lives seem to have persisted and in many cases to have worsened.

This negative product resulting from deliberate intervention by governments on the basis of well considered, theoretical prescriptions has led to much soul searching within the academic community. As the present decade has progressed, the volume of literature re-examining the premises and theoretical underpinning of attempts at development in the poor countries of the world has increased. Besides economics, other disciplines in the social sciences, whose possible contributions have not received much attention in the past, now find greater receptivity to their points of view and to the alternative perspectives on the problems of development which they provide.

One of these disciplines is geography with its traditional concern with regions and spatial organization. Initially, the geographical interest in development was conditioned by its regional preoccupation and much of its theoretical contribution was to the emerging field of urban and regional planning.[1*] Indeed, as long as the subject shied

*Superior figures refer to the Notes and references on pages 347–72.

away from confronting the problems of development in their totality, its conceptual contribution came to be limited largely to seeking ways and means of correcting at the urban and regional levels distortions arising from decisions taken at the national level. In the last ten years it has become increasingly obvious that such attempts at correction were either impractical or counter-productive, since they tend to worsen existing distortions. Concepts such as those of growth centre strategies of regional development came to be seen as attempts at, as it were, pouring new wine into old wineskins with all the consequential disastrous effects this entails.

In these circumstances an alternative viewpoint has been evolving within geography which sees development as an attempt by organized human communities to define new spatial relationships among their members and between them and their environment. This viewpoint takes the nation state as its unit of interest, in terms of both its internal social relationships and its external relations with other states. Thus, while it is conscious of regional inequalities in countries at different levels of development, it is even more mindful of the ever widening gap in the opportunities for meaningful human existence between advanced industrial countries and those referred to as underdeveloped.[2]

The extension of the spatial perspective to problems of underdevelopment is a very recent event. Yet it is already yielding insights into the process of development which need to be given greater attention than hitherto. In the present volume, although this perspective has been largely generalized, some attention has been paid to the variety of forms and levels of spatial organization found among the underdeveloped countries. The existence of such variety must not, however, be allowed to blur the appreciation of the real essence of the problem of underdevelopment. The need constantly to distinguish between the essential character and the differing surface manifestations of underdevelopment is one which remains paramount throughout this book.

The heterogeneity of underdeveloped countries

As of 1975, some 108 countries (excluding China) were regarded, according to Abdalla,[3] as constituting the Third World (see Figure 1). The term 'Third World' is often used as synonymous with underdeveloped countries and together these countries accounted for some 1958 million people or 63 per cent of world population, excluding China. The most significant fact about all of these countries is their

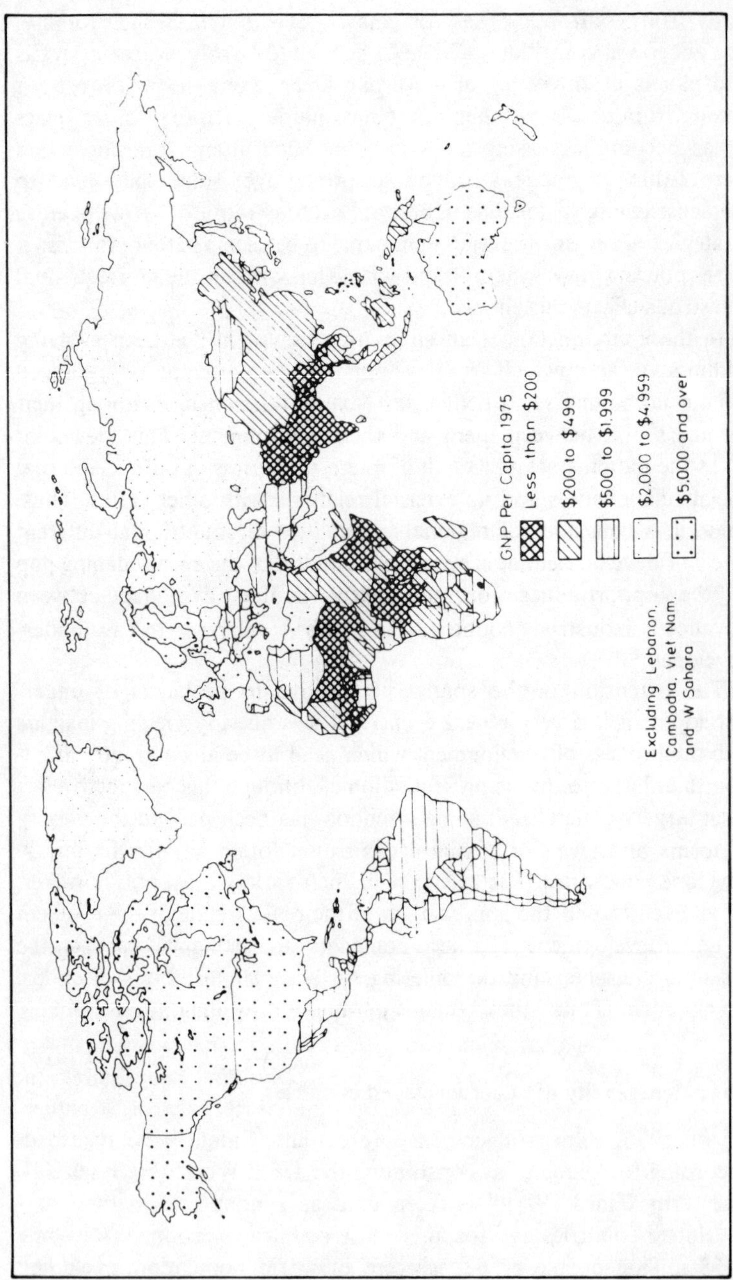

Figure 1 *The distribution of underdeveloped countries*

integration into a global economy where their major role has been that of producers of industrial raw materials. This integration has been largely the product of an industrial revolution which began in England in the eighteenth century and had become well established in Western Europe and North America by the second half of the nineteenth century.[4] This revolution brought about an international division of labour whereby the countries of Western Europe and North America came to concentrate on production and activities which benefited most from technological progress, as well as from the fuller and more rational utilization of the abundant resources of labour and land found particularly in other areas of the world. The increased overall global activity was accompanied by the accentuation of interdependence among all the countries of the world. Within the framework of this interdependence, two patterns of development emerged. The first was in those countries of early industrialization whose economies came to be based on technological progress and a rapid accumulation of capital. These countries constituted the core or centre from where the volume and trend in global economic activities came to be determined and directed. Their development entailed increasingly complicated production processes which constantly required not only a change in the relative quantities of production factors, particularly an increasing amount of capital per unit of labour, but also a qualitative transformation of these factors through a progressive improvement in the characteristics of human labour.

The second emerging pattern of development took place in countries peripheral to the central industrial region in response to changes in overall demand emanating from and usually effected through agencies based in the core. This second type of development was characterized by its essential dependence on circumstances at the centre. In general, it gave rise to economies which were always extensive in character and whose productivity could be increased without significant changes in the focus or mode of production. Thus, the replacement of a subsistence crop such as yam by an export crop such as cocoa brought about an increase in overall output but required no major changes in production organization or techniques. In other instances, such as mining, this peripheral development took the form of assimilating modern techniques and intensifying the input of capital in a production sector which lacked the capacity to transmit its growth to the economy as a whole and was strictly geared to export. In either case, peripheral development had little capacity to transform traditional techniques and organization of production. Nevertheless, by requiring the modernization of infrastructures and some part of the state apparatus, this type of development set in motion an historical process which opened up important new possibilities in countries of the periphery.

The peripheral and dependent nature of their economies is perhaps the single most important factor influencing conditions and events in underdeveloped countries today. However, among themselves, these countries are so heterogeneous that the surface manifestations of these influences give the appearance of tremendous variation. The heterogeneity of underdeveloped countries can be examined under three broad headings: population, natural resources and current levels of development.

Population

With regard to population, the most significant factor differentiating underdeveloped countries from one another is to what extent their peoples are indigenous. In virtually all the countries, the initiation of the current pattern of dependent development followed from the migration of substantial numbers of people from the central industrial countries. In the countries of Africa and Asia, the size of this foreign population was never considerable and the prospect of their complete assimilation into the local population was limited. Even where, as in southern Africa, they formed a settler community, they remained a distinct minority whose survival within the country was based on the exercise of brute force. By contrast, in Central and South America, the immigrant groups came to dominate the local population so thoroughly that the latter became the minority struggling to preserve its cultural identity. In development terms, the significance of this distinction has been, at least until recently, a greater ease of penetration of the countries of Central and South America by forces from the central industrial region. Such penetration, either by individuals or enterprises, does not present serious problems of racial and cultural differences such as are noticeable in the case of Asia and Africa. In consequence, one notices a higher level of foreign investment and development in this group of countries.

Another important feature of differentiation is the size of the countries. Underdeveloped countries show tremendous variation in size due to the intense struggle for colonies among the countries of the centre during the nineteenth century. As a result, the three continents of Africa, Asia, Central and South America have between them 133 out

Table 1 *Size distribution of countries, 1984*

Size classes (millions)	*Europe*	*North America*	*Oceania*	*Central and South America*	*Asia*	*Africa*	*Total*
Not available	–	–	–	–	3	–	3
under 1	9	–	11	18	5	11	54
1–5	4	–	2	9	10	14	39
5–10	7	–	–	7	3	15	32
10–50	10	1	1	5	9	12	38
50–100	4	–	–	1	5	1	11
over 100	1	1	–	1	4	–	7
Total	35	2	14	41	39	53	184

Source: *World Bank Atlas* (1986)

Table 2 *Density distribution of countries, 1984*

Density intervals people per km²	*Europe*	*North America*	*Oceania*	*Central and South America*	*Asia*	*Africa*	*Total*
Not available	–	–	–	–	3	–	3
under 5	2	1	1	3	2	7	16
5–9	1	–	4	3	2	7	17
10–24	4	–	1	8	3	12	28
25–49	1	1	2	6	7	11	28
50–99	9	–	2	3	5	8	27
100–199	12	–	4	6	7	6	35
200 and over	6	–	–	12	10	2	30
Total	35	2	14	41	39	53	184

Source: *World Bank Atlas* (1986)

of the 184 national units (or 72 per cent) making up the world. Table 1 shows, however, that just over half of these nation-states have fewer than 5 million people. Sixty-seven out of 93 of such countries are to be found in the underdeveloped regions of Africa, Asia, Central and South America. At the same time, the Third World has countries such as India with nearly 750 million, Indonesia with 159 million, Brazil with 133 million and Nigeria with 97 million. This wide variation in poulation greatly affects the capacity for independent action by many of these countries as they attempt to become more self-reliant in the course of their development.

The density of population is also a major factor of differentiation. Most African countries have densities of less than 50 persons per square kilometre, with only the two small countries of Mauritius and Rwanda having densities of over 200. By contrast, most countries of Asia have densities of over 50 per square kilometre and more than a quarter of them have more than 200 persons per square kilometre. Central and South American countries are intermediate between the two with most of the mainland countries having densities of less than 50 per square kilometre whilst the islands have more than 100 per square kilometre. Again, the fact of density is significant since it provides a crude index of available land resources for future development. This index, it must be emphasized, is particularly crude, since in most of the countries actual cultivable land is limited and density per unit area of cultivated land is considerably higher than depicted.

Natural resources

The heterogeneity of underdeveloped countries is further compounded by variations in their natural resource endowments, especially as viewed from the perspective of the central industrial countries. This perspective is crucial to the extent that it has determined the position of each of the countries within the system of the international division of labour. On this basis, underdeveloped countries fall into three broad groups. The first are the temperate countries of South America whose climatic and vegetal resources made them eminently suitable for the production of those agricultural commodities, notably beef, requiring an extensive use of land. Countries such as Argentina and Uruguay became part of the expanding frontier of the industrializing European economy to which European agricultural techniques were transplanted at an early stage. The very extensiveness of the agriculture practised and the sheer volume of freight involved, necessitated the creation of a widespread transportation network in these countries which indirectly led to the rapid unification of their domestic market, focusing on the major ports of shipment.

The second group comprise countries in the tropical zone whose agricultural exports were needed mainly as industrial raw material.

Most of the underdeveloped countries of Asia, Africa and parts of Latin America fall into this category, their role being to supply the central regions with such commodities as sugar, tobacco, coffee, tea, cocoa, groundnuts, palm produce, cotton, rubber, sisal and timber. The rapid expansion of demand for these commodities in countries of the centre during the second half of the nineteenth century was an important factor in their colonization during this period and resulted in the integration of the economies of the tropical underdeveloped countries into the world market system. None the less, tropical commodities were of limited significance as a factor in the development of countries of the centre, although they involved the opening up of large areas for settlement. Consequently, tropical countries did not attract much capital for the creation of a complex infrastructure and production was not infrequently allowed to continue within the framework of traditional organization and techniques.

The third group of countries are those whose economy came to be dominated by the export of mineral products. In Central and South America such countries included Mexico, Chile, Peru, Bolivia and later Venezuela. In Asia, most of the Middle Eastern countries, notably Iran, Iraq, Saudi Arabia, Kuwait, as well as Malaya, are in this category. In Africa, Zambia, Zaïre, Liberia and Mauritania, and more recently Libya, Algeria and Nigeria, are of this type. The advance technology which mining required by the nineteenth century, and its high capital intensity, soon set these activities apart from the rest of the economy and made them function as a separate economic system. Foreign control was a particularly conspicuous feature of the mining economy and this meant that the major share of the income generated was deflected from the domestic economy. As a result, the value of mining activity as a factor for inducing direct change in the domestic economy has been minimal. Moreover, since the infrastructure created to serve the needs of mining tended to be highly specialized, the resultant external economies were also minimal or non-existent for the economy as a whole. Finally, since this activity used inputs of industrial origin acquired outside the country and created only a limited flow of wage income, it made no significant contribution to the creation of domestic markets. Its potential as a dynamic factor became evident only when the state intervened, obliging mining companies to acquire part of their inputs locally and collecting, in the form of royalties and tax revenue, a significant share of the flow of income traditionally remitted abroad.

Levels of development

Given these variations in the evaluation by countries of the centre of the natural resource endowments of underdeveloped countries, it is not surprising that their levels of development differ significantly. Although various indices and socio-economic indicators can be provided to depict heterogeneity in the levels of development among underdeveloped countries, it will suffice in the present context to use the gross national product per head as a crude index for this purpose. Table 3, for instance, shows that in 1984 some 60 per cent of African and just over 25 per cent of Asian countries had a per caput GNP of less than $500. Most of these countries belong to the category which the United Nations has designated as low income economies. Together, these countries account for some 2.3 billion people or roughly 48 per cent of the world population. These percentages quickly rise to 83 per cent in the case of Africa and about 50 per cent in the case of Asia, if the cut-off point is $1000 and less. In terms of the population involved, the proportion rises to 62 per cent of total world population of 4714 million in 1984.

It is, however, worth noting that two groups of underdeveloped countries are not accounted for by these crude indices of per caput GNP. These are the OPEC countries and the so-called rapidly industrializing underdeveloped countries. Eight OPEC countries, namely Kuwait, United Arab Emirates, Qatar, Libya, Saudi Arabia,

Table 3 *Gross national product per caput by country, 1984 (US $)*

GNP per caput	*Europe*	*North America*	*Oceania*	*Central and South America*	*Asia*	*Africa*	*Total*
Not available	7	–	4	5	13	7	36
Less than 200	–	–	–	–	3	6	9
200–499	–	–	1	2	4	22	29
500–999	–	–	1	8	5	9	23
1000–1999	2	–	2	14	3	4	25
2000–4999	8	–	1	9	1	4	23
5000 and over	18	2	5	3	10	1	39
Total	35	2	14	41	39	53	184

Source: *World Bank Atlas* (1986)

Gabon, Oman and Venezuela are among the wealthiest countries in the world in terms of their exceptionally high per caput GNP. All of these countries have a per caput GNP of over $2000. Indeed, the first three countries in the list have a caput GNP of over $10,000 and are easily the richest countries in the world in these terms. However, all these countries are small in area and population, accounting together for less than 38 million people. Also their wealth is based on petroleum which is a non-renewable natural resource whose extraction and exportation by foreign firms yield most of the revenue. The question of whether this wealth can be equated with development depends on how well income from this source can be converted into long term productive investment. The present tendency in many of these countries is to expend this income on short-term conspicuous consumption, on the acquisition of sophisticated weaponry, on building prestige projects and engaging in similar irrelevant investment. Certainly, as Table 4 shows, in terms of such social indications as birth and death rates, life expectancy at birth, adult literacy and labour force participation in agriculture, these countries remain essentially underdeveloped.

The list of rapidly industrializing countries includes Brazil, Mexico, Argentina, Iran, Taiwan, and South Korea. These are all countries where the per caput GNP is generally between $1000 and $2000 and where many industrial plants have been established. This has tended to create the illusion of escaping from underdevelopment although questions concerning the ownership of these plants, their products, their markets, their direct and indirect effects on other sectors of the national economy, as well as their impact on the physical and cultural environment, soon reveal their limited contribution to the task of transforming the lives of a substantial proportion of the population. Again, as with the OPEC countries, Table 4 shows that in terms of birth and death rates, particularly the resulting rate of natural increase, many of these countries share characteristics with other underdeveloped countries. The same goes for their labour force participation in agriculture. Argentina stands out as exceptional with regard to most of these indicators, yet in a very real sense, she represents no more than the most advanced of the underdeveloped countries.

The changing paradigm of analysis

This attempt to indicate the high degree of variety to be encountered

Table 4 *Social indicators of selected 'well-to-do' underdeveloped countries, 1984–5*

	Total population (millions)	*Birth rate (annual live births per 1000 population)*	*Death rate (annual deaths per 1000 population)*	*Life expectancy at birth (years)*	*Labour in agriculture (per cent of total labour force)*	*Primary school enrolment (all ages as percentage of primary school age population)*	*Adult literacy (percentage of total population over 14 years)*
OPEC countries							
Kuwait	1.8	35	3	71	2	91	70
Libya	3.6	45	11	58	19	n.a.	67
Saudi Arabia	10.8	43	12	56	61	67	n.a.
Gabon	0.8	35	17	50	77	n.a.	62
Venezuela	17.8	35	6	68	18	105	87
Rapidly industrializing countries							
Brazil	132.6	30	8	64	30	96	78
Mexico	76.9	34	7	66	36	121	90
Argentina	30.1	24	9	70	13	119	95
Iran	43.8	40	10	60	39	97	51
S. Korea	40.6	23	6	68	34	100	98

Sources: *United Nations Yearbook*, *UNESCO Yearbook* and the *Yearbook of Labour Statistics*

among countries of the Third World does not significantly detract from the fact of their essential underdevelopment. This appreciation has come about through a change in the paradigm of analysis used in the social sciences to examine and evaluate the complex manifestations of the phenomenon of underdevelopment. There was, for instance, a time when underdevelopment of these countries was equated with the 'backwardness' of their peoples and the reason for this was sought in broad environmental conditions. This paradigm of with the 'backwardness' of their peoples and the reason for this was sought in the broad environmental conditions. This paradigm of 'environmental determinism' had lost most of its appeal by the second half of the twentieth century and although there were still a few publications on such topics as climate and development, they were more cirumspect in their assessment of the relative significance of environment factors *vis-à-vis* socio-cultural conditions.[5]

A more important paradigm that came to dominate thinking about development for nearly three decades derived from inductive generalizations of certain aspects of the past of advanced Western industrialized countries. This paradigm would appear to have been inspired by a somewhat perceptive comment by Karl Marx towards the end of the nineteenth century that 'the country that is more developed industrially only shows to the less developed the image of its own future'.[6] Underdeveloped countries were thus seen as countries which were still chiefly agricultural and rural and whose populations were held in thrall by archaic social organizations and outmoded attitudes not compatible with sustained growth in production. For such countries, development would occur with industrialization which would bring about a massive shift of the labour force out of agriculture and other primary production sectors into processing and manufacturing.[7] Furthermore, it was envisaged that industrialization, because it required high capital investment and a shift from animate to inanimate sources of energy, would bring about higher productivity which would translate directly into higher incomes for workers. This, in turn, would make possible capital accumulation through savings (deferred consumption) which would again be reinvested in factories, machines and infrastructural facilities. This would produce a spiral effect to move the economy progressively to higher and higher levels of development. Imported capital may be needed but its main role would be to serve as a 'pump primer' for the process of development, particularly where local savings were inadequate. Moreover, with industrialization, significant economies of scale would be achieved through centralization of production and markets and this would give rise to the mutually reinforcing tendencies of development and urbanization.

On the social level, because industrialization, unlike traditional agriculture, requires a population with higher skills, literacy and creativity, a built-in requirement of this pattern of development is universal education. Along with this would go the inculcation of a new system of social values which is supportive of industrial growth, namely: personal discipline, punctuality, hard work, and responsiveness to monetary incentives by working longer and harder. Similar emphasis would need to be placed on those characteristics necessary for the emergence of an entrepreneurial class, notably rationality, sobriety and a willingness to take risks.

An important element of this paradigm was the benign role it conferred on developed countries with regard to the underdevelopment of the rest of the world. Underdevelopment was seen as the product of causes internal to the various countries concerned. To help these countries out of the vicious circle of poverty, ignorance and disease, the developed countries, out of a feeling of common humanity or 'enlightened self-interest', were inveigled to extend assistance in the form of technical expertise, loans and credits and to train local entrepreneurs, technicians, managers and industrial workers either *in situ* or abroad. Such assistance was important since development was conceived of as resulting from innovations and changes generated in the developed countries and diffused to the underdeveloped. These innovations have been introduced into the metropolitan centres of underdeveloped countries and from there are further diffused into the rural areas. Such diffusion is assumed to be beneficial in its effect, serving to modernize the whole country and to raise the level of the more backward periphery to that of the central region.

This paradigm of analysis has come under serious criticism in recent years. For one thing, many of the solutions that derived from it have been attempted and, almost without exception, have led to a worsening of conditions for a sizeable proportion of the population of underdeveloped countries. For another thing, the criticisms have gone hand-in-hand with the emergence of another paradigm for analysing the phenomenon of underdevelopment. This alternative frame derives much intellectual direction from Marxian dialectics and sees underdevelopment as intrinsically part of the same process that brought about development in the present advanced industrialized countries. Specifically, this new paradigm criticized the neo-classical economic development analysis on three major counts. First, that it lacks historical specificity. It treats underdevelopment without any awareness of the historical reality of the world capitalist economy out of which these conditions have emerged. As such, the neo-classical paradigm evades the discussion of the specific historical position that underdeveloped countries occupy within the polarized structure of the world capitalist economy.[8] Yet, without an understanding of this position, no analysis of an underdeveloped society can hope to advance beyond a superficial and mechanistic analysis of forms and appearances. Furthermore, the neo-classical development paradigm overlooks the global expansion of metropolitan capital, ignores its inner dynamics during the imperialist stage and pays little attention to its impact on the space economy of peripheral societies found in underdeveloped countries. In consequence, one of the major problems with the paradigm is that it is divorced from the very process of history and the specific stage of development reached by these societies. The result is that development strategies are framed in terms of prices, wages, savings and investment ratios for societies within which, as Ranis observed:

> Prices do not respond to changing relative scarcities, or entrepreneurs . . . to changing prices; immobilities of resources are such as to prevent necessary movements of factors, or some other characteristic of the general social milieu interferes with the economic process working in such a manner as to generate continuing economic growth. . . . It is apparent that such barriers must be eliminated before a society can play host to the kind of cumulative development process with which advanced countries have become familiar.[9]

A second criticism levelled against the neo-classical development paradigm is its lack of class analysis. This paradigm, it is alleged, assumes that the development and organization of an economy can take place in the setting of an implicitly harmonious social order where there would seem to be no internal structural contradictions. Where the analysis refers to inequalities in the distribution of income, or to the existence of certain social and political elites, no explanation is given concerning the origins and reproduction of these groups. Yet such explanations are important, for even among the small-scale peasant farmers who tend to be regarded as a homogeneous group, there invariably exists differentiation, albeit of varying degrees of importance, under differing social conditions.[10] This differentiation bears within it many implications for the regional pattern of underdevelopment encountered in many countries. Consequently, the neo-classical development paradigm ignores the process whereby specific classes or particular fractions of classes acquire political consciousness, begin actively to challenge the structures of the existing socio-economic order and exert influence on the whole development of a space-economy.

A third criticism of the neo-classical development paradigm is its failure to consider the importance of spatial structure in the process of development. This criticism has two aspects to it. On the one hand, advance beyond a superficial and mechanistic analysis of forms and appearances. Furthermore, the neo-classical development paradigm overlooks the global expansion of metropolitan capital, ignores its inner dynamics during the imperialist stage and pays little attention to its impact on the space economy of peripheral societies found in underdeveloped countries. In consequence, one of the major problems with the paradigm is that it is divorced from the very process of history and the specific stage of development reached by these societies. The result is that development strategies are framed in terms of prices, wages, savings and investment ratios for societies within which, as Ranis observed:

> Prices do not respond to changing relative scarcities, or entrepreneurs ... to changing prices: immobilities of resources are such as to prevent necessary movements of factors, or some other characteristic of the general social milieu interferes with the economic process working in such a manner as to generate continuing economic growth. ... It is apparent that such barriers must be eliminated before a society can play host to the kind of cumulative development process with which advanced countries have become familiar.[9]

A second criticism levelled against the neo-classical development paradigm is its lack of class analysis. This paradigm, it is alleged, assumes that the development and organization of an economy can take place in the setting of an implicitly harmonious social order where there would seem to be no internal structural contradictions. Where the analysis refers to inequalities in the distribution of income, or to the existence of certain social and political elites, no explanation is given concerning the origins and reproduction of these groups. Yet such explanations are important, for even among the small-scale peasant farmers who tend to be regarded as a homogeneous group, there invariably exists differentiation, albeit of varying degrees of importance, under differing social conditions.[10] This differentiation bears within it many implications for the regional pattern of underdevelopment encountered in many countries. Consequently, the neo-development planners in many countries and on international discussions about development. Dudley Seers explains this state of affairs as due to cultural lags. According to him:

Cultural lags protect paradigms long after they have lost relevance. The neoclassical growth paradigm has been remarkably tenacious – in fact it still survives in places. It has suited so many interests. It has been highly acceptable to governments that want to slur over internal ethnic or social problems. It has offered (not only in the hands of Walt Rostow) a basis for aid policies to inhibit the spread of communism. . . . It has provided international and national agencies with an 'objective' basis for project evaluation, and goals for what should be called the Second Growth Decade. It has generated almost endless academic research projects and stimulated theorists to construct elaborate models. It has not been fundamentally unacceptable to economistic modernizers across a broad political spectrum, including Marxists as well as members of the Chicago school. Above all, as a paradigm it is very simple.[12]

Conclusion: the significance of a spatial perspective

The Marxian paradigm of development analysis is, none the less, of increasing importance for many reasons, not least of which is the new and serious appreciation of the spatial perspective which it encourages for a full understanding of the development process. This spatial perspective is predicated on one fundamental premise, namely that every underdeveloped country has within its boundaries the two most important resources necessary for its development: productive land and the labour of its population. The development process is consequently the application of rational thought to the mobilization and utilization of these two fundamental resources to improving the material conditions of the people as a whole. The weakness of the older paradigm is that by focusing on capital, it limits the concern to what can be done for people, usually a small segment of the population, defined by the scope of individual projects, rather than what the people can do *together* for themselves.

The position is perhaps best illustrated by the following reported comments of two Campeche peasants on the current paradigm of development as exemplified in the small irrigation projects jointly financed by the Mexican Government and the inter-American Development Bank:

These projects belong to the Government. They are the *patrons*. They employ *campesinos* like us as *peons* so that we can earn some money and not starve. But they have not paid us any wages yet. Perhaps they were lying again.

But even if the projects succeed what benefit will they be to the Camino Real? . . .How many people are members? Very few. We need some help for all the *campesinos* right now. *What good is it to make a few* campesinos *rich men in ten years, if all the rest of us are still stuck in the mud?*[13]

The objective of the present volume is thus to present an alternative, strongly spatial perspective for looking at the development process. In an era in which underdeveloped countries are seeking self-reliant ways to their own development, this perspective provides ideas as to how they can set about the full mobilization and utilization of their internal resources for that purpose. The perspective is rooted in the growing realization that concern with the spatial reorganization of a country can induce the release of tremendous physical and mental energies whose practical outcome is certain to give rise to the socio-economic transformation necessary to launch a country on to a path of self-centred, self-reliant and self-sustaining development.

The book is divided into four parts. Part One is concerned with orientation. It examines in detail what should be understood by the term 'development' and considers its close relation with geographical space and the structural arrangements and organization of that space in every country. That arrangement and organization is conceived of in two main forms: rural and urban.

Part Two examines the rural structures of underdeveloped countries. Using mainly African examples, it shows the collapse of traditional structures and discusses strategies which could bring about the establishment of new structures more consistent with the current goals of development of underdeveloped countries.

Part Three turns attention to the urban situation in these countries, particularly the crisis which cities represent for most of their governments. It shows how this situation is closely related to the breakdown of conditions in the rural areas. It then considers ways of establishing a new system of cities which would, in particular, achieve a more beneficial relation between urban and rural areas and a highly rationalized interpenetration of industry and agriculture.

Neither of these two types of restructuring and reorganization can take place independently of each other nor be unmindful of other critical factors in the circumstances of particular countries. An emphasis on a new type of national integration is thus called for.

In Part Four different aspects of this integration are considered, involving new forms of regional specialization and division of labour, greater geographical mobility of the population, enhanced flow of information and increased internal trade. An important chapter in this section examines the implication of this new type of national integration for the external relations of underdeveloped countries and discusses the importance of a strategy of selective closure for incubating a self-reliant process of development. A concluding chapter raises the issue of ideology as it relates to a spatial perspective on the development process. It emphasizes that while the process requires an articulated vision of the type of society it seeks to create, such a vision, in order to be realizable, must be grounded in the realities of the cultural and physical environment of each country and to that extent must give rise to an authentic ideology of development different in character, though not in essence, from one country to another.

Part One

Orientation

2 Defining development

One of the major factors in the current concern about what has been achieved from the three or four decades of central planning in many underdeveloped countries derives from the ambiguity surrounding the word 'development'. In the literature, the primary role of economic forces in bringing about the development of a society has often been taken as axiomatic, so that development and economic development have come to be regarded as synonymous. For some time also, it was unexceptionable to use economic development interchangeably with economic growth. This gave rise to a situation where a small and backward country could overnight become rich and 'developed' if it was fortunate in having a raw material such as petroleum which, as a result of a sudden upward change of price, became a tremendous earner of foreign exchange. Indeed, so confused had the situation become that Dudley Seers wondered whether, instead of worrying about brushing aside the web of fantasy and slipshoddedness surrounding the word 'development', we shouldn't simply abolish its use and look for a better and less debased word.[1] None the less, for various reasons, he argued that a redefinition rather than an abolition was what was called for.

Part of the reason why it is necessary to define the word clearly, is also the change in its real conceptual meaning over the last ten years. This is perhaps best illustrated by the shift in the conceptual use of the word in two seminal papers of Dudley Seers himself. In 1969, Seers conceived of development as involving not only economic growth but also conditions in which people in a country have adequate food and jobs and the income inequality among them is greatly reduced. As he puts it:

> The questions to ask about a country's development are three: What has been happening to poverty? What has been happening to unemployment? What has been happening to inequality? If all three of these have declined from high levels, then beyond doubt this has been a period of development for the country concerned.[2]

Eight years later, in 1977, in reviewing this conception of development, Seers asserts that he has left out one essential element which must now be added. This is self-reliance. According to him, the addition of this new element entails that the main emphasis in development would no longer be on overall growth rates or on patterns of distribution. The crucial targets from now on would be:

> Ownership as well as output in the leading economic sectors; consumption patterns that economise on foreign exchange (including imports such as cereals and oil); institutional capacity for research and negotiation; and cultural goals of the country.[3]

The possibility of such a remarkable conceptual shift strongly underlines the importance of making clear from the outset the sense in which the word 'development' will be used in the following pages. In order to do this, an attempt has been made to discuss four major ways in which the word has been used in the literature and to indicate how the preferred interpretation in this volume embraces most of the others but goes well beyond them to ensure that due emphasis is given to the spatial dimension in the development process.

Development as economic growth

Especially in the economic literature of the early, post-Second World War period, development was defined as 'a rapid and sustained rise in real output per head and attendant shifts in the technological, economic, and demographic characteristics of a society'.[4] This definition of development by itself is unexceptionable since, in developed societies, real output per head rises on a self-sustained basis with certain changes in the technology available to the society, in the forms of economic organization and in a certain reduction in the burden of dependents each individual has to support.

However, in practice, this interpretation came to be applied in a gross, macro-structural sense in which the role of the individuals involved in the process became completely unimportant compared to the total volume of commodity produced and the proportion put aside as savings for further investment. It is from this perspective that Lewis asserts that 'the central problem in the theory of economic growth is to understand the process by which a community is converted from being a 5 per cent to a 12 per cent saver – with all the changes in attitudes, in institutions and in techniques which accompany this conversion'.[5]

This conceptualization certainly gives priority in the development process to increased commodity output rather than to the human beings involved in the production. Indeed, it makes it possible to think of development as easier to achieve under a tyrant or dictator who, with coercion and violence, can make the life of his people unbearable while taking away most of their production as savings and investment surpluses. This approach was certainly important in the prescription given to developing countries to concentrate on export production, either of agricultural raw materials or minerals, as a means of raising real output per head and hopefully generating more surpluses to be invested in more export production and consumption goods.

The result is the well known phenomenon of a dual economy found in most underdeveloped countries. The export sector engages only a small proportion of the population but gets all the advantages of technological improvement and economic reorganization which induces a process of demographic transition, particularly in that portion of the population resident in urban centres. The traditional sector, on the other hand, encompasses the vast majority of the people but remains dependent on the age-old and inefficient technology and outmoded economic organization. Moreover, the demographic behaviour of this majority exacerbates the fine balance between resources and the human population.

The dual character of the economy in these countries becomes more pronounced in the era of industrialization. Many of the industries established were concerned with producing locally many of the durable and non-durable consumer goods formerly imported. They were in fact owned in many cases by the same foreign firms that used to import the goods and tended to depend on automated or semi-automated machinery. The raw material needs of these machines were such that they could not be supplied from within the country, with the result that very little interdependence developed between the industrial and the primary production sectors of the economy. The discrete, almost isolated functioning of these two sectors put in to bolder relief the dual character of the economy.

None the less, increases in the output of such industries are recorded as growth in the economy and development for the country. Yet, it is clear that the role of the country in all this has been no more than providing a location for the local assembly of foreign machines, technical know-how and raw materials in return for wages and employment for a little skilled and unskilled labour.

Development as modernization

Such an excessively narrow economic interpretation of development came later to be tempered by the need to engender changes of a social, psychological and political nature in the process. Development, still in the sense of economic growth, came to be seen as part of a much wider process of social change described as modernization. According to Lerner:

> Modernization is the process of social change in which development is the economic component. Modernization produces the societal environment in which rising output per head is effectively incorporated. For effective incorporation, the heads that produce (and consume) rising output must understand and accept the new rules of the game deeply enough to improve their own productive behaviour and to diffuse it throughout their society.... This transformation in perceiving and achieving wealth-oriented behaviour entails nothing less than the ultimate reshaping and resharing of all social values, such as power, respect, rectitude, affection, well-being, skill and enlightenment.[6]

The emphasis in development as modernization is thus on how to inculcate wealth-oriented behaviour and values in individuals. This, in a sense, represents a shift from a commodity to a human approach. It involves principally how to make the population of a country 'understand and accept the new rules of the economic growth game'. Investment in education, or more broadly in human resources, came to be seen as a major and critical basis of societal change. Indeed, in many countries a new epithet 'social' came to be added to development, to counter the previously almost synonymous association with economic development.

Development as modernization saw a new concentration in many developing countries on the building of schools and colleges, expansion of enrolment at all levels of education and in adult education, extension of the coverage of mass media particularly through radio and television, growth in the number of health centres and medical establishments, provision of better housing and recreation facilities, and new interest in youth and youth activities. This form of development was, however, less critical of content than of form. In a situation of changing social conditions, it continued to educate, to inform, and to minister to health needs through processes reminiscent of the period of colonial tutelage or procedures borrowed directly from the advanced industrial countries.

The 'new rules of the game' also had a consumption dimension. To be modern meant to endeavour to consume goods and services of the type usually manufactured in advanced industrial countries. This fact made it easy to accept the logic of a strategy of industrialization based on import substitution, since this meant that a colonial country continued to enjoy a style of consumption not based on its own internal resources but on a less obvious form of imports. Such a country also gives an appearance of modernizing by having rows of industrial factories in its capital city whose technology of production is well beyond the appreciation of the people.

Modernization thus bred its own needs and desires in those exposed to it. In particular, the enhanced appetite for the consumption of certain 'modern' goods and services which it stimulated, created considerable disaffection with traditional conditions especially in the rural areas. The result has been a tremendous exodus of people from such areas to the urban industrial centres. Shortage of agricultural manpower in the rural areas came to exist side by side with large urban unemployment. The disequilibrium arising from this state of affairs had far-reaching consequences. Not least of these is the progressive pauperization of those left behind in the rural areas or those caught in the throes of chronic urban unemployment. Equally important has been the increasing inability of these so-called 'modernizing' countries to feed their population. More than ever before, development as modernization has exacerbated income inequalities between individuals, between regions of a country and between urban and rural areas. Most underdeveloped countries came to exhibit more sharply the picture of a small minority of extremely wealthy individuals living off, as it were, the backs of a large, poverty-stricken and destitute majority.

Development as distributive justice

By the end of the 1960s, it was becoming clear that neither development as economic growth nor as modernization was having the expected wide-ranging effect on the standards and conditions of living of the majority of individuals in many Third World countries. If anything, the relative position of the masses worsened *vis-à-vis* that of the elite. Widespread poverty and destitution became visible, tangible and compelling. Development came to be seen not simply as raising per caput income but more important, of reducing the poverty level among the masses or, as it was more picturesquely put, satisfying their 'basic' needs.

This situation in underdeveloped countries coincided with the anxiety in the advanced industrial countries about environmental pollution and the fear of a rapid depletion of global natural resources. The United Nations Conference on the Human Environment held in Stockholm, Sweden, in the summer of 1972 dramatized worldwide concern for equity and social justice in the distribution of national and international resources and for the negative externalities that sometimes result from their mindless exploitation. For the underdeveloped countries, it was accepted that their environmental problems were largely those resulting from poverty which could be removed only through new strategies of development.

Interest in development as social justice thus brought to the forefront three major issues: the nature of goods and services provided by governments for their populations (otherwise referred to as public goods); the question of the accessibility of these public goods to different social classes; and the problem of how the burden of development (defined as externalities) can be shared among these classes.

Public goods as the product of collective action range from such intangibles as defence and police protection to concrete objects such as hospitals, schools and recreation centres. This wide variation permits a basic distinction between what have been termed (perhaps not very felicitously) 'pure' and 'impure' public goods.[7] Pure public goods refer to collective goods which, once produced, are freely available to every citizen. Defence is the best example, although one can think also of modern electronic mass media products such as radio news and communication. Impure public goods are localized so that citizens do not enjoy homogeneous quality and quantity in their consumption.

A new factor, that of accessibility, thus becomes critical in determining the amount and quality of such goods consumed by the individual citizen. Accessibility to public goods such as employment opportunities, resources and welfare services, can be defined both in physical and social terms. In a physical sense, it relates to the distance to be covered by an individual in an attempt to secure the good; socially, it relates to barriers of class, status or recognition which he may also have to overcome in the process. In either case, a price has to be paid either for transportation or for social betterment. The factor of accessibility, especially in its physical forms, means that individuals in certain localities or regions benefit less than others from the distribution of the product of social co-operation and development.

The situation would be less objectionable if the underprivileged population did not have to bear some of the externality effects of development. Externality effects can be defined as 'the by-products, wanted or unwanted, of other people's activities that immediately or indirectly affect the welfare of individuals'.[8] They arise when certain effects of production and welfare go wholly or partially unpriced. One of the most obvious cases of externality effect is that of air or water pollution resulting from the activities of industries. One can also think of the effect of urban development on rural areas or of economic growth in certain regions of a country on conditions in others. In each case, individuals who do not in any way profit from the presence of such activities in the particular area do, none the less, have to suffer from the effects of the unwanted by-products.

For some considerable period, the issue of social or distributive justice was seen largely in the light of consumption and attempts to cope with it were conceived purely in terms of the transfer of resources from the privileged to the disadvantaged groups in society. This was undertaken largely through progressive taxation or government subsidies. In underdeveloped countries, this approach to the problem meant giving primacy and priority to issues of economic growth while treating problems of distribution as a residual matter of only subsequent interest. The situation was well depicted in the usual metaphor of associating development with the baking of the national cake and the issue of justice in sharing it out. Hence, to have enough to share around, one must be concerned first with how to bake a large cake.

It was only later that it was appreciated that there was a production or asset distribution side to the issue of social justice which a country neglects at a cost to itself. Indeed, without paying adequate attention to this, growth in an economy can reach only a certain level and will then be stymied. In the circumstances, new approaches to development planning began to receive some interest, at least in the literature. As Ahluwalia and Chenery put it:

> The relations between distribution and growth and the importance of asset concentrations, lead to a basic change in the terms in which development objectives are formulated. For one thing, the allocation of investment cannot be separated from the distribution of its product. They should be thought of as different dimensions of a single development strategy. . . . If it is provided in an appropriate mix of education, public facilities, access to credit, land reform and so forth, investment in the poor can produce benefits in the form of higher productivity and wages in the organized sectors as well as greater output and income for self-employed poor.[9]

The new development strategy paid special attention to disadvantaged target groups such as small, peasant farmers, landless labourers and submarginal farmers, urban under-employed and urban unemployed. But perhaps, more important, the new strategy underlined the importance of regional development planning in seeking to even out, or at least narrow the gap in, the life chances, employment opportunities and real income of citizens irrespective of the region of the country in which they live. New conceptual tools such as growth poles and industrial complexes were put forward as vital to achieving the goals of more even development and better distributive justice in a country.

Development as socio-economic transformation

Scholars of a Marxist philosophical persuasion argue that the questions of distribution and social justice cannot be considered or resolved independently of the prevailing mechanisms governing production and distribution. In the absence of these radical considerations, they insist, any development strategy based on such notions and conceptual framework is sure to be self-defeating. They point to numerous countries, mostly in the advanced capitalist industrial world, which have attempted various forms of redistribution of real income or undertaken anti-poverty programmes, with rather indifferent results. The reason for failure, they claim, is obvious: 'programmes which seek to alter distribution without altering the capitalist market structure within which income and wealth are generated and distributed, are doomed to failure'.[10]

Against this background, the development of a society or country is seen as essentially a transformation of its 'mode of production'. Mode of production refers to those elements, activities and social relationships which are necessary to produce and reproduce real (material) life.[11] The elements comprise the raw materials existing in nature, productive equipment and infrastructure, and human labour. Activities relate to the process of bringing these elements together in the production of goods and services and in the context of available technology, social division of labour and social tastes and patterns of demand. Social relationships, on the other hand, define the social basis for co-ordinating the productive activities of the numerous individuals involved in these processes. They determine the nature of the social structure and are maintained through political, legal and other means. These means are, in turn, related to the prevailing objectives of the society or at least of the leaders of the society at a particular historical epoch. Such objectives are not always written out or consciously proclaimed and professed. But they can be gleaned from the nature of production and distribution organizations and the social relations that regulate their functioning.

It is easy to appreciate the fact that societal objectives have changed from one historical period to another. Such fundamental changes, as distinct from random variations in fashion or tastes, have enabled us to characterize societies as primitive, communal, feudal, capitalist or socialist. Each of these societies has not only a distinctive set of societal objectives but also different patterns of social relations which determine how various activities are undertaken and co-ordinated, and the value placed on the various elements involved. In many African countries, for instance, until the colonial period, societies were organized largely on a communalist basis. Social relationships were defined largely on the basis of kinship and this provided a framework within which activities were co-ordinated and various elements brought into productive use. Access to land was predicated on kinship affiliation and so was participation in various crafts and trades. By contrast, in present day modern society, social relations are largely determined through the economic value set on each individual, and activities are co-ordinated through the medium of self-regulating markets. Access to various factors of production is mediated through the market which also determines the purchasing power each individual can command.

Basic shifts in any of the aspects of the mode of production can trigger off wide-ranging changes which may culminate not only in the transformation of the mode but also in changes in the relative importance of social classes. It is such a socio-economic transformation that really constitutes development. However, it hardly ever happens that the productive relations at one societal stage are completely transformed before a new mode of production is established. In other words, no one historical epoch is the exclusive domain of one mode of production, even though a particular mode may be clearly dominant. As Lukács puts it:

> A particular mode of production does not develop and play an historic role only when the mode superseded by it has already everywhere completed the social transformation appropriate to it. The modes of production and the corresponding social forms and class stratifications which succeed and supersede one another, tend in fact to appear in history much more as intersecting and opposing forces.[12]

The manner in which these intersecting and opposing forces operate within a given country can be said to determine whether the country 'develops' or becomes 'underdeveloped'.

This is particularly important in the present context because of the critical role of the capitalist mode of production for understanding the phenomenon of development and underdevelopment in the world today. According to the Marxian interpretation of history, the capitalist mode of production depends on the existence of wage labour whose labour power produces a surplus (that is, product not needed for purposes of sheer subsistence) which, however, is appropriated and accumulated by the employer – or capitalist class – within a system of exploitative social relations. This inherent exploitative relation results in the emergence of class consciousness and class conflict in capitalist societies.

The Marxian interpretation of history also insists that everywhere in the world the capitalist mode of production will eventually succeed all pre-capitalist modes. How this would happen, however, is the subject of considerable current theoretical controversy whose details need not concern us.[13] What is perhaps of greater relevance in the viewpoint is its discussion of the interrelations likely to develop in the confrontation between capitalist and pre-capitalist modes of production. This discussion has given rise to other sets of theories which seek to explain the imperialist expansion of the capitalist industrial nations of Europe in the nineteenth century. One of the best sustained of these theories is the Hobson–Lenin thesis.[14] According to it, imperialism was a response both to a crisis of overproduction in the industrial nations, for which colonial territories provided guaranteed markets for excess manufactured goods, and to changes arising from the increasing domination of the system by corporate investors, such as banks, which needed new zones of investment opportunity to compensate for the declining rate of profit in their home regions.

Whatever the merit of this thesis, the reality of imperial expansion showed the opposite tendencies. Colonial territories proved to be net exporters rather than net receivers of capital. So-called 'colonial development' was characterized more by asset stripping than investment.[15] It is this which has provoked the rise of a new school of thought on the global pattern of development and underdevelopment. This school – the 'dependency school' – suggests that the capitalist 'exploitation' of underdeveloped countries occurs more in the realm of acquisition and exchange (as evidenced in the generally adverse terms of trade for Third World countries) than of production.[16] According to this school, the appropriation of raw materials and agricultural commodities on extremely favourable terms to the industrial countries, rather than the direct appropriation of the surplus of the workers' labour power, characterizes the development or, more correctly, the 'underdevelopment' process in most of Africa, Asia and Latin America. Consequently, class conflict was replaced by the equivalent conflict between the enriched and 'developed' metropolitan centres and the plundered and 'underdeveloped' peripheral regions.

It is important to stress this interrelation between development and underdevelopment so as to prevent the attempt to see the two phenomena as different and separate. In the modern period, capitalism as a mode of production has intruded into social life in virtually all countries of the world. Some countries have resolved the resultant intersection and opposition of forces which it generated in regard to the preceding social forms in a manner that has encouraged the improvement of social life and the enhancement of productive capacity. Such countries are regarded as 'developed'. Others have not been able to undertake such a resolution either because, as colonies, the decision apparatus was not under their control or because, even when they ceased to be colonies, they failed correctly to assess the nature of the situation confronting them. Thus, as Frank emphasized, 'in the colonial period, it was not isolation but integration into the Western capitalist system which created the reality of [Brazilian] underdevelopment'.[17]

This close interrelation between development and underdevelopment, so clearly underlined in the Marxian conceptualization of development as socio-economic transformation, has three important implications. First, it emphasizes that development is essentially a human issue, a concern with the capacity of individuals to realize their inherent potential and effectively to cope with the changing circumstances of their lives. Or as Cairncross puts it, 'the key to development lies in men's minds, in the institutions in which their thinking finds expression and in the play of opportunity on ideas and institutions'.[18] In other words, the growth of goods and services or the diffusion of the material products of other cultures cannot be regarded as development itself; at best they are no more than necessary aspects of development. In the modern period, they do not constitute a sufficient index of socio-economic transformation.

Second, development involves the total and full mobilization of a society. The task of changing the 'institutions in which the thinking (of individuals) finds expression' cannot be undertaken in an *ad hoc*, piecemeal fashion. It has to be comprehensive and to invoke total political commitment. Hence Myrdal's definition of development as 'the movement of the whole social system upwards'.[19]

Third, development represents a redefinition of a country's international relations. It involves a shift from an outward-oriented, dependent status to a self-centred and self-reliant position with regard not only to the processes of decision-making, but more importantly the pattern and style of production and consumption.

All of these three dimensions – individual, societal and international – are criticial for an appreciation of the complexity and multi-faceted nature of the development process. More importantly, they are vital for an understanding of the problem of underdevelopment in many Third World countries. They help to undermine the implicit teleological attitude to the development issue – an ethnocentric approach which believes that some peoples are incapable of development. It is an attitude found not so much in the writings of scholars but one which appears to underline the actions and activities of their governments and multinational corporations. They thus underscore the fact that development or underdevelopment must be seen within a 'world system' context since the contemporaneous existence of different modes of production in the same geographical world means that boundary conditions also exist and make critical the unit and scale of any analysis.[20] This, however, should not blind us to the caveat by Roxborough that given the present state of knowledge and theory concerning underdevelopment, there is need to strive for a modest and historically specific approach which attempts to further understand shifts in modes of production and in the role of different classes and class fractions at particular points in time and space.[21] Notwithstanding, this new approach enables a reassessment of the nature of underdevelopment and provides a new framework for appreciating what is involved in an effective strategy for moving out of the state.

The nature of underdevelopment

The concept of underdevelopment is relative or, more accurately, relational. It is a state of societal well-being which, in relation to conditions elsewhere, is far from satisfactory. The state reflects certain absolute and concrete conditions, notable among which is the loss of self-reliance and the inability to be the master of one's own fortunes. This loss of self-reliance especially, in countries described as underdeveloped, can be shown to be an inevitable concomitant of the undermining of their traditional socio-economic formation by the international capitalist system.

In Asia and Latin America, the period of undermining dates back to the eighteenth century, but in Africa it has been ascribed to the period between 1830 and 1885, that is, preceding the colonial domination of the continent. Specifically, with regard to Africa, this undermining involved not only the expansion of European trade into the continent but the forcible subjugation of its people when it became necessary to raise the profitability of their intercourse with Europe. This subjugation effectively removed the independence with which the African trader related to his European counterpart up to then, and left him powerless to resist blatant exploitation and depredation. This transition has been described by Brunschwig as follows:

> Black Africa was shaken and changed, just as Europe had been by the coming of the inventions and discoveries which brought it out of the Middle Ages. . . . The evolution took place at the pace of the Black who was free to accept or refuse the novelties. The African did not feel dominated or constrained. In general, he dealt on equal terms with the foreigners and did not feel carried away in spite of himself on to a path which was alien to him. This evolution could have continued. It was interrupted in the last quarter of the nineteenth century. The European conquest (or colonization) did not give a different direction to the path on which the African had now started. The break did not come from a change of direction, but from a brutal thrust which took away from the Africans control over their progress.[22]

What gave the Europeans the power to achieve this 'brutal thrust' not only in Africa but in the other underdeveloped regions of the world, has often been attributed to their technological supremacy deriving from the industrial revolution. While technology provided the means for making violent thrusts all over the world, it certainly did not account for the motivation. That motivation was the compelling need to extend the global sway of capitalism which, as a new mode of production, had by the middle of the nineteenth century engulfed much of Europe. For efficiency, capitalism must treat labour as just another commodity brought to the market. Its price or wage must be determined by its supply and the demand for it. Yet, it must be obvious that labour is not a discrete commodity. It is an attribute of human beings who cannot be shoved about, used indiscriminately, or even left unused, without affecting also the human individual who happens to be the bearer of this peculiar commodity. This was why at its emergence in Europe, the impact of this new economic order on the masses of the people was to reduce them to the very depth of abject and humiliating poverty.[23]

In commenting on this system as it took root in England in the early half of the nineteenth century, Polanyi, for instance, noted as follows:

> The system, in disposing of a man's labour power, would incidentally dispose of the physical, psychological and moral entity, 'man', attached to that tag. Robbed of the protective covering of cultural institutions, human beings would perish from the effects of social exposure. They would die as victims of acute social dislocation through vice, perversion, crime and starvation. Nature would be reduced to its elements, neighbourhood and landscapes defiled, rivers polluted, military safety jeopardized, the power to produce food and raw materials destroyed. Finally, the market administration of purchasing power would periodically liquidate business enterprises, for shortages and surfeits of money would prove as disastrous as floods and droughts in primitive society. Undoubtedly, labour, land and money markets are essential to a market economy. But no society could stand the effects of such a system of crude fictions even for the shortest stretch of time, unless its human natural substance as well as its business organization was protected against the ravages of this satanic mill.[24]

In the face of such social consequences, it was no wonder that societies in Western Europe and later in North America found themselves from the middle of the nineteenth century up till today, forced to protect their members from full exposure to the ravages of capitalism. Such protection took the form not only of the growth of the trade union movement but also of the rise of central and representative governments committed to restraining the socially disruptive potency of capitalism through such regulations as tariff laws, factory laws, social security and pension laws, labour codes and a host of other legislation.

These restraints came to have grave and negative consequences on profit margins. According to the Hobson–Lenin thesis, the capitalist economic system was thus forced to seek new fields where labour could be treated as a cheap commodity so as to ensure a higher rate of profitability than was possible in the home region and the history of global colonialism in the last quarter of the nineteenth century derives largely from this single fact. Although it is equally true that colonial territories were sought both for the raw materials they provided and the market for cheap manufactured goods which they constituted, they were viable and greatly desired because of the enormous profit derivable from the easy exploitation of the labour of their populations.

Colonies thus became areas where the indigenous society was

unable or rendered incapable of protecting its members against full exposure to the ravages of capitalism. The situation in southern Africa today provides the most blatant and vivid example of the socially disruptive potency of capitalism. But even in countries where the disruption was not so extreme and where colonialism could not fully incapacitate the society, the damage done was not inconsiderable. In place of traditional self-confidence, the people were reduced to a state of imitative dependence, a highly degraded state associated not only with an inability to provide themselves adequately with the material means of sustenance but also with the loss of cultural and psychological integrity. As Polanyi again perceptively observes:

> Not economic exploitation, as often assumed, but the disintegration of the cultural environment of the victim is the cause of the degradation. The economic process may, naturally, supply the vehicle of the destruction, and almost invariably economic inferiority will make the weaker yield, but the immediate cause of his undoing is not for that reason economic; it lies in the lethal injury to the institutions in which his social existence is embodied. The result is loss of self-respect and standards, whether the unit is a people or a class, whether the process springs from so-called 'culture conflict' or from a change in the position of a class within the confines of a society.[25]

It is thus no wonder that in any nationalist struggle against colonialism, the initial strategy is an appeal to cultural revival, to a new effort to assert the fidelity of cultural norms and values. Unfortunately, such appeals tend to focus on the superficial elements of a culture and often ignore a deeper appreciation of the need for institutional adaptation and reconstitution especially in the dynamic context of a modernizing society. The result is that development planning is made to appear as involving no more than a simple-minded preoccupation with the allocation of capital investment funds and the importation of various sophisticated equipments and technologies from advanced industrial countries.

Conclusion

In what follows, therefore, an attempt will be made to present another view of development and development planning. This view recognizes that any theory of development reflects the social, historical and national background of its authors. Certainly most of the theorizing on this issue to date shares a metropolitan world view that is preoccupied chiefly with the comprehensive elegance and logical consistency of the intellectual construct. By contrast, the present volume is concerned with reflecting a view from the periphery in which questions as to 'what is to be done' outweigh, though do not overwhelm, the attention to theoretical comprehensiveness.

The vantage point for this new perspective is provided by the consideration of the role of space in the development strategy of societies in different countries and at different times. The basic thesis is to show that efficiency in spatial organization arising from an ability to transform spatial structures in a manner consistent with a particular mode of production, is a critical if not a major factor in the development of a country. It is this intricate relation between forms of spatial organization and development that provides the *raison d'être* for this book. More than this, this relationship focuses attention on the nature of geographic space and the manner in which it intrudes into every conception of development such that, in truth, one can speak of development as essentially a socio-spatial process.

3 Geographic space and development

A country, says Vidal de la Blache, 'is like a metal struck in the likeness of a people'.[1] This view of geographic space emphasizes its inert quality and gives the impression that the image or likeness struck is essentially a function of the differentiating characteristics of the people. It is a view of space that sees it as a container, providing a framework or perhaps serving as the field of human action. This perception of the role of space is reflected in the scant attention given to the spatial dimension in the literature on development. Indeed, it is responsible for the lack of serious regard paid to it in the social sciences in general, a situation which geography has been striving quite strenuously to correct.

Yet, a little reflection would show how very superficial such a conception of space is. Geographic space refers essentially to the earth's surface and the varying attributes of that space. These attributes comprise not only the more easily visible features due either to natural processes or human construction, but also non-visible valuation due to social signification. This is why, for instance, two acres of flat land, one in the centre and the other at the edge of a town, can generate widely different streams of social consequence. An acre of land in the centre of a town has all the advantages of easy accessibility produced in consequence of the existence of an urban community. Such an acre is thus the object of considerable competition among members of the community and so its market value rises very high. To own this acre is to have an asset which confers tremendous economic power to an individual. An acre at the edge of town is also socially important, especially for residential development, but it does not have the same economic potential as a central location. Hence, while social significance can place different values on similar units of space, space in turn confers new and differential values on various members of the society.

Space as part of socio-economic formation

It is this capacity of space both to serve as an intransitive object and yet operate as a transitive medium of socio-economic valuations that has forced a growing re-evaluation of its role in the development process. In this regard, it is important to appreciate the three ways in which space can be conceptualized.[2] First, space can be regarded in an absolute sense as a thing in itself with a specific existence which is uniquely determined. This is the surveyor's and the cartographer's space, identified through a conventional grid reference, especially of latitudes and longitudes. Second, there is relative space which emphasizes relationships between objects and which exists only because those objects exist and are related to each other. Thus, if we have three settlements, A, B, C, in which the first two are close together in a physical sense whilst C is farther away but has a better transportation link with A, it is possible to speak in relative spatial terms that settlements A and C are nearer to each other than A is to B. Third, there is relational space in which space is perceived as containing and representing within itself other types of relationships which exist between objects. For example, the space in the centre of a town contains and represents within itself relationships deriving from the valuation which society chooses to place on particular locations within its occupied space. These three conceptions of space are not mutually exclusive. Any given space can become one or all simultaneously depending on the circumstances.

In terms of human activity, therefore, the second and third conceptualizations are clearly the more significant. The location of objects and the perception and social evaluation both of the objects and their location, do influence the pattern of behaviour of members of a given society. It is in this sense that space can be regarded as shaped matter, acquiring specific forms and styles of disposition depending to a large extent on the prevailing mode of production. For example, the space where people meet to buy and sell is often referred to as a market. Yet, in many African countries, it is possible to see the difference in spatial form of a traditional market and a modern market. The former is usually an open space, dotted here and there by trees, containing sheds which serve as stalls for numerous small scale traders, occupied largely on a periodic basis and still retaining in many communities its traditional social and ritual significance; the latter is almost invariably a built-up area, with much of the selling conducted on a daily basis within substantial buildings owned and used by relatively large scale, highly capitalized commercial enterprises. It is in this sense that Santos argues that 'spatial forms . . . constitute a language of the modes of production'.[3] Indeed, Santos insists on the need to distinguish between a mode of production as a pure concept of social relations, and as concretized spatial forms within an historically determined territorial base. This concretization is referred to as the socio-economic formation. To put it in his words:

> A mode of production . . . organizes the process of production into a particular form in order to have an effect on nature and obtain from it the elements necessary for the satisfaction of society's needs; that society and 'its nature' – that is, the portion of 'nature' from which it extracts its production – are indivisible, and together are called the 'social formation'.[4]

This indivisibility between society and the natural environment which it must transform for its continued existence and survival, has helped to highlight the close interrelation between spatial forms and social processes represented by the concept of a mode of production. Harvey argues that spatial forms must be seen 'not as inanimate objects within which the social process unfolds but as things which "contain" social processes in the same manner that social processes are spatial'.[5] It is in the same sense that Vieille contends that 'space is certainly a constitutive category of the mode of production. Genetically, the process of creation of space and that of the mode of production are inseparable. The latter cannot be understood if an abstraction is made of the former'.[6]

Much about development can thus be said to be concerned with the creation and organization of spatial forms or structures. Such creative acts take place on different scales varying from the national level right down to the level of individual household space, and representing differing orders of power-sharing and decision-making. Yet it is an interesting fact of history that it is those decisions of direct and immediate impact at the level of household space that are the most difficult to make. It is, for instance, relatively easier to create new administrative units or states within a country, than to bring about a new system of family land-holdings. The recent example of state creation in Nigeria, whereby between 1967 and 1976 the country changed its pattern of spatial organization from four to twelve (1967) to nineteen (1976) states simply by fiats of military governments, is a case in point. It is equally relatively easy, as again happened in Nigeria, to go below the level of each of these states to create new districts or local government areas. But to go below this level to define and create forms appropriate for decision-making at the level of households or groups of households, presents infinite and complex problems and difficulties.

Yet, it can hardly be denied that until decision-making at all the relevant spatial levels is made convergent in terms of the prevailing preoccupations and objectives of a given society, opportunities for frustration and contradictory tendencies are multiplied. In other words, certain types of spatial form and arrangement are more conducive to the achievment of certain objectives than others.

Although spatial forms and arrangements at the household level are a direct result of the actions of individuals, they none the less reflect the guiding hands of society as a whole. No single individual, for instance, can establish his own land tenure system. Even when possessed of all the weapons of violence and coercion, no individual can establish a claim to any unit of space on a permanent basis without the concurrence of and validation by society. It is in this sense that spatial forms reflect social processes and enable us to perceive in a concrete fashion how efficient, or otherwise, particular processes have been or are operating.

Thus, development, to the extent that it is a social process, is a creator and organizer of space. In turn, spatial reality provides us with a dimension not only for evaluating the nature and efficacy of the process, but also for influencing and directing it in a desired manner. It is this possibility of spatial transitivity that enables geography to provide new insights into the development process. This new insight starts from the analysis of spatial structures and recoils back on the processes that gave rise to them with a view that, through appreciation of the need to create certain forms, new social processes may be set in motion. Such a perspective gives an active role to space and spatial considerations that goes far beyond any traditional conception of space as a passive element or a conditioning or limiting factor. For any country, therefore, space is thus not only like the 'metal on which the likeness of a people is struck' but, more importantly, it is an active factor in the process of determining what type of image to strike.

This propulsive role of space poses not so much a philosophical as a methodological problem for thinking about development. It requires greater conceptual concern *a priori* with the spatial consequences of particular social decisions as well as with the social consequences of specific spatial arrangements. Such conceptualizations are not derivable through deductive reasoning but are essentially the product of human activities in different countries and at different historical periods. They result from processes of induction, of trying to distil out the essence and common denominator in the past activities of societies which have made the transition to self-centred development. This inductive approach to the conceptualization of development concentrates on articulating the processes whereby entire societies are mobilized and the total national space is restructured and reorganized to facilitate increased social and economic productivity and on establishing a more efficient system of participating in the creation and distribution of the greatly enhanced national product. Spatial structure and spatial organization are thus two important conceptual building blocks of this perspective to development. As such, it is important to elaborate briefly on each of them.

Defining spatial structure

By spatial structure is meant the ordered relations which exist between individual spatial forms or elements and the whole of which they are a part. These forms are concrete physical manifestations such as farmlands, hamlets and villages, fences and roads, factories and shops, towns and cities, railways and waterways and a host of other shaped matters serving human societies in establishing a complex set of beneficial relations with the natural environment. As structures, they are realized as landholding systems, rural settlement patterns, systems of central places, sets of urban firms, communication networks and so on. It is, of course, the relations between the individual forms which define the nature of the structure and determine the order of its complexity. Such structural relations tend to show a high degree of constancy and continuity over time, largely because of the reflexive interactions between spatial and social structures.

This reflexive interaction is most easily perceived in the effect of spatial structure on the behaviour of members of a given society. Nystuen, for instance, uses the interior of a mosque to illustrate the relationship between structure and behaviour.[7] A mosque is usually devoid of furniture and has a flat, highly polished tile floor. Imagine the entrance into such a space of a teacher and his pupils. The teacher settles himself on the polished floor, choosing no place in particular, nor does he face any particular direction. Yet once he has chosen his specific location, his decision not only determines the form of arrangement of his pupils but also the pattern of their behaviour and interaction over that space. One finds, for instance, that the pupils do not choose random positions but set about arranging themselves in a very determinate fashion. They settle close in front of the teacher – but not too close. The first row forms a semi-circle. Before the first row extends very far around the teacher a second row begins to form. It is more desirable to be in the second row directly in front of the teacher than in the first row far off to one side. The second row is probably staggered so that pupils in this position can see past the heads of those in the first row. The arc of the students and the number of rows grow in an interdependent fashion. This relationship will depend upon the teacher's voice, how loud it is, how well he enunciates, whether there is an advantage in seeing his mouth as he speaks, and so on.

Furthermore, there will be a tendency for some distance to separate each pupil from another to prevent crowding, However, as with the number of rows, this separation will depend upon the number in the class. Crowding will occur if the class is very large. There is probably a greater tendency for crowding near the front than at the rear of the class. At some distance, the teacher's voice no longer carries the message distinctly. A certain amount of shuffling around occurs, some pupils look around or sit facing other directions. They might even start talking among themselves. Certain pupils might even seek out a location in the last row so as to be beyond the reach of the teacher's eye.

One can go on expanding on the other locational and behavioural possibilities that follow on the decision of the teacher to determine his own position on this otherwise featureless plain. But what has been said so far should help to underscore the point and to enable us to explore further the concept of spatial structure. In the illustration provided, one of the more important, though implicit, aspects of spatial structure that help to give form to the arrangement of pupils around their teacher, is the expectation of what will be done or ought to be done by the other pupils. In other words, for any society to work effectively and to have what may be called a coherent spatial structure, its members must have some idea of what to expect. Without such a pattern of expectations and a scheme of ideas about what people think others ought to do, it would not be possible for members of a society to order their lives and their behaviour in geographic space. Spatial structure thus tends to reflect ideal patterns of behaviour which are socially set and which may in fact be formally determined by rules, regulations and legislation.

However, to see spatial structure purely in terms of a set of ideal relations is to miss its more important aspect arising from the dialectics entailed in its actual realization as concrete realities. So, any concept of spatial structure must also see it as resulting from acts of individuals in response to other influences, which in consequence have an infinite capacity recognizably or imperceptibly to transform or change some essential characteristics of the structure. Two such influences are of particular significance. The first arises from the membership of individuals of a persistent group within a society, such as clans, castes, age-sets, secret societies or classes or is due to position in a kinship system or to status with regard to a political system. This helps to determine how individuals define their relation to or their interaction within the structure. The second influence derives from technological developments which may impinge directly on the physical principles or social rules on which the structure has been constructed.

The interaction and sometimes the contradiction between the idealized and realized relations concretized in specific spatial structures are, in a sense, the grit from which change and transformation are fashioned. Outmoded spatial structures generate as much conflict as outmoded social structures and to that extent contain the seeds of their own destruction. For example, small, scattered intermixed farm holdings reflecting a pre-capitalist mode of production and a dependence on somewhat simple agricultural technology, become no longer viable when agricultural production is organized on the basis of a capitalist mode. Such a farm holding system must in consequence undergo tremendous restructuring and rearrangement. The same thing can be said of settlement patterns that reflect certain idealized expectations in their origin, expectations which over time come to be no longer relevant or viable and consequently are discarded in favour of new and more appropriate ones. The fact that some individual settlements survive over many centuries should not blind us to the fact that the relations which they concretize at different periods have changed.

Elements of spatial organization

The idea of change and transformation of spatial structures leads inevitably to a consideration of spatial processes, of the forces represented by those acts of individuals which help to define, and are defined by, given spatial forms. Spatial organization embraces this concept of spatial processes. It refers to the spatial arrangement of actions in sequences conforming with selected social ends. These ends have some elements of common significance for the set of individuals concerned in the action. The significance need not be identical, or even similar for all the persons; it may be opposed between some of them. But in all instances, the result is to create a structure or situation which is regarded as socially desirable.

The three most important processes of spatial organization are those of competition, integration and diffusion.

Spatial competition

Spatial competition refers to processes of spatial organization which in part results in the resolution of the opposition among spatial elements, by action which allows one or other of them to come to final expression. Such action succeeds by reason of the element concerned possessing an uncommon attribute or advantage. The broad climatic zones of the world, for instance, can be seen as the product of spatial competition between air masses of differing temperature and moisture characteristics. Similarly, in a woodland, plants which can strike their roots down deepest or respond most actively to photosynthetic stimuli, end up displacing most others and eventually dominating a particular landscape. Attributes or advantages of certain plants or animals to be competitively successful can also depend a lot on social valuation or usages. Thus, domesticated plants and animals have come to occupy significant portions of land where their wild counterparts are being progressively squeezed out.

In human terms, aspects of spatial competition are reflected in location theories relating to agricultural, industrial or commercial activities. Tremendous emphasis is placed in these theories on the role of transport development, especially in so far as this determines not only the relative cost of moving commodities from the production to the consumption sites but also the profit and therefore the competitive advantages of the particular processes of production in occupying a given site. Some importance is also placed on issues such as labour costs in the production centre, on the issue of internal (scale) and external economies, on agglomeration cost advantages as well as on the size of the market.

Competitiveness in this context involves specialization or the narrowing of the scope of activities so as to ensure greater effectiveness or dexterity in the discharge of the particular range of activities or the production of particular commodities or services. Such greater effectiveness implies increased productivity and, in economic terms, tremendous cost saving. Specialization of functions in turn encourages areal specialization. It leads to the domination of a given area of land by one form of activities. In agricultural terms, such specialization gives rise in its extreme form to monoculture, although more often than not there is simply a dominant crop. At any rate, areal specialization has been responsible for one view of geography as 'the study of the areal differentiation of the earth's surface'.[8]

One aspect of spatial competition which is seldom given due attention in the literature is the social relations of the various human actors caught up in the process. Social relations refer not only to ownership rights and privileges but also to rules governing the general access of individuals to resources. It is easy to see that in a subsistence economy the adoption of an income-generating export crop would be easier for an individual with access to fifty hectares of land, than for one with access to only one hectare. Yet this simple fact at the beginning of a process of change could lead to the smaller producer being pushed out completely from the economic scene, particularly if the advantage of large holdings also involved easier access to improved technology, labour and credit facilities.

Over the long term, spatial competition gives rise not only to areal differentiation but, more importantly, to sharp disparities in regional development within a country. The principle involved here is that identified by Myrdal as one of 'circular and cumulative causation'.[9] This principle states that once spatial competition gives advantage to one area over another 'a cumulative process of mutual interaction sets in whereby change in one factor would continuously be supported by the reaction of the other factor, and so on in a circular way. The whole system would move in the direction of the primary change but much further. Even if the original push or pull were to cease after a time, both factors would be permanently changed, or the process of interacting changes would continue without a neutralization in sight.'[10]

The effect of circular and cumulative causation is to deepen the differences between various areas of a country and to give rise to spatial structures reflecting different patterns and rates of development. It is these differences in structures which have fed and supported the dualistic model of development. Clearly, in most underdeveloped countries areas which were competitively advantaged because they were able to produce exportable commodities during the colonial period, came to receive most of the investment in transportation and social development. The higher income-earning capacity of the population made these regions more attractive for later industrial development which again gave the regions greater competitive advantage for future development. The disadvantaged regions did not stand still throughout this set of changes. Every development change in the advantaged region was marked by a corresponding worsening of their conditions, especially with the out-migration of the younger, more energetic population to the growing region and the consequent loss of the labour, entrepreneurship and capital which they represent.

In development terms, therefore, what is important is the fact that the forces of spatial competition are not neutral. They can have, for a country as a whole, progressive or potentially disruptive consequences. For this reason, they need to be watched closely in the process of development. The operation of the self-regulating market from which these forces derive may, if unfettered, increase rather than decrease differences, especially of income and opportunities, between regions in a country. As Myrdal so eloquently puts it:

> If things were left to market forces unhampered by any policy interference, industrial production, commerce, banking, insurance, shipping and, indeed, almost all those activities which in a developing economy tend to give a better than average return – and, in addition, science, art, literature, education and higher culture generally – would cluster in certain localities and regions, leaving the rest of the country more or less in a backwater.[11]

The development process must not only be concerned with the present organizational features resulting from the operation of forces of spatial competition, but also appreciate and evaluate their long term tendencies. It is for this reason that direct policy intervention is often advocated in the process of development.The main objective of this intervention is to correct disparities in the spatial competitive capabilities of individuals, groups or regions of a country. Such corrections are not always easy to accomplish, not only because they touch on very critical areas of social relations but also because of the problems of leakage arising from interdependency among spatial units in a country. None the less, one must note that while, theoretically, regional disparities can be expected to be reduced as development matures in a country, the evidence that this will happen irrespective of the strategy or style of development is far from conclusive.[12]

Spatial integration

Within a national framework, spatial integration is a necessary corollary of spatial competition. It involves the co-ordination of spatial activities to achieve a harmonious effect. However, it goes beyond mere co-ordination to imply structural interrelationships so that activities do not take place in space as discrete events in themselves but are closely determined and influenced by all other activities. The structural elements in a spatial integrative process include not only nodes or the human settlement system but also various channels or linkages through which exchanges or interactions within the system take place. Of the nodal structures, the urban system is perhaps the most decisive, while linkages relate largely to transportation and communication networks, as well as to all media through which information, goods and people move between the various nodes. In structural terms, it is possible to conceive of spatial integration occurring at different levels such as those of the city, the district, the region or the total national space.

In the developmental context, while spatial integration at each of these levels is important, it is the national level that is perhaps most crucial. This is why, in the literature, there has been perceived, for instance, some intricate relation between the degree of development in a country and its level of urbanization. The nature of this relationship has always been elusive and has not infrequently been demonstrated in terms of a simple correlation between urbanization ratios and various indices of economic development. In some cases where the relationship has been conceptually elaborated in detail, it has not been easy to operationalize the variables or find appropriate and adequate data to test meaningful hypotheses in this field.

Yet it must be obvious that the core of spatial integrative relations comprises the effectiveness of the linkages that exist between different places and activities in a given territory. This is a function of both the geographic flows of people, material and information and the vertical relations in the production process especially between primary, processing and distributive activities. One can thus identify two dimensions in the spatial integrative process: the interactive and the functional. The former relates to movement between places, usually with no transformation in the elements being moved, while the latter involves varying degrees of transformation in the process of being transposed. Functional spatial integration, to the extent that it involves the valorization of the products of given areas of a country, is of critical importance in the process of development in that it increases, in real terms, the stock available to a given society.

As with spatial competition, the process of spatial integration is not neutral. Indeed, it is possible to evaluate the process in terms of the goal to which it is directed. Since this goal can vary, the form in which a territory is spatially integrated can also vary. It is, for instance, easy to show that the goal of spatial integration in a colonial economy must be to facilitate the generation and evacuation of exportable surpluses from the dependent territory. By contrast, the goal for an independent country could embrace not only increased production but also better social distribution and enhanced national unity.

In developmental terms, therefore, an aspect of the concept of spatial integration is probably best conveyed by the complementary concept of integrity. A spatially integrated economy has a certain wholeness to it which enables policies applied to any part of it to have system-wide effects or repercussions. Such an economy is intrinsically centred on itself and is not a composite of parts, some of which are autochthonous while others are more effectively involved in another and exterior spatial system. A colonial economy lacks this wholeness. And it is not for nothing that colonial administrations in many African countries resisted the development of a national road network which might compete with their investment in the railway. It is equally noteworthy that the history of colonialism everywhere shows a certain resistance to the development of manufacturing industries in dependent territories. In recent times, even though manufacturing has now spread to most underdeveloped countries, the import substitution strategy behind its present development constitutes in itself a major obstacle towards its full integrative role.

Spatial integration, whether in its interactive or functional form, cannot be seen as a process leading to the attainment of a static goal. The state of that goal is one whose quality is constantly being modified as a result of the impact of innovations. Such innovations affect not only the production processes but also the consumption patterns and thus determine both the quantity and the varied character of output. A critical aspect of development therefore concerns the nature of the mechanism for the spatial diffusion of innovations within a country.

Spatial diffusion

The mechanism for the spatial diffusion of innovations has received considerable attention in the geographical literature, starting particularly with the work of Torsten Hägerstrand in Sweden in 1952.[13] Within the development context, two types of innovations are recognized: consumer and entrepreneurial innovations.[14] The diffusion of consumer innovations takes place at the level of the household and is, to a large extent, concerned with the acceptance of new consumption goods and services or the adoption of new household technologies. As such, its spatial diffusion often gives the impression of wave propagation or the expansion of ripples from a given source. By contrast, the diffusion of entrepreneurial innovations tends to appear spatially as terminals of a set of spokes from a common hub because of the rather selective nature of the adoption.

Particularly with regard to entrepreneurial innovations, the role of interpersonal, face-to-face communication is regarded as critical. Part of the reason for this is the complexity of practice that often goes with the adoption of even a simple innovation such as a new type of seedling. When should it be planted; how should it be planted; how long should it stay in the soil; what is the nature of its moisture requirements? Although all these questions can be answered in a publication, face-to-face interaction is often still necessary to cope with any reservations which a potential adopter may have. It is equally noteworthy that in spite of the existence of numerous trade journals and technical reports, contact analysis underscores the fact that even among large, multinational corporations, many innovations continue to be diffused through interpersonal contacts at meetings and trade fairs.

The diffusion of consumer innovations is propagated through various strategies besides face-to-face interaction. Mass media communication, especially using suggestive and persuasive advertising as well as various techniques of sales promotion, are veritable methods of creating favourable attitudes to the adoption of various consumer goods. Yet, even here, particularly in underdeveloped countries, interpersonal contacts have a vital importance for the diffusion process. They serve to put a premium on the identification of those locations where face-to-face interactions are maximized and where people are likely to be more receptive to new ideas. They also emphasize the importance of community leaders and migrants as vital foci in the process of diffusion.

Productive contacts resulting in the diffusion of innovations are facilitated by certain types of spatial conformation in the human geography of a country. These include, for instance, market places and the hierarchical system of urban centres. Thus, urban centres, apart from their role in ensuring the spatial integration of a country, serve the equally important function of facilitating the diffusion of innovations. Of comparable significance in this regard is also the transportation network. This has the effect not only of improving accessibility but also of increasing the contact and information fields of individuals and enterprises. It is partly for this reason that the establishment of a transportation network has been regarded as a most important factor in the development of a country.

Spatial diffusion can thus be appreciated as a vital mechanism for keeping the momentum of development going. Its effectiveness within a country depends on the national network of social communication. Social communication, in this context, is defined as 'a comprehensive term which includes all aspects of interpersonal contact, ranging from a weekly social meeting to the growth of markets and a money economy. Anything which broadens the information field of an individual promotes social communication'.[15] This includes, for instance, schools and voluntary and political organizations. Spatial diffusion studies reveal, however, the existence in every community or country of areas of resistance or actual barriers to the spread of specific types of innovation. Resistance is often the product of social attributes which make it difficult for potential adopters to appraise correctly the value or importance of an innovation. Such attributes can be educational, religious, ethnic, class or income group in origin. Resistance, however, can be broken under the influence of persistent communications. Where the resistance is so formidable that it not only defies all interactive relationships but also makes it difficult for the innovation to spread beyond the resisting area, a barrier is created. Barriers to the spatial diffusion of innovations need not be limited only to those of social origins. Physical features such as mountains, deserts and large bodies of water, to the extent that they greatly limit interpersonal contacts among people, often act as veritable barriers to spatial diffusion processes.

Resistance and barriers should, however, not be thought of only in negative terms since spatial diffusion is itself not always a neutral process. Just as innovations may be beneficial when they are diffused into an area when the time and conditions for them are right and appropriate, so they may have adverse consequences in a country and a society if these requirements are not met. There is, for instance, increasing concern about the wisdom of the diffusion of certain consumption styles into underdeveloped areas of the world, especially where the effect has been to raise expectations which cannot be met or sustained. Sometimes, such openness to diffusion processes undermines the cultural integrity of a people, with wide ranging consequences for their self-confidence and sense of identity. Moreover, the potency of spatial diffusion processes for creating large-scale uniformity in preference patterns has provoked a countervailing reaction in favour of diversity and the social preservation of a wider range of life style choices for individuals and groups. All of this has meant a willingness overtly and formally to create resistances and barriers against these processes.

Development as spatial reorganization

It is possible from the perspective presented to conceptualize more clearly the development process as one of spatial reorganization. The reorganization arises as a result of the fact that development implies the articulation of a new set of social goals. The nature of those goals has been clearly spelt out in the preceding chapter. What is being argued here is that for these goals to be attainable, a new set of spatial actions is called for. This clearly involves a deliberate exercise of choice in the making of decisions. It rests on the dialectics of societal evaluation of the general ends towards which activities are to be directed and particularly on the importance which the given society wishes to confer on the individual in the scheme of things. For instance, since for most underdeveloped countries, land is the primary source of income and wealth and the basis of social relations for over 80 per cent of the population, development cannot be said to have begun without some form of land reform with consequences for other elements of the spatial structure in such countries. Such redefinition of social relations not only affects the form and manner of operation of spatial competitive processes but also introduces new strands into the processes of spatial integration and spatial diffusion.

Like all social processes, spatial reorganization involves a new system of resource allocation which in turn implies some notion of efficiency. In other words, certain types of spatial arrangement can be expected to make a relatively better contribution to the attainment of specified goals than others. In consequence, the patterns of resource allocation can be ranked or evaluated in terms of what type of spatial arrangement they make possible. This concept of efficiency calls attention to two other aspects of the spatial reorganization process. The first is the importance attached to magnitudes. This aspect, for instance, emphasizes that it is not enough to know how many people live in a country. It is equally necessary to have an idea of the number and sizes of the settlements they occupy, the number and quality of routeways that link these settlements, the number, size and ownership distribution of land as well as other assets, and so on. All these are important in determining the efficacy of the new system of social relations and its effect on the other processes of change. The more attention is paid to matters of magnitude and their changing complexity, the greater the chances of development in a country being well managed and directed.

The other aspect of spatial reorganization deriving from the concept of efficiency is the recognition of the time factor. Time is at once a denominational and an evaluative factor in spatial reorganization. In its denominational role, time operates as a boundary condition; inasmuch as efficiency can be measured as the number of events occurring per given unit of time. Such events, in so far as they relate to human spatial behaviour, depend for their orderly qualities on common definitions, assumptions and actions with regard to their location in time. The orderly qualities, on the other hand, relate to two aspects of denominational time. The first is synchronization, the occurrence of an event due to the carrying out of certain activities simultaneously by a number of people; the other is sequence or the occurrence of an event arising from actions following one another in a prescribed order. Temporal strategies involving both synchronized and sequential changes thus become an important aspect of spatial reorganization. For instance, certain fundamental structural changes, because of the high degree of interdependent relations which they entail, may require a great measure of synchronization in their execution. Thus, no country can successfully transfer from driving on one side of the road to the other side without an attempt to synchronize the activities of drivers throughout its length and breadth. The whole process of change, of course, could be seen as sequential, starting with the preparations and arrangements for the change, through the change itself and then to activities connected with dealing with the consequences of the change. In this manner, efficiency in spatial reorganization may depend on knowing what type of temporal strategy is appropriate for what set of actions. Or, as Moore puts it, 'time thus becomes, along with space, a way of locating human behaviour, a mode of fixing the action that is peculiarly appropriate to circumstances'.[16]

In its evaluative role, time or the frequency of events during a period of time is denoted by the concept of rate and has acquired special significance in a modernizing or industrializing society. The ability to increase the frequency of producing individual units of goods within a given time period, for instance, is regarded as an index of efficiency. The improvement of such ability, either through practice or enhanced knowledge, becomes a desirable aspect of social behaviour and leads to the phenomenon of specialization. Such improvement when it becomes societal in its extent, can itself be the cause of other structural spatial changes. Hence, technological changes especially in the speed of communication and transportation have brought about tremendous reorganization in spatial activities and in the social, economic and political life of a country. None the less, speed is not always socially significant and its desirability depends on the events concerned and the circumstances in which these are being considered.

The temporal dimension of spatial reorganization, enables us to characterize development on a linear graph (Figure 2) showing sharp discontinuities. The abscissa of the graph is identified as time while the co-ordinate represents the levels of societal change. Change in this context encompasses both development and growth. Development is shown as an intense concentration of activities requiring a high degree of synchronization and sequential ordering over a short period of time during which the whole social fabric of a country is transformed and its spatial structure reorganized. This allows the populace to operate and function at a higher level of efficiency in the production, distribution and consumption of both material and non-material goods. Growth, on the other hand, is a process involving different rates of incremental change during which the possibilities of a given organiza-

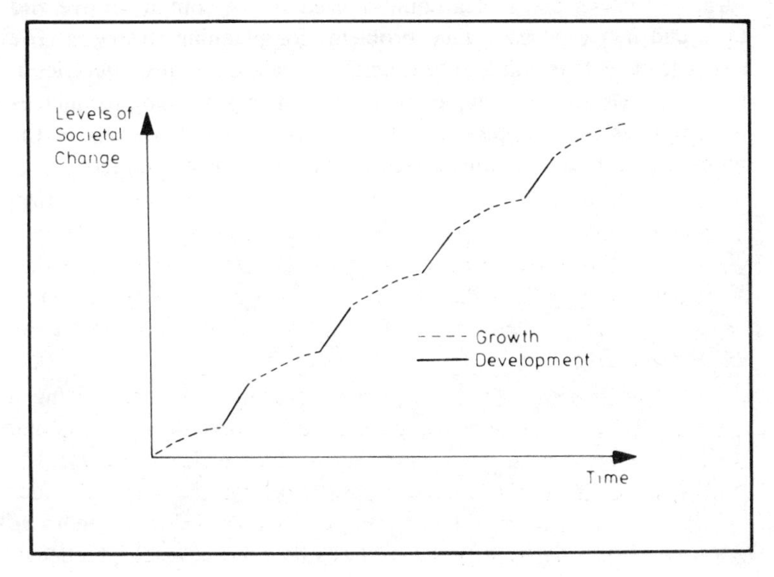

Figure 2 *Temporal aspects of development and growth*

tion are fully exploited. What the graph emphasizes, therefore, is that any organization has the capacity for growth but only up to a point. To go beyond this point involves a concern with the development of the organization. This entails structural transformation to make the system better able to respond to the new demand for a high level of efficiency and equity.

Conclusion

In the present context, spatial reorganization is seen as synonymous with development in the sense that spatial forms represent physical realizations of patterns of social relations. The need for a pattern of social relations which can incubate new processes of production thus requires the reconstruction of spatial structures both in the rural and urban areas of a country. Clearly, certain types of rural spatial structures, more than others, must make it easier to launch a country into an era of high agricultural productivity. When such transformation of the rural spatial structure is being designed, it will be found that its success is also contingent on a new structure of urban places. A new system of towns and cities is, however, not only a necessary complement for rural transformation but also an important component of spatial reorganization in its own right. This task of transforming the rural and urban areas of a country needs to be comprehensive and total and hence poses special problems for planning strategies. The next decade is thus going to be crucial for underdeveloped countries in terms of their ability to move away from current planning techniques to new ones more appropriate to the task in hand. Details of the nature of this task constitute the content of subsequent chapters.

Part Two

Rural Development

4 Collapse of traditional rural structures

Traditional rural structures refer to elements in the rural production system which existed prior to the recent period of modernization and development. Being traditional, however, does not mean being static and, apart from various changes during pre-colonial times, we must add the massive adjustments which the structures have had to make to the various distortions of colonialism. In most underdeveloped countries, the traditional rural structures embrace the spatial production system together with the technology, institutions, organizations and value systems that have developed around it. The social relation of land is central to this production system and is reflected not only in the land-holding and rural settlement pattern but also in the relation of the latter to the urban centres within the countries concerned.

Among underdeveloped countries, however, traditional rural structures are found in such great variety that it is difficult to trace for each of them their growing inappropriateness for the development of their respective countries. For instance, in India traditional rural structures evolved from the collective tribal ownership of ancient times to a peasant–proprietor system (the *ryotwari*) which is still prevalent in a large part of western and southern India. In the rest of the country, however, the system was overlaid by the caste institution. This recognized certain families as belonging to the land-owning castes and, along with the Brahmins who performed ritual functions, these owned most rural lands. However, members of such families were ritually prohibited from actually labouring in the fields and so these families became the patrons and often absentee landlords for a group of client families who tilled the land. The situation became complicated in the late Moghul period by the *zamindar* system based on revenue collection from land and this was further distorted during the British colonial period. The result was that by 1960, 42 per cent of rural households in India had farm holdings of less than one acre each and operated only 1.3 per cent of the total cultivated acreage; about 45 per cent of rural households with holdings of one to ten acres, operated about 39 per cent of all cultivated acreage; 13 per cent of households with over ten acres each operated almost 60 per cent of all acreage.[1]

The same colonial distortion is noticeable in Indonesia where the traditional communal ownership of land was replaced under Dutch colonial administration by individualized forms of holding such that by 1960, 51 per cent of farmers had holdings of less than half a hectare and 28 per cent held between a half and one hectare.[2] Thus, 80 per cent of all proprietors held only 50 per cent of the cultivated areas in Java, with the remaining 50 per cent held by the 20 per cent of the proprietors with holdings of more than one hectare. Even more significant is the fact that by 1960, 60 per cent of the rural population were already landless labourers.

In Latin America, the traditional rural structures derive largely from the system of large land grants made to the Spanish and Portuguese *conquistadores* in the sixteenth century. These large land holdings, or *latifundia,* have survived in most Latin American countries up till today. They provide the basis for a patron–client relationship in which the majority of the rural population work as tenants or share-croppers on the land of largely absentee landlords. Indeed, as of 1950, it was claimed that some 1.5 per cent of agricultural enterprises had an average of more than 1000 hectares per unit and accounted for over 65 per cent of agricultural land while at the opposite extreme 73 per cent of the enterprises had an average of no more than twenty hectares and represented only 3.7 per cent of the area.[3] Although some of these large land holdings have been turned to highly productive modern plantations, by far the majority have remained traditional and inefficient, and today constitute the greatest impediment to modernization and development in most Latin American countries.

By contrast to the Asian and Latin American situation, the traditional rural structures in most African countries are still based largely on communal ownership with distortion towards individualization during the half century of colonial rule. In very few instances does one find structures developed around a patron–client relationship. Yet, one of the more notable facts is that these structures share one common denominator with those in other regions of the underdeveloped world, namely that they also impede easy access of the vast majority of the rural population to various factors of production and supporting services. It is this fact, of the incapacitation of the rural masses, that accounts for growing dissatisfaction with rural conditions and that is resulting in some countries in the collapse of the production system, and in others in the increasing pauperization of the rural population. The situation in Africa enables us to trace the path of this decline to stagnation and best illustrates the contradiction posed by traditional rural structures to incubating the development process. In this chapter emphasis is placed on the situation in Africa, although references will be made to conditions in Asia and Latin America which are similar in their consequences even though different in their origin.

Traditional rural structures in Africa

To understand the nature of traditional rural structures in Africa, it is useful to begin with a consideration of the value system that surrounds land as the most critical production factor in traditional society. This value system derives essentially from what one might call broad societal objectives. For most traditional societies, the primary societal concern was survival. This was perhaps understandable given the great odds against any individual surviving childhood. Once that threshold had been crossed, society tries to ensure that the individual does not lack the wherewithal for keeping body and soul together. Since the basic unit of societal organization was the family, the mechanism for achieving this objective was to be found at this level. The concept of family varied from one society to another, being closely related to the accepted practice of marriage. A nuclear family usually implies a family based on a monogamous marriage and comprising a man, his wife and their children. Technically, however, it could also be extended to cover a family based on a polygamous marriage and comprising a man, his wives and their children. On the other hand, an extended family would include not only a man's wives and children, but his brothers, their wives and children, and the wives and children of his grown up children. There is also the concept of lineage which relates all those families which can trace their descent from the same ancestor.

In Africa, the functional basic unit of society is the extended family. This is the unit that owns land and whose members relate to this land on the basis of values and rules sanctioned by the society. These values have both ethical and philosophical aspects to them. On the ethical side, societal values insist that nobody shall be allowed to be destitute. Since land was a key factor of production every member of society must have access to land. This was ensured through an inheritance law which guarantees that all the children or sons of a man shall have a part of all his movable and immovable property. This is the principle of partible inheritance which is found among virtually all African peoples. It stands out in sharp contrast to the inheritance principle of primogeniture, whereby it is the first male child who inherits the land and may or may not take responsibility for the survival of his siblings.

The philosophical aspect of the social values attached to land is evident in the attitude to ownership and right of alienation. In most African societies, life is regarded as a passage and the living are thus no more than a link between previous and future generations. Family land belongs to all of these – the ancestors, the present generation and posterity. The relation of the living to the land is thus no more than that of trustee with rights of beneficial use (or usufruct) as long as he lives, but with no power to alienate the land from the ownership of the family. Thus, no African society recognizes the right of sale of land. Unallocated, unused and relinquished land can be allocated to strangers or to families growing faster than their resources of land but the land cannot be sold. The head of the community who exercises this allocative power on behalf of the community cannot unilaterally appropriate land to the use of his own family.

Communal ownership, however, did not mean that individual rights to a piece of land were not recognized. Although such rights were usufructary, they were clearly delineated and could often be inherited. Hence, most households frequently made use of common land and individual holdings simultaneously. Indeed, there were a few communities such as in Bornu (Nigeria) and among the Diola of Mali, where such individual holdings were treated virtually as freehold.[4] Nor was the system of land tenure without its own internal dynamics within traditional societies although detailed study of this is yet to be made. For example, Gueye noted that the establishment of Muslim rule in the Fouta-Toro area of northern Senegal towards the end of the eighteenth century, by insisting on the equality of all male heirs, led to the fragmentation of holdings, encouraged a greater degree of individual exploitation of land and led to the migration of heirs whose inheritance was too small to provide them with a living.[5]

None the less, it is important to stress that it was labour that was the scarce factor of production in traditional African society rather than land. Hence, while rights to labour were rather closely defined, that to land, which was in general an abundant resource, was less specific.[6] As a result, social rank, in most African societies, was not seen as giving an individual any undue privilege to the land beyond what he was entitled to within his extended family. This is one reason why no landed aristocracy has been established in any African society. Most of these societies are what have been called 'rank societies', defined by Fried as

> one in which positions of valued status are somehow limited so that not all of those of sufficient talent to occupy such statuses actually achieve them. Such a society may or may not be stratified. That is, a society may sharply limit its position of prestige without affecting the access of its entire membership to the basic resources upon which life depends . . . Accumulation of signs of prestige does not convey any privileged claim to the strategic resources on which the society is based.[7]

This value system, in so far as it relates to land, was similar to what obtained in many developed countries at an earlier stage in their evolution. Stenton describes the Anglo-Saxon English attitude:

> Medieval practice suggests very strongly that the holding of the pre-Conquest Ceorl had been partible among his sons, or among his daughters if he had no son, and that the conception of the holding as 'family land' was so firmly established that its possessor had no right to alienate any part of it to the disadvantage of his expectant heirs.[8]

The Norman Conquest and the growth of feudalism were to alter this greatly in the direction of primogeniture but even then it survived in various areas right up to the nineteenth century. The records of many other European countries reveal similar situations, among the masses of the people if not among the aristocracy.

This value system has implications for some of the production organizations and institutions. Kinship relation is a primary factor not only in determining access to land but also in the organization of labour and of residence. The supply of labour for agricultural production is generally provided by members of the extended family. During periods of peak agricultural activity such as land preparation or harvesting, a farmer may also call on the services of his in-laws or those of his age mates (when he is young) or a group of younger relations and their friends (when he is old). For this form of communal labour, the farmer is expected to pay in kind, usually by providing a sumptuous meal and drink. The question of labour specialization between the sexes varies from group to group. Among the Yoruba in Nigeria and Benin, women are generally exempt from farming and spend most of their time trading. In Bamenda, Cameroon, on the other hand, they are very important cultivators.[9] By and large most agricultural work is done by adult males although women and children do help on the farm during the period of weeding and of harvesting.

Apart from organizations for the supply of labour to individual family farms, there are also organizations for undertaking communal tasks such as the repair of pathways and roads, maintenance of the compound of the village headman and sometimes the policing of the community. Usually, such organization is based on age grades although a number of age grades can be made to collaborate on a particular task. It has been claimed that the demand of irrigation agriculture for a high degree of labour co-ordination to maintain the flow of water, construct and repair the channels and the dams and ensure timeliness in planting and harvesting, tends to encourage a more authoritarian organization of society.[10] This may be so for large scale irrigation although this claim has been challenged. It is certainly not manifest in the many examples of rudimentary irrigation farming found among many African groups south of the Sahara, especially among the Mandingo and Diola of the West African savanna and the Sonjo of Tanzania.[11]

However, a certain degree of authoritarianism is implied in the institution of slavery, which in many societies, particularly in West Africa, provided a substantial labour force for rural production. Such slave labour was present in West Africa long before the rise of the trans-Atlantic slave trade and constituted a major factor in the rise of many of the medieval empires and kingdoms such as Mali and Songhai in the savanna and Ashanti and Dahomey in the forest belts. Not every farmer, however, could depend for his additional labour supply on slaves who were not always cheap to own. More often than not, large numbers of slaves were owned by leading state officials or chiefs to provide foodstuffs for them and their immediate circle of dependents as well as for the army. The situation in Songhai in the fifteenth century and in the rural areas of Sokoto (Nigeria) in the nineteenth century are very typical in this respect.[12] Seldom were slaves used in producing export surplus, the striking exception in this regard being in Dahomey where, frustrated in selling his slaves to the Europeans in the mid-nineteenth century, the king of Abomey was encouraged to use them to grow his oil palm plantations for export production.[13]

Of the various institutions operative in rural communities, perhaps the most relevant from the point of view of production is that concerning the provision of credit. In virtually every rural community, some form of credit has always been needed by individuals to cope with their problems of production or domestic responsibilities. Various means have been devised to meet this. A man may secure credit on his personal recognition as a scion of a well-known family. He may do so by 'pawning' his own labour or that of his child over a given period of time or he may join a rotating savings group, known as *esusu* among the Yoruba, whose members contribute periodically and for each period one of them receives the total collection. All of these various ways of securing credit are predicated on personal knowledge of a type which is guaranteed by kinship relation. Sanctions of various kinds were none the less employed against defaulting and, as Hopkins observed, most African societies had rules which defined various types of loan and laid down regulations for recovering debt.[14] This was true both of Muslim states in western Sudan and of the states of the forest zone such as the Yoruba kingdoms.

Four major aspects of the technological component of traditional rural structures are relevant to our purpose. These are:

(a) the crops and animals that are the object of productive activities;
(b) the instrumental means of cultivating or rearing them;
(c) the instrumental means of storing, preserving and processing them; and
(d) the ideas or conception of ensuring sustained yield from the land over a relatively long period of time.

In most traditional African societies, the crops grown are those whose history of domestication is lost in the distant past or those which have been introduced as a result of contact with other societies.[15] These include grains such as millet (mainly *Sorghum* and *Pennisetum*), maize, rice and funio (hungry rice); roots such as yams, cocoyams and cassava; plantains and a variety of legumes, bulbs and fruits. Animal husbandry in Africa is as old as agriculture with cattle, goats and sheep as the major livestock. We know little of the tradition of systematic development of new species or varieties with special characteristics, although different breeds and various other animals have been introduced at different times in the past. The implements of production are distinguished by being dependent mainly on animate power. In many African societies, for instance, this power is provided by human beings using relatively simple implements such as hoes, cutlasses and axes for clearing the forest, preparing the land, sowing, weeding and harvesting. In some societies such as in the Gambia, a simple type of plough known as the *kedabo,* is used for the task of preparing the land.[16] Storage containers for the harvest vary from pots and similar containers to special buildings such as granaries or barns. For purposes of preservation, various simple processes are observed.

Drying and smoking are used in many societies and processing and conversion are common means of extending the time-utility of agricultural products.

The idea of ensuring a sustained yield from the land over a relatively long period of time is perhaps the most critical aspect of man's technological capability with regard to agricultural production. For many societies, it is generally recognized that yields can be maintained by the application of domestic refuse or animal droppings to the land. But there is never enough of these materials for the total area cultivated. Hence, the tendency in most rural communities to divide the farm land into two categories: one near the homestead, usually small but manured intensively and cultivated permanently, the other and larger area farther away from the homestead and alternating a period of cultivation with a period of fallow during which the land is allowed to recuperate naturally. Virtually all known agricultural communities in the world recognize these two categories of land and, with regard to development, it is the use of the larger field that has been critical.

This brief description of the value systems, organization, institutions and technology of agricultural production should facilitate an appreciation of the spatial manifestation of traditional rural structures in Africa. Central to this is the residential location of the extended family which takes the form of a compound comprising a number of individual huts, or a continuous building of many rooms. Such compounds may be isolated or grouped into hamlets or villages. Usually, the areas around the compounds are manured, intensively cultivated and cropped every year.[17] The main fields are divided up among the families. Within each piece of family land, various individuals have right of use of various plots, which tend to be small and scattered and intermixed with those of others as a result of the inheritance law of the society. Footpaths or tracts of various width link the farms to the villages and hamlets and these, in turn, to the market towns.

The impact of colonial exploitation

One of the more common characteristics of underdeveloped countries today is the fact that, at one time or another, they have all been subjected to colonial exploitation. The significance of this form of exploitation in the geography of development is that it not only undermined and hastened the collapse of the traditional rural structures but also, in the process, it distorted certain basic relationships in the society. For African countries in particular the undermining was achieved through the extension into and penetration of the rural structure by the capitalist mode of production. In more common parlance, this penetration is referred to as the monetization of the economy.

The implication of this penetration as indicated in the introductory chapter is the conversion of all the traditional factors of production, notably land and labour, into commodities which can be bought and sold in a self-regulating market. Yet, as Polanyi again indicated:

> What we call land is an element of nature inextricably interwoven with man's institutions. To isolate it and form a market out of it was perhaps the weirdest of all undertakings of our ancestors. . . . Land is tied up with the organization of kinship, neighbourhood, craft, and creed – with tribe and temple, village, gild, and church. . . . The economic function is but one of many vital functions of land. It invests man's life with stability; it is the site of his habitation; it is a condition of his physical safety; it is the landscape and the seasons. We might as well imagine his being born without hands and feet as carrying on his life without land. And yet to separate land from man and organize society in such a way as to satisfy the requirements of a real-estate market was a vital part of the concept of a market economy.[18]

The attempt to separate or alienate land from man took two forms. The first was to individualize its ownership. Individualization of land tends to eliminate all claims on land originating in kinship or neighbourhood organizations. In this way, all social restraints on land are removed and individuals can give away, mortgage, sell or alienate a piece of land at will. Correspondingly, this means that society must recognize a new and contradictory phenomenon: landless peasants. The second form which overlaps with the first, arose out of the consequent subordination of alienable land to the needs of a swiftly expanding urban population created by the market economy. Although land cannot be physically mobilized, its produce can, if transportation facilities and the law permit. Thus, as Ohlin observed, 'The mobility of goods to some extent compensates the lack of interregional mobility of the factors; or (what is really the same thing) trade mitigates the disadvantages of the unsuitable geographical distribution of productive facilities'.[19] Such a notion, according to Polanyi, was entirely foreign to the traditional outlook.[20] The goods of everyday life were not supposed to be regularly bought and sold. Surpluses were expected to provide the neighbourhood, especially the local town. They were not meant to be taken hundreds of miles away to feed the population of a growing metropolis with which the peasants have no common affinity. The agglomeration of population in industrial towns in Europe from the second half of the eighteenth century changed this situation completely, first on a national, then on a world scale. Farmers in regions as far away as the tropical and subtropical zones of the world were drawn into the vortex of this change although its origins were obscure to them.

Hopkins described the resulting distortion, and the beginning of income inequalities in the rural areas of West Africa, in the following words:

> Little is known about the elements of luck, skill and necessity which led to the development of this 'kulak' class. Nevertheless, the evidence points to the growth of inequalities among farming communities in the second half of the colonial period. On the Gold Coast, a small group of wealthy farmers, who were responsible for a disproportionately large amount of the cocoa shipped from the colony, had already appeared by 1930. In Nigeria, about a quarter of the farmers in Oyo Province depended primarily on the production of cocoa in the 1930s and no longer grew all the foodstuffs they needed. By the 1950s about 10 per cent of cocoa farmers held 41 per cent of the land under cocoa in Nigeria and over half the total volume of cocoa was grown by a minority of producers, each of whose aggregate holdings exceeded six acres. A parallel trend appeared on the Ivory Coast following the rapid expansion of cocoa farming after the Second World War, though there it was often 'strangers' who accumulated large holdings. Evidence for the regions exporting palm oil and kernels is harder to find because the economic history of these products in the twentieth century has been unjustly neglected, but there are indications at least, in eastern Nigeria, that a similar process of differentiation occurred. The groundnut producing areas have been better documented. In northern Nigeria, inequalities in one rural community have been studied in depth while in Senegal it is clear that a relatively small number of wealthy Mourides managed to perpetuate the dominance they had achieved in the late nineteenth century. [In short], the expansion of export activities speeded the commercialization of land. The twentieth century saw an increase in the amount of land in the export producing regions held virtually as freehold.[21]

This mobilization of land in the interest of capitalist accumulation provoked countervailing social forces in the Europe of the eighteenth and nineteenth centuries to modulate the disastrous effects. Common law in land and agrarian tariffs and protectionism helped to stabilize the European countryside and weaken the total dissolution of rural life. Yet such protective social mechanisms were precisely what colonialism denied the peoples in the dependent territories, so their exposure to the full operation of the capitalist market was to lay the basis for later deterioration of conditions in the rural areas in different parts of the continent.

The penetration of the capitalist mode of production into peripheral rural areas of underdeveloped countries required the creation of a new and more appropriate spatial order in those countries. Fundamental to such an order is the need to facilitate the enforcement of law and order and to establish stable conditions for capitalist accumulation. For instance, Prescott describes the process of spatially organizing the territory of northern Nigeria for colonial exploitation. According to him, two stages can be recognized: multiplication and integration.

> The stage of multiplication was concerned with the pacification of the territory. Boundaries were traced in bold lines on sketch maps and indicated the limits of military jurisdiction, within which the indigenous population was contacted and pacified if necessary. As the area over which the Government exerted direct authority was increased, new provinces were created. Once the Government had succeeded in establishing effective control over the whole territory the second stage commenced. Colonial administrations were encouraged to be thrifty, and one way of reducing expenditure was to have an efficient administrative system with the fewest possible provinces and provincial offices. Accordingly, some of the earlier provinces were amalgamated, and the boundaries were drawn with a view to assisting administrative efficiency and economy.[22]

The efficiency of the new administrative organization was, however, only in terms of the objectives of colonial exploitation. Certainly, the administration paid little regard to the desires and preferences of the local population except where these were too strong to be ignored. Often, however, once the administrative structure had been established some minimum modicum of social services, especially in the field of education and health, was usually provided. Part of the effort at economy implied training some of the local population to take on most of the drudgery of administration. The need to guarantee the well-being of the few colonial administrators, in turn required that attention be paid to environmental sanitation and that some medical facilities be provided.

Once the broad outline of the administrative structure had been established, integration continued and, in fact, came to be influenced by decisions on installing a spatial order for direct capitalist accumulation. The two main elements of this are the transport network and the urban system. Both of these are discussed in greater detail later in this book. Their importance here is that they represent the means of concentrating the rural surplus of colonial territories and drawing the mass of peasants unwittingly into the vortex of global market forces about which they know next to nothing. A colonial administration does this, of course, through identifying the produce of local cultivation which can find a market outside the country. Where no produce fits the bill, new crops are introduced which can serve this purpose. In colonial territories such as those in Africa, this situation immediately created a three-fold category of farm produce where only one existed before:

(a) newly introduced crops grown mainly for export, for example cocoa, coffee and tea;
(b) local crops grown mainly for export, for example groundnuts, palm oil and palm kernels, cotton;
(c) local crops with no export potential, for example yam, millet, cassava.

It is significant that in most countries, the first two categories came to be referred to as cash crops. The ability of land in a particular district to produce such crops immediately changed its economic status and prospects. This was further enhanced by research directed at raising overall productivity. Research stations were established to develop better varieties; extension services were set up to teach the farmers better methods of production; farmers were organized into co-operatives for better production or processing; credit facilities were made easily available; standardization processes were established; channels for effective disposal were instituted and price stabilization policies were formulated. In short, everything was done to stimulate the production of those crops that had export value. By contrast, there was a high degree of indifference or ineffectual attention paid to the food crops which make the third category above. None the less, the implied specialization on export crop production in some areas of the country meant that their demand for food crops must be made up through purchases from the non-exporting regions which were thus indirectly brought into the new global economic system.

The establishment of a colonial spatial order generated, within the traditional society, new tendencies which have had tremendous implications for the integrity of the rural structures. Perhaps the most potent of these tendencies is that concerning population dynamics. Although both birth and death rates have remained relatively high, the effect of the more stable social conditions was an increasing excess of births over deaths. A gradual rise in the population growth rate became noticeable, with not unexpected regional differences. Equally noticeable were the movements of people generally, from those regions with few export opportunities to those with high export potentials. Most of these movements were from one rural area to another and comprised individuals in search of money wages. Although a small proportion went to the new towns of the colonial administrators and traders, the period of massive rural to urban migrations came much later.

Effects on the traditional rural structure

The colonial domination of African countries as part of the process of expansion of global capitalism has had at least six effects on their traditional structures:

The effect on the land tenure system

As indicated above, the direct effect of colonialism on land as a factor of capitalist production was to encourage its individualization and commercialization in such a way that all the legal, administrative and commercial rules already developed in European countries came increasingly to be applied to land in African countries. This was relatively easy in those countries with a large settler population where extensive tracts of land could be set aside specifically for European occupation. It was more difficult in areas of dense indigenous population where no settler population had become established. In these latter areas, various means were used to achieve the same end. First, land was compulsorily acquired by the government for settlement and resettlement schemes, and the land tenure practice was made to conform with an individualized system. On this matter, and with regard to British colonial territories in Africa as a whole, Lord Hailey observed:

> . . . a regime of individual holdings has for long been the ideal of the official Agricultural Departments. They have never failed to impress on the Administration the implications of the classic observation made by Arthur Young when he studied the results of the cultivation of the seemingly barren sand dunes near Dunkirk. 'The magic of property has turned sand into gold.' Hitherto, however, there has been little in the way of direct legislation, though the individual holding has been the practice followed in establishing some of the projects of agricultural settlement.[23]

Second, for the vast area still under indigenous proprietorship,

English common law justice was dispensed in such a way as to emphasize the colonial preference for individual holdings.[24] In the forested areas of West Africa, for instance, it came to be the law that if a person is allowed to use a plot of land for cultivating food crops and inadvertently was not restrained from planting tree crops on the land, the land passed to him in individual ownership. Litigation in land became a common preoccupation and encouraged ruthless individuals to dispossess even their own less educated kinsmen from their rights to family land.

Other factors came to reinforce the trend to the individualization of land in the rural areas. But since this trend has not been institutionalized and made all-pervasive, the land tenure situation in many underdeveloped countries in Africa is characterized by extreme complexity which leaves the majority of the rural population in uncertain bewilderment. Thus, even though the rural areas of underdeveloped countries in Africa were penetrated by the capitalist mode of production, there was little attempt to transfer the rural economy as a whole along capitalist lines. The result was a confused situation in which farmers could neither go forward to full modernization nor retreat to the security of survival offered by the pre-colonial system.

Effect on rural land holding

The increase in population which accompanied the establishment of the colonial spatial order reacted on the inheritance law to change conditions in the rural areas. The improvements in the survival prospects of children meant that family land was having to be cultivated by many more individuals. Fragmentation became the order of the day and was particularly serious in areas of dense agricultural settlement, as among the Ibo of Nigeria. None the less, fragmentation of holdings could also be produced by other than inheritance law although the operation of this law tends to exacerbate its consequence. Thus, de Wilde noted that fragmentation may be due to the pattern of land use, such that a farmer may have three different parcels of land – one around his house, another adjoining the village, and still another in the bush.[25] He may also have a piece of bottom land used for growing a particular crop such as rice. Fragmentation may also result from special clearings undertaken at different times particularly for export crops such as cocoa or coffee. Such fragmentation, noted de Wilde, does not in many cases interfere seriously with farming, but where fragmentation through successive subdivision of land among heirs becomes such as to make rational use of land impossible at the very time when population pressure dictates the need for greater intensification and optimum land use, intervention through consolidation becomes urgent and imperative. Even with consolidation, as the example from Kenya shows, if no deliberate attention is paid to the inheritance law, the cycle of fragmentation may begin all over again.[26]

Moreover, as fragmentation makes it more and more difficult for farmers to practise meaningful cultivation, new solutions are found to their problems of survival. Some take on a number of ancillary occupations such as sale of water, trading or porterage. Others simply mortgage away or sell their various parcels of land, become landless labourers or migrate away from the village.[27] Land fragmentation thus encourages increasing inequality in the access to land in the rural areas and the incipient development of a class structure. Hill, for instance, documents the position in northern Nigeria and emphasizes the increasing differentiation occurring within the farming community.[28]

Effect on land use pattern

The fragmentation of agricultural land is an event that follows the demise of a farmer when his children take over the cultivation of his land. But long before this stage is reached, the farmer himself would have to resolve the crisis of expanding the output from his farm to feed the increased size of his family. Farmers respond to this situation of population pressure in various ways. Some seek out unused land to cultivate, either as a permanent addition to their former holding or on a leasehold or rental basis from their neighbours. Others try to increase the productivity of their farm land, for instance through mixed cropping. But almost invariably all of them are forced progressively to reduce the length of fallow so that the same land can be used more often than previously. Greater frequency of usage in the absence of artificial fertilizers leads to rapid diminution of yields. In the Kikuyu area of Kenya, in the period preceding the Mau Mau emergency and the Swynnerton Plan for land consolidation, District Commissioners' reports were full of references to the influence of fragmentation: 'the fertility of the land decreases, the population increases and the fragmentation never ceases so that the economic return gets smaller to the family each year.'[29] Inevitably for some farmers, this trend ends with land abandonment. In the case of the Kikuyu it led to serious political agitation. More often, however, what it entails is that families caught in this situation and without access to a reserve of land, cannot afford to set aside land for the innovative adoption of new crops and are condemned to depending for their sustenance on traditional food crops. Mbagwu, for instance, notes the difficulties of farmers in Ngwaland to provide land for growing new varieties of oil palm as a means of increasing their income.[30]

Effect on rural indebtedness

The fragmentation of land and declining productivity invariably mean that certain households can no longer meet all their needs from the returns of their labour. The first sign of crisis is the tendency to 'pledge' the product of the farm as security for credit. Because of the vagaries of climate, the purveyor of such credit, usually the village moneylender, invariably insists on a high interest rate. The combined loan and interest rate are such that the net income from succeeding harvests is usually inadequate. A new round of borrowing takes place, leading ultimately to a total alienation of farm land to the moneylender.

Although indebtedness is more usually regarded as a serious problem in the rural areas of South Asia, particularly in India and Indonesia, it is increasingly becoming a fact of rural life in many parts of Africa. In a longitudinal study of the Gambia covering the period 1947 to 1974, Haswell noted the increasing pauperization of a significant proportion of the rural population due to the excessively usurious nature of rural loans and the rise of a small class of relatively wealthy rural merchants and moneylenders.[31] Similarly, Beckett in his study of the cocoa village of Akokoaso in Ghana noted that of 201 families in the village, some 125 were indebted in various degrees. According to him, 'the eradication of this family debt is a problem of major importance and of extreme difficulty, as many families have undoubtedly fallen below the point at which they can by their efforts liquidate their debt. Some might do so at the expense of their status as independent farmers, by relinquishing their farms to the mortgagor and becoming share-croppers or labourers on their former farms.'[32]

In short, in many underdeveloped countries, an important feature of the current situation in their rural areas is the amount of land being appropriated by moneylenders from the impoverished villagers. Sometimes, when the moneylenders are strangers to the particular area, local reaction can take the form of organized violence to drive them away. More often, farmers, on losing their land, stay on to sell their labour to their creditor for what it will fetch.

Effect on rural settlement

The breakdown in the joint family relation to land and production has led to a weakening in the need to stay together in those parts of Africa where nucleated rural settlement used to be the rule rather than the exception. Particularly in West Africa, the establishment of a stable condition of law and order under colonial administration also rendered nucleation less important than it used to be, when it also served some defensive purposes. So one consequence of colonial domination has been the disintegration of rural settlements, with farmers moving out to settle right on or close by their farmland.[33] This dispersal of rural settlement was not always recognized by the colonial administration as concomitant with processes which they had set in motion, especially as it tended to aggravate the problems of effective administration. An example of this difficulty was reported by one administrator in the Lagos Colony of Nigeria in 1900, who observed that the people of Irele (or Ikale) had scattered among their farms deserting Irele, formerly a large town.[34] 'Major Ewart and I have endeavoured to persuade them to rebuild their town, they did so and straightway left it again. I can understand their liking to be near their work, but it makes it very difficult to discuss any question with them.'[35]

A similar experience is reported from East Africa from the land of the Lugbara in the West Nile District of Uganda. Here, Middleton and Greenland noted that the Lugbara traditionally live in villages with each homestead having a homefield near it, and an outside field beyond the village.[36] As a result of increased population pressure, however, members of family groups moved out to occupy the outside field and open their own homefields. The result is that today a large part of the Lugbara area shows a succession of homesteads, each with its homefield but without much of an outside field. Also, the staple millet mixture which traditionally was grown in the outside field is now grown on the homefields only.

Effect on rural labour

The disintegration of rural settlements is often the physical manifestation of the collapse of the control traditionally exercised by the extended family. This collapse means that while it still may be possible to muster labour for peak agricultural activities on a traditional basis, the sanctions against non-compliance have weakened considerably. In Hausaland, for instance, the tendency has been noted for traditional farm labour organization (*gandu*), particularly the paternal type, to collapse with the death of the father. The *gandu* is a voluntary, mutually advantageous agreement between father and married son, under which the son works in a subordinate capacity on his father's farm in return for a great variety of benefits. Sometimes, a *gandu* can comprise mainly cousins related through a common grandfather or even great-grandfather. In either case, fragmentation of land holding and the increasing scarcity of land has been a major factor in the collapse of such organizations. As Hill observed, 'the hiatus which occurs on the death of a *gandu*-head is a period of disarray, rather than of reorganization according to accepted procedures. Sons often exert their newly found freedom to sell their father's farmland, perhaps to finance their migration as farmers.'[37]

The collapse of traditional labour organization thus means the creation of a reserve of labour which, through migration, has been able to provide additional hired hands, especially in areas of export production. In the areas from which such labour originates, however, there is a real shortage which is not fully made good by the labour of the women and old men left behind, who often cannot cope with timely land preparation and bush clearance. So even in such areas, able-bodied men sometimes have to resort to wage labour. As long as local competition for such labour was not great, rural farmers could depend on it. But policies in the urban areas and in the social fields gradually came to make this situation untenable in the long term, thereby hastening the total collapse of the traditional rural structures.

Post-colonial aggravation of rural conditions

Two major elements in the post-colonial development strategy followed by many underdeveloped countries were to have the unintended effect of aggravating problems in the rural areas and accelerating the collapse of traditional rural structures. The first of these was the emphasis on social development. It was generally advocated that one of the obligations of a government in a politically independent country was to rescue the mass of its population from the shackles of disease, ignorance and poverty. To deal with disease and ignorance many African countries embarked on massive programmes to increase and improve available health and educational facilities. The result was the second element, an acceleration in the rate of population growth in most of these countries.

This new development had an immediate impact on the situation in the rural areas. The rapid growth of population meant an intensification of the process of land fragmentation with all of the consequences outlined above. The type of education being offered to rural children made things even worse. Since this was essentially oriented to fit them for urban employment, there was introduced into the rural areas an active disdain of and disaffection with the rural way of life. This situation was already manifest in India in the 1950s but became strikingly noticeable in many African countries only in the 1960s. In both cases the result has been a massive out-migration of youth, usually those with primary education, from rural areas to urban centres, particularly to the metropolitan areas of their countries. While these massive movements have not reduced the number of mouths a country has to feed, they have certainly diminished the number of hands available to produce the food, especially when technological changes are not compensating for labour loss. A common feature of rural production in some underdeveloped countries today is thus the acute shortage of farm labour and the higher age of the farming population. In a study of peasant agriculture in the cocoa belt of western Nigeria, for instance, Olayemi noted that 78 per cent of the farmers interviewed claim they now experience some non-seasonal labour shortage.[38]

Within the context of development plans, a fashionable strategy for dealing with the problem of poverty was import substitution industrialization. For the rural areas, this strategy simply deepened the alienation from the urban centres. Since most of the raw materials utilized were themselves often imported, there was little productive relation between activities in rural and urban areas. Hence, there was hardly any means of making growth in productivity in urban centres stimulate corresponding growth in the rural areas.[39] Rather, the effect of such growth was immediately transferred abroad in terms of demand for more imports. Consequently, in at least three ways import substitution industrialization aggravated the already parlous conditions in the rural areas of underdeveloped countries.

First, in its effect on prices. For the industries to survive foreign competition most governments had to erect quite a high customs tariff against foreign goods and, because of the recency of the industrial culture, there is a serious problem of quality control, so locally produced manufactured goods tend to be priced above their real value. Invariably, urban wages are adjusted to reflect these artificial prices but there is no corresponding mechanism for adjusting the prices paid to farmers to the level of that for manufactured goods. Thus, the tendency has been for farmers to pay high prices for manufactured goods from the city and to receive relatively low prices for agricultural produce sold to the city.

Second, the urban wages which are set to reflect the highly inflated prices of manufactured products become a source of distraction for the rural labour market. Unskilled and skilled farm hands are lured away from the farms by expectations of urban employment.[40] Even where the prospects of securing urban jobs are not bright, the effect is to encourage a demand for rural wages far above what is economic in the overall farm budget. The result is to aggravate the labour shortage situation.

Third, given these two situations, it is clear that the terms of trade between rural and urban areas in most underdeveloped countries are generally against the rural areas. In other words, what import substitution industrialization has done is to accelerate the rate of capital transfer from the rural areas and, in the process, to intensify the pauperization of the rural populace.

Conclusion

The preceding description of the collapse of traditional rural structures may appear to paint too gloomy a picture of the situation in the rural areas of underdeveloped countries. It may also give the impression of uniformity of conditions between the various countries, or even between regions within the same country. Such an impression must, however, be kept in perspective. As mentioned earlier, rural structures vary considerably within and between underdeveloped countries. What is remarkable, however, is the similarity in the processes – the colonial exploitation, population growth and outward-oriented industrialization to which they have all been exposed in the last century or so, and the countries' almost identical responses – which are making discussion of rural problems throughout the underdeveloped world sound so familiar and repetitive. Even in areas where, because of abundant land resources, these responses are not as yet starkly evident, the tendencies are there for anyone to perceive.

Furthermore, the nature of the rural problem in the process of development may be misconstrued if the word 'collapse' is interpreted only in the sense of a breakdown of structures. A more useful interpretation is in terms of disorganization and increasing inappropriateness of the structures. This is why in most underdeveloped countries, and particularly among a large proportion of their small peasant farmers, knowledge and practice of modern farming has made so little headway. The overall result is well represented in Table 5. This table indicates that total agricultural production showed a remarkable increase in the period 1948-58, immediately after the Second World War, which was also the period of very rapid population growth. The fact that such increases were not marked by any significant rise in per caput production, implies that much of the growth in total volume was not the result of rising productivity. Since that time, the position in the underdeveloped world as a whole has been one of steady decline in the rate of both total and per caput production.

This situation constitutes the most serious challenge in the development process of most underdeveloped countries. The issues involved are not just those related to the adequate production of food for the particular country, although this is not unimportant. Rather, they impinge directly on how to improve the access to national resources of the vast majority of the population of a country who live in the rural

Table 5 *Annual rate of growth of total and per caput agricultural production (per cent)*

	1934/38 to 1948/52	*1948/52 to 1958/62*	*1958/62 to 1968/72*	*1934/38 to 1968/72*
Total				
Africa	1.7	3.1	2.5	2.4
Far East	0.3	3.3	2.5	1.9
Latin America	1.2	3.5	2.5	2.2
Middle East	1.0	4.0	3.1	2.5
Underdeveloped countries	0.9	3.4	2.7	2.1
Per caput				
Africa	–0.2	0.8	–*	0.1
Far East	–1.0	1.2	–0.2	–0.1
Latin America	–1.0	0.7	–0.4	–0.3
Middle East	–*	1.4	0.3	0.5
Underdeveloped countries	–0.9	1.1	0.1	–*

Source: Paul Bairoch, *The Economic Development of the Third World Since 1900*, Cynthia Postan (trans.) (London, 1975), p. 18.

* Less than 0.5 per cent

Table 6 *Share of agriculture in selected African development plans (per cent)*

Country	*Plan period*	*Share of total investment*	*Agriculture, public investment*	*Planned annual increase of GDP*	*Agricultural production*
Cameroon	1966–71	19	18	5.8	3.3
Ethiopia	1963–67	21	15	4.3	2.3
Ivory Coast	1967–70	–	30	7.7	–*
Kenya	1966–70	4	26	5.2	4.8
Mauritania	1963–66	9	17	9.2	2.0
Nigeria	1962–68	–	14	4.0	–
Senegal	1966–69	20	42	6.0	5.4
Sudan	1962–71	21	27	5.2	4.0
Tanzania	1965–69	15	28	6.7	4.0
Togo	1966–70	23	26	5.6	3.5
Uganda	1966–70	13	27	6.1	5.2
Zambia	1966–70	10	15	–	4.7

Source: United Nations Economic Commission for Africa, *A Survey of Economic Conditions in Africa 1967* (New York, 1969), p. 56.

* Annual rate of growth of 2.8 per cent for traditional agriculture, 5.3 per cent for industrial agriculture and 20 per cent for forestry.

areas, and to ensure their participation in those areas of decision-making which directly affect their lives.

These issues are not new. They have always been recognized as the basic ones which a government must tackle if it is to achieve any development in agricultural production. The development plans of virtually all underdeveloped countries are full of declarations of government intent to increase agricultural output and improve the lot of the rural population. Indeed, as Table 6 shows for a select number of African countries, during the 1960s most of the governments planned to invest between 14 and 42 per cent of their capital fund in agriculture, in the expectation of a growth rate of between 2.0 and 5.4 per cent per annum. Yet, as Lewis observed, most of these plans recognize that 'it is difficult to raise the productivity of small farmers . . . but the truth is rather that the appropriate effort has seldom been made. Few underdeveloped countries have come near to setting the scene as it needs to be set.'[41] In the next chapter, therefore, emphasis will be placed on outlining what sort of scene needs to be set and what conditions need to be created to ensure cumulative development in the rural areas of underdeveloped countries. The scene needs to be not just for improvement in agricultural production by itself, but for raising the overall quality of life of the rural population, hence the emphasis on rural development rather than simply on growth in the output from agricultural activities.

5 The nature of rural development

Rural development is concerned with the improvement of the living standards of the low-income population living in rural areas on a self-sustaining basis, through transforming the socio-spatial structures of their productive activities. It should be distinguished from agricultural development, which it entails and transcends, for that is concerned with only one aspect of their productive life. In essence, rural development implies a broad-based reorganization and mobilization of the rural masses so as to enhance their capacity to cope effectively with the daily tasks of their lives and with changes consequent upon this. Since land is basic to the viability of rural life, it is the contention of this study that comprehensive spatial reorganization is central to the attainment of this objective and that much of the failure of past attempts at rural development has been due to the relatively scant attention paid to the spatial dimension of rural development planning.

After many years of inadequate emphasis, during which pride of place was given to industrialization, it is now becoming generally recognized that rural development represents perhaps the only logical way of stimulating overall development. For one thing, industrial production, because of its current capital intensiveness and its dependence largely on imported inputs, tends to cater for a small market of relatively high-income domestic urban consumers. For this market to expand so as to give industrial activities the opportunity to grow and further transform the whole economy, the vast subsistence rural sector must be developed. Moreover, there is the welfare aspect in which rural development is the most effective means of improving the well-being of the vast majority of a country's population.

A critique of rural development projects

Even before the wide recognition of the primacy of rural development, governments in many underdeveloped countries launched various programmes and projects aimed at rural development. Uma Lele offers a five-fold classification of these programmes.[1]

(a) *Commodity programmes* These are concerned with the relatively straightforward objective of increasing the production of export crops among smallholder farmers. These programmes are usually financed by European commercial companies, development corporations, or the governments of former colonies. Examples of such programmes include the Kenya Tea Development Authority established by the Kenya Government in 1960; the tobacco project of the Tanganyika Agricultural Corporation in the Urambo region of Tanganyika, now managed by the Urambo Farmers Co-operative Society; the Mali Cotton Production Project undertaken in 1952 with the assistance of the Compagnie Française pour le Développement des Fibres et Textiles.

(b) *Functional programmes* These are undertaken to remove a single constraint that is considered to be particularly critical for getting rural development under way. Investment may be directed to the development of a national network of adaptive agricultural research, to training extension services, to the construction of feeder roads, to the provision of agricultural credit or to improving the agricultural marketing network. Examples of such programmes include the agricultural credit administered by Kenya's Agricultural Finance Corporation.

(c) *Subsectoral development programmes* These involve the provision of a number of services related to the development of a specific subsector or region. The Kenya Livestock Development Project involving the development of commercial (company, individual and group) ranches in Kenya's Central, Eastern and Rift Valley Provinces, is an example. So is the Minimum Package Programme of the Ethiopian Government, begun in 1971 on the initiative of Swedish technical experts, involving the provision of services considered critical for rural development such as agricultural extension, credit, co-operative development and feeder roads.

(d) *Rehabilitated spontaneous efforts* These are programmes on squatter settlements established by disadvantaged social groups of unemployed or landless, usually on land to which they have no title, which eventually compels a degree of government attention for the establishment of schools and health clinics. Examples of such squatter development are the Kibwezi-Mtito-Chylulu Hill Zone settlement in south-east Kenya and the Muka Maku Co-operative Society in Central Kenya.

(e) *Regional rural development programmes* These are undertaken with substantial initiative and participation by national governments.

A subcategory of this comprises the integrated rural development programmes undertaken mainly on the initiative of donor agencies, planned and administered by expatriates. Of the first type are the Rural Development Programme in Kenya and the Ujamaa Programme in Tanzania. The former was conceived in 1966 and involved the initiation of a variety of rural development schemes such as the master farmer project, the development of hybrid maize and cotton, a cattle development programme, youth training, road construction and home economics extension. The programme in Tanzania was an attempt to bring about the devolution of administrative power by fostering participation of the peasant masses in the planning and implementation of rural development programmes. Programmes in this subcategory include the Chilalo and Wolamo Agricultural Development Projects both in Ethiopia, the Lilongwe Land Development Programme in Malawi and the Zones d'Action Prioritaires Intégrées and the Société de Développement du Nkam settlement scheme, both in Cameroon.

This five-fold classification covers the types of agricultural and rural development schemes found to a significant extent in the underdeveloped regions of the world. Table 7 shows a select list of projects representative of these approaches to rural development in Africa. It emphasizes three important aspects of current efforts, namely, their high capitalization, the important role of foreign donors and the limited number of people involved in each project. Indeed, as indicated in the footnote, the population figures refer to persons living within the project area, many of whom are not directly involved in the activities of the project, but it must be assumed that they none the less 'benefit' somehow from the projects' interventions.

Moreover, although not so obvious from the table, most of these programmes and many others like them elsewhere in Africa, place more emphasis on enhanced agricultural exports than on food crop production. The result, as already indicated, is that while farmers, including those within specific projects, learn a lot about the cultivation and husbandry of the former, their techniques and organization for producing the latter have hardly changed. The establishment of research centres such as the International Institute for Tropical Agriculture in Ibadan, Nigeria, as well as earlier ones in Manila, the Philippines and Los Baños, Mexico, which concentrate specifically on food crops and farming systems, represents a major shift in a desirable direction. The interest in farming systems is particularly noteworthy since, with regard to Africa, this has encouraged a re-evaluation of the

Table 7 *Select list of rural development projects in Africa*

Country	*Project*	*First year of implementation*	*Project cost*	*Financing agencies*	*Project population**
Type A: Commodity programmes					
Kenya	Special Crops Development Authority (SCDA): Kenya Tea Development Authority (KTDA)	1960–4 1964	$15.6 million (to 1971)	IDA, CDC FDR and Government of Kenya	66 500 farmers (1972)
Mali	Compagnie Française pour le Développement des Fibres et Textiles (CFDT)	1952	$9.8 million (to 1971/72)	EDF FAC	Population of 1 million within cotton zone (1972)
Mali	Bureau pour le Développement de Production Agricole (BDPA) – Opération Arachide	1967	$19.7 million (to 1971/72)	FAC World Bank	0.75 million within the operation (1975)
Tanzania	Urambo Settlement Scheme, Tumbi Settlement Scheme, World Bank Flue-Cured Tobacco Project	1951 1954 1970	$0.26 million $0.10 million $14.7 million (by 1978)	TAC BATC World Bank, IDA and Tanzania Government	2400 farmers 4600 farmers 15 000 to 30 000 farmers
Tanzania	Sukumaland Cotton Development	1950	Not available	Government of Tanzania and IDA	Population of Sukumaland 2.5 million
Type B: Functional programmes					
Kenya	Agricultural Finance Corporation (AFC)	1963	$33 million (total asset 1971)	IDA, SIDA, FDR, BLTP and Kenya Government	Covers the entire country

Country	Project	First year of implementation	Project cost	Financing agencies	Project population*
Type C: Subsectoral programmes					
Ethiopia	Minimum Package Programme (MPP)	1970	$30 million (to 1976)	IDA, SIDA, FAO and Ethiopian Government	Approx. 5 million people living in MPP areas in 1974
Kenya	Kenya Livestock Development Project (KLDP)	1970	$11.3 million (to 1974)	IDA, SIDA, USAID and Government of Kenya	2500 – 3000 ranchers or pastoralists
Nigeria	Small-scale Rural Industries: Industrial Development Centre (IDC) and Small Industry Credit Schemes (SIC)	IDC, 1962 SIC, 1966	Not available	USAID with Nigerian Government	IDC, northern and eastern regions SIC, northern region
Type D: Rehabilitated spontaneous programmes					
Kenya	Spontaneous Land Settlement	Not available	Not available	Not available	Approx. 300 000 (1969)
Type E: Regional rural development programmes					
Cameroon	Zones d'Action Prioritaires Intégrées (ZAPI)	1967	$3.18 million (1972)	FAC and Cameroon Government	Approx. 175 000 ZAPIs
Cameroon	Société de Développement du Nkam (SODENKAM) Land Settlement Scheme	1966	$0.5 million (1972/73)	Government of Cameroon	18 000 to 22 500 inhabitants by 1981
Ethiopia	Chilalo Agricultural Development Unit (CADU)	1967	$19.3 million (to 1975)	SIDA with Ethiopian Government	Population within Chilalo Awraja is 400 000

Ethiopia	Wolamo Agricultural Development Unit (WADU)	1970	$5.1 million (to 1974)	IDA, UK, WFP and Ethiopian Government	Population within Wolamo Awraja is 240 000
Kenya	Special Rural Development Programme (SRDP)	1971	$1.6 million (in 1972/73)	USAID, SIDA, UK, NORAD, Netherlands and Kenya Government	Six districts with population 826 000 (1971)
Malawi	Lilongwe Land Development Programme (LLDP)	1967	$14.6 million (to 1975)	IDA and Malawi Government	Population of 550 000 within district

Source: Uma Lele, *The Design of Rural Development* (London 1975), pp. 8–11.

* Except for Sodenkam, KLDP, KTDA and Smallholder Tobacco Development, the population figures refer to the number of persons living within the project area, many or most of whom are not direct participants in programme activities.

IDA: International Development Association
CDC: Commonwealth Development Corporation
FDR: Federal Republic of Germany
EDF: European Development Fund
FAC: Fonds d'Aide et de Coopération (France)
TAC: Tanganyika Agriculture Corporation
BATC: British American Tobacco Company
SIDA: Swedish International Development Agency
BLTP: British Land Transfer Programme
FAO: Food and Agriculture Organization
USAID: United States Agency for International Development
WFP: World Food Programme
NORAD: Norwegian Agency for International Development

traditional bush-fallow and mixed-cropping system which is prevalent in much of the continent. It is now recognized, for instance, that in terms of energy input-output ratios, subsistence agriculture is many times more efficient than a modern American farm with its heavy consumption of fossil fuel.[2] However, energy efficiency is only one criterion for judging farming systems; productive efficiency is another. And there is no gainsaying the fact that modern agriculture (with its emphasis on appropriate organization and technology) produces more than enough food to feed a large and growing population.

The point to emphasize with regard to agriculture, therefore, is that it is simply one of the variables in the process of rural development. The type of agriculture pursued by a country, particularly with regard to its level of technological sophistication and energy efficiency, will be dependent on a number of contingent factors, not least of which is the overall level of development of the country and the size of population that now has to be fed from the productivity of its agriculture. More than this, there is the very important issue of the social relations which a country wishes to see established in the process of agricultural development. It is quite possible to ensure a high level of agricultural production in a country – allow a few individuals to take control of most of the land resources, give them easy access to capital, and treat the majority of the rural population as a reservoir of cheap labour. On the other hand, if the idea is to increase the standard of living of the masses, then the improvement in agricultural production is taken as given, and the emphasis shifts to the conditions and circumstances under which the majority of farmers have to conduct their productive activities.

It is from this perspective that various criticisms have been launched against the present set of rural development programmes. The issue here is not whether these programmes have succeeded or failed in achieving their objectives, but whether those objectives are themselves the most appropriate in the circumstances. For example, the commodity programme for tea production undertaken by the Kenya government had, by 1972, brought some 25,000 hectares in ten districts under tea cultivation and produced almost 313,000 quintals of green leaf.[3] The net return from one hectare of tea (1970–71) was approximately $543. This has meant an average per caput annual net cash income from tea production of about $41 compared to £7 in the pre-programme era.

Clearly, a programme which brings about a six-fold increase in per caput income cannot be said to have failed. The same thing is true of the Sukumaland Cotton Development Project in Tanzania where annual per caput cash income increased from less than $2 in 1950 to about $22 in 1966, although this was partly due to the world market price for cotton which rose from $0.09 to $0.21 per kilogramme between 1950 and 1952. Similar income increases have been noted for commodity programmes – cocoa, coffee, rubber, oil palm – for small-holder farmers in many underdeveloped countries. Indeed, such programmes account for a large part of what is regarded as agricultural and rural development in many of these countries. Yet, as Oluwasanmi emphasized with regard to Nigeria:

> Changes in productive techniques and organization are a more accurate measure of agricultural progress than the mere acceptance of production for the market by peasants. While the Nigerian farmer has readily taken to the cultivation of new crops for the export market, he has done so within the framework of the traditional system of farming and in almost complete ignorance of modern techniques of cultivating and processing these crops. To the extent that the Nigerian farmer continues to use primitive techniques in the cultivation and processing of his export crops, to that extent is his response to the new economic situation incomplete and inadequate. This phenomenon of primitive techniques of production within a social and economic environment otherwise regarded as 'revolutionary' or 'dynamic' constitutes in Nigeria the very essence of backwardness.[4]

With regard to functional programmes, again successes are as notable as distortions. This is perhaps most obvious with regard to credit provision. The function of rural credit is not merely one of meeting exigent credit needs; its aim should be to finance those techniques and organization of production which improve rural income, savings and investment. The purchase of an animal or tractor-drawn plough, the hire of spraying equipment or of seasonal labour, are a few examples of the dynamic role of credit. Yet, in spite of repeated efforts in many countries, normal banking institutions have had difficulty in extending credit to those small-holder farmers who have no legal title to the land they cultivate. In Kenya, for instance, the position improved considerably once land was consolidated and registered and farmers could obtain mortgages on the title to their land.[5] Yet, even there, most banks who have lent considerable sums to African farmers on mortgages emphasize that procedures for realizing on such security are time-consuming and costly, and that in the last analysis it is difficult, if not impossible, to sell land when a court judgment has been obtained, owing to widespread popular antagonism to the foreclosure of mortgages.

Yet, since the need for credit is periodical on many of these farms, failure to pay back is usually the symptom of a more serious problem So even when the government sets up agricultural finance agencies to replace or supplement the efforts of commercial banks, they too run into the same difficulty with small-holder farmers. This has the result that they tend to concentrate on the bigger farmers thereby setting in motion processes causing heightened inequalities in the rural areas. The Kenya Agricultural Finance Corporation project started in 1965, for instance, with the aim of assisting the small-holder farmers with a per caput cash income of less than $50. Yet by 1972, it allowed only about 30 per cent of its total disbursement to this category of farmers.[6] Similarly, the credit programme of the Chilalo Agricultural Development Unit was directed at those tenants and land owners with less than 25 hectares of land. Yet, when credit statistics were analysed, it turned out that in 1968-69, 32 per cent of the funds had gone to farmers holding more than 40 hectares, a percentage which rose to 34.5 in 1969-70 but fell to 2.0 in the following year when a check led to a restatement of policy to concentrate on the target population.[7] Similar distortions have been reported from various Asian and Latin American countries where the implementation of credit programmes has often worked in favour of larger land owners and against small-holder farmers.[8]

Even with regard to adaptive agricultural research, there have been many instances where the overall effect has been against the majority of small-holder farmers and in favour of a minority of big farmers. This is particularly so where national agricultural research and development policies focus on capital-intensive production methodologies and new varieties of seed, especially when based on large-scale irrigation or mechanization.[9] Given local resource availability, such innovations can often be inappropriate, especially for small farms. The case of many high yielding crops which need more water and other inputs has often been mentioned. In Mexico, Pakistan, the Philippines and India, such new varieties have been most widely adopted in the irrigated and well-watered regions and by the larger farmers who are already prosperous. Small farmers without adequate credit or training have been left out.[10]

The subsectoral development programmes also have the characteristic that, while they can succeed in terms of agricultural productivity, they often run the risk of further distorting social relations and encouraging increased inequalities. By contrast, the governmental rehabilitation of spontaneous efforts by disadvantaged social groups is generally in the right direction, but is too limited in scope and paternalistic in operation to be of lasting benefit.

Regional rural development programmes, as currently conceived, are often no more than the juxtaposition within one limited area of the other programmes, with the emphasis still largely on immediate increased productivity. The result is a substantial level of capital investment, usually on a scale which cannot be replicated in other parts of the country.

In reviewing the success that has attended these various types of development projects over the last two decades, Lele identified various factors which have made most of them 'less than fully effective in making the process of development of the low-income sector self-sustaining'. To quote her:

> Their limited effectiveness cannot be attributed to the inadequate or inappropriate specification of target groups but rather to a combination of factors. First, the objectives of rural development have changed considerably over time. Many of the projects reviewed were designed with what now appear to be limited objectives, as for example, increasing export crop production among smallholders. [Second], the projects were also based on more limited knowledge than is now available of broad sector and policy questions and of their possible impact on the performance of the individual programs. [Third], frequently, despite the fact that the likely impact of domestic policies and institutions were anticipated, for a variety of reasons analyzed in this study, national policies could not be changed to improve project performance. [Fourth] the programs were often based on inadequate knowledge of technological possibilities and of their suitability to small-farm conditions. [Fifth], experience with regard to the appropriate forms of administrative institutions and their transferability was limited when many of these programs were planned . . . [Sixth], they also suffered from poor knowledge of the sociocultural and institutional environment in which they were to be implemented. Consequently, the programs were rarely designed with a view to anticipating the effect of sociopolitical factors on the response to interventions or with an intention to introduce modifications in plans in the course of implementation to achieve maximum effectiveness. *Finally* and most importantly the programs often experienced extreme scarcity of trained local manpower.[11]

These criticisms of rural development projects to date is predicated on the assumption that they could be improved and made more effective as a vehicle for stimulating development in rural areas. A more fundamental criticism, however, at present coming mainly from Latin America, questions the whole notion of conceiving such projects as having any 'developmental' rationale. According to this view, such projects are seen as part of a new technocratic ideology or *proyestismo* which assumes that production in the inefficient antiquated agrarian sector in underdeveloped countries can be stimulated simply through modernization efforts involving massive injections of advanced technology and outside capital, without any fundamental change in the existing agrarian structure.[12] The ideology of *proyestismo*, literally translated as 'project fever', is further seen as a strategy for modernization which is truly western in origin and which is based on the prevailing technocracy cult in the developed world. Although this viewpoint recognizes that many of these projects are genuinely undertaken to improve the welfare of the rural masses, ironically, as pointed out above, their overall effect has often been to maintain the *status quo* rather than promote real social change.

The position is perhaps better appreciated when one takes into account the proportion of the rural population in underdeveloped countries living in poverty and we can redefine the problem of rural development in terms of how speedily their life chances can be improved. Table 8 shows the position with regard to the major regions of the underdeveloped world. It emphasizes that while most rural development projects calculate their target population in thousands, the enormity of the development task must be appreciated in millions. The table, moreover, has defined poverty in terms of people with an income below $75 per caput. Except for America the proportion of the rural population so defined is more than half. If the threshold value were to be raised to $100 per caput, it is likely that the proportion could rise to between two-thirds and three-quarters of the total rural population.

It is this fact that challenges the rationale and long term advantage of the present 'project by project' strategy of rural development. Clearly, for rural development to be effective, long lasting and self-sustaining, it must be comprehensive in its scope and must address itself without equivocation to the task of redefining the social relations embedded in land as the primary asset of the rural population. Such redefinition basically involves the spatial reorganization of the rural areas of a country. Although recognizing the variety of existing conditions in different countries, the following is an attempt to give a broad description of what such spatial reorganization would entail.

The nature of rural spatial reorganization

Spatial reorganization involves the rearrangement of all spatial

Table 8 *Extent of rural poverty in underdeveloped countries*

Region (1)	*Total population 1969 (2)*	*Rural population 1969 (3)*	*Population with income below $75 per caput (4)*	*Population with income below 1/3 national average per caput income or below $50 per caput (5)*	*Rural population in poverty as per cent of rural population*	
					(4) as per cent of (3)	*(5) as per cent of (3)*
	—— figures in millions ——					
Underdeveloped countries in:						
Africa	360	280	140	115	50	41
America	260	120	30	45	25	38
Asia	1080	855	525	370	61	43
Total	1700	1255	695	530	55	42

Source: World Bank, Sector Policy Paper, *Rural Development* (February, 1975), p. 80.

elements in the rural areas so as to improve the physical and social access of farmers to all national resources. Although land reform is central to its achievement, it entails far more than is usually denoted by that programme. Its objective is to rescue the farming population from the contradictions of operating in a modern industrialized economy with the means, institutions, organizations and rules of a traditional pre-industrial society. What such rearrangement entails is two interrelated operations. The first concerns the rural settlement and its community of farmers; the other the land that goes with these settlements.

Land reform

It is perhaps necessary to deal first with the land and then to consider the appropriate settlement and community system that goes with it. In actual operation, this may or may not be the order in which changes take place. None the less, the spatial reorganization of land must aim, first, at ensuring that rural lands enter into the stream of current societal assets and can be used and treated as capital in all senses of that word; second, at doing away with small, scattered holdings of farmers so as to facilitate the introduction and adoption of more appropriate systems of agricultural production and management; and, third, at preventing small-holder cultivators from being easily dispossessed by the emerging class of rural capitalists. This three-fold process of valorization, consolidation and conservation is critical for rural development and thus needs to be further elaborated.

The problem of valorization, that is, of giving new and added value to land, arises from the economic disadvantage in which the contradiction between traditional and modern conceptions of what is an asset places the majority of the rural population. This problem has already been indicated with regard to the difficulties of small-holder farmers in securing credit as often as they need it, in so far as they have no secure title or tenure to the land they cultivate. The consequence of this, as already discussed, is that each time governments establish credit institutions to ease the provision of credit, it is the wealthier, sometimes absentee, farmers who have the necessary security, that receive the most. So a major task in the spatial reorganization of most developing countries is a massive cadastral survey of property and the establishment of institutions for registering and offering title deeds to the individual or family owners.

The necessity for consolidation is perhaps sufficiently obvious. In an age when efficiency in operation is part of the new societal objective, the loss of valuable working time in moving between scattered holdings makes consolidation a necessity. Moreover, the likely adoption of mechanization at one stage or another in the development process and to a lesser or greater degree in the farming operation, makes consolidation critical for progressive agriculture in these countries. Land consolidation, however, raises two other problems of the spatial reorganization process. These relate to the question of individualization of holdings and the appropriate form of inheritance law. The problem of whether or not to consolidate land into individual farms depends on the type of social relations which a country wants to establish in its path to development. This is considered in greater detail later. The inheritance laws are considered as an aspect of conservation.

Conservation requires that over the long term, land is kept in good heart. The more conventional concept is usually defined in terms of ecological balance but often ignores the social situation which makes this possible. For instance, it has been emphasized in the previous chapter that current inheritance laws in many African countries require that land be shared out among a man's surviving sons or children. Furthermore, that as population has been increasing rather rapidly, this subdivision has reached a stage where it is now difficult for many of the inheritors to observe the traditional length of fallow needed for the land to recuperate. The progressive shortening of the fallow period increases the exposure of the soil to climatic forces, encourages the incidence of sheet and gully erosion and leads eventually to the destruction of valuable farming land. One way of checking this chain reaction could be by redefining the inheritance laws in such a way as to discourage fragmentation. This again raises issues of social relations which are dealt with later.

Land valorization, consolidation and conservation define the scope of land reform in many countries. In a number of cases where countries have embarked on such reform, there is evidence that after an initial period of adjustment to the transformation, which is marked by a slight decline in production, marked increases both in total production and in productivity become evident. Table 9 compares agricultural production in Iran for periods before and after the Land Reform Act of 1962. It shows that by the fifth year after the reform, the production of wheat, rice, cotton and sugar beet had increased by over 35, 60, 17 and 250 per cent respectively compared with the last year before the reform. As Ajami noted, the average annual rate of growth in the agricultural sector of the country has been around 3.8 per cent for the decade 1960–70.[13] Similarly, it has been observed that in countries such as Bolivia, Chile, China, Mexico, South Korea, Taiwan, Venezuela and former North Vietnam, land reform programmes have substantially improved income, security and equity of small-holder farmers.[14] At the same time, it must be mentioned that in some other countries, notably Argentina, Bangladesh, Brazil, Colombia, Ecuador, Guatemala, India, Indonesia, Iraq, Ivory Coast, Pakistan, Peru, Philippines and former South Vietnam, land reform programmes have not had the effect of substantially raising peasant incomes. Griffin suggests that one reason why land reform has not worked so well in the latter group of countries is that the necessary infrastructure, credit, advice and other supports were not made available to the new small landowners. None the less, he argues that land reform is necessary in an underdeveloped country, not only on the grounds of equity but also of efficiency. According to him:

Factor markets in underdeveloped countries are highly imperfect. . . . The imperfections are such that large landowners face a very different set of

Table 9 *Comparison of agricultural production in Iran for periods before and after land reform*

Year	*Estimated value of total agricultural sector production (million rials* at fixed price)*	*Wheat*	*Rice*	*Cotton*	*Sugar beet*
			(thousand tons)		
Before land reform					
1959	85 119	2900	810	265	706
1960	86 744	2924	709	328	707
1961	86 984	2803	576	348	810
Average 1959–61	86 282	2875	698	313	741
After land reform					
1962	88 315	2700	700	276	860
1963	89 893	3000	860	346	1191
1964	92 159	2600	800	363	1028
1965	99 020	3000	845	420	1411
1966	102 750	3190	875	339	2280
1967	110 853	3800	930	405	2857
Average 1962–67	97 165	3048	835	348	1638

Source: Central Bank of Iran, *National Income of Iran, 1962–67,* Teheran, 1969 (in Persian).
* 76 rials = 1 US dollar

relative prices from those confronted by small-holders and tenants. Land and capital are abundant and cheap relative to labour for the big farmers, whereas the reverse is true for the small. Thus, the big farmers tend to adopt techniques of production with relatively high land–labour ratios and small cultivators and tenants adopt very labour intensive techniques. These differences in techniques are a reflection of allocative inefficiency, namely, a failure of the large farmers to economise on the scarce factor of production – land, and to use intensively the plentiful factor – labour. Moreover, it has been shown, from Guatemala to India that the misallocation of resources is due to the fact that different types or classes of farmers face different relative factor prices.[15]

Rural settlements and communities

If, for both equity and efficiency reasons, the emphasis of rural development is on the fuller utilization of rural labour, then attention must turn to the form in which this factor of production is distributed throughout the rural areas, that is, to the community and rural settlement structure. Rural settlements represent not only an aggregation of the farming population but also the physical manifestation of both the social relations of land and the ecological, technological and organizational basis of its utilization. In general, one recognizes two types of rural settlement: dispersed and nucleated. It is not easy to be categorical as to why one form or the other tends to be prevalent among certain communities. Part of the difficulty lies with the historical persistence of a specific settlement type long after the factors of their formation have changed. None the less, it is assumed that dispersed isolated farmsteads are the characteristic settlement type among communities with little development of either central administration or trade.[16] Moreover, as soon as some exchange relations begin to develop, the need for some management of this interaction encourages agglomeration into hamlets or villages. Defence may also come to be a factor of agglomeration. Whatever the factor, the juxtaposition of people soon creates conditions of differentiation, especially in terms of access to land and to social position.

In recent times, the introduction of elements of modern technology has been a factor in changing the patterns of rural settlements in many parts of the underdeveloped world. Modern transportation facilities, for instance, have encouraged the evolution of nucleated, usually linear-type rural settlements along rural roads, while the security provided by a colonial administration as well as the increased income from exported agricultural production, have facilitated wider dispersal.[17] None the less, whatever the prevailing form of rural settlement whether nucleated or dispersed, the critical element in spatial reorganization is the second-order unit of rural territorial organization, within which these settlements are incorporated through a network structure that facilitates their access to all kinds of national resources. It is remarkable that in traditional societies such units, generally below the level of district, were always a critical element of administration. In the republic of Sudan, for instance, Barbour noted the *omodia* to be the basic unit of territorial organization above the level of the household, usually comprising ten to forty or more households under the control of an *omda*.[18] The *omda* supervises the heads of households (*sheikhs*) under him in the collection of taxes and the maintenance of order. It is instructive that in their colonization of Britain, the Anglo-Saxons organized the territory on much the same basis, in what were called 'hundreds' (or a hundred peasant households) for the maintenance of peace and order, the adjustment of taxation and the settlement of local pleas.[19] These were later transformed into parishes as the Christian religion came to dominate much of the social life in the period after the Norman Conquest. It is equally instructive to note that, because overall development was never the objective of a colonial administration, the traditional units of rural territorial organization in many underdeveloped countries of Africa were allowed to disappear or fall into disuse, even where they remained of topographical significance. In the Yoruba areas of Nigeria, for instance, rural and urban wards fell into disuse during the colonial period. On the other hand, the nature of the indirect rule system in northern Nigeria which superimposed the British over the earlier Fulani hegemony of Hausaland, enabled the system of village areas and wards (*ungwa*) to survive till today, but generally at a level of realization below formal official recognition.[20] In such circumstances, it is understandable that in many countries current strategies of rural development are not oriented to utilizing these units and their critical network structure and in consequence have been indifferent to their revival or re-establishment.

The unit of rural territorial organization must, however, be distinguished from the administrative structure of a country. Although both can be closely related, it is more common for these units to operate below the district or county level, without being regarded as a formal tier of authority in the administrative system. Where the two are formally integrated, the role of these primary units tends to be confined to the maintenance of law and order and the collection of taxes. Yet, in development terms, these units of territorial organization in traditional societies reflect the nature of prevailing societal preoccupations. Given that these preoccupations have changed in recent years towards increased emphasis on production, social equity and enhanced national integration, this unit must be defined and re-established so as to facilitate the achievement of these goals. Emphasis must be placed not only on the settlements that make up the unit but also on their route linkage with each other and with the higher order settlements which provide them with marketing, storage, social and cultural services.

The unit of rural territorial organization relates closely to the primary or lowest level of the central place system in a country. It may be conceived of as the market or tributary area of the first-order central place within the national system. But its function is more than marketing. In fact, its production organizational role needs to be more strongly stressed in a developmental context. In this regard, it is instructive to note that this is precisely what the Chinese have done with the commune which they define as 'the basic unit of the social structure . . . combining industry, agriculture, trade, education and the military . . . [while being] at the same time the basic organization of social power'.[21]

The unit of rural territorial organization thus comes to determine the layout of farm and market roads, the location of schools, health clinics, recreation centres, market place and storage facilities, repair stores and so on. It also strongly influences the whole system of settlement hierarchy in a country. The form this influence takes depends, for instance, on whether the behavioural assumption underlying it is that of individual or corporate decision-making of the type represented by well-organized co-operatives. This consideration is important since, as the Israeli example amply demonstrates, societal behaviour based on the decisions of well-organized co-operatives needs fewer levels of central places to satisfy their varied demands.[22] Such considerations are matters of choice and can change in the process of a nation's development. But it is important to bear the matter in mind both because it relates to issues of social relations and because it can affect the efficiency of the operation of the settlement system.

Social relations in spatial reorganization

The social relational element in the process of spatial reorganization of a country emphasizes the point made in Chapter 3 of the close interrelation between spatial structure and social systems and the reflexive nature of these relations. For a social system based on capitalist relations, for instance, it is clear that land, like other factors of production, must be treated as a discrete, individualizable commodity which is negotiable and subject to the rules of a self-regulating market. At least at the extreme, such a system requires that no social constraint be placed on the individual, or the market, in terms of how much land he can buy and own as long as he has the money to bid for it. Furthermore, given the finite nature of the land resources (at least the cultivable portion of a country), the system accepts the consequence that a large number of smallholders can be bought out, dispossessed, turned into landless peasants and be without employment.

Where a country accepts such a system of social relations, the rural landscape will reveal large individual farms with a tendency towards isolated farmsteads owned by wealthy and prosperous farmers able to purchase the most sophisticated farm equipment, who can utilize to a high degree of effectiveness the results of modern scientific research. Where such a class of farmers belongs to the same class as the people in government or the urban bureaucracy (and they usually do), the appropriate network of roads, telephone lines and repair centres will be developed for such farms. However, since these large landowners will still need labourers, one feature of such a landscape will be labourers' cottages tucked away somewhere on the farms. Such labourers will appear fortunate compared with the mass of landless or dispossessed peasants who must eke out an existence by squatting on the land of the large landowners under constant threat of eviction and arrest. The picture depicted, of course, calls to mind the situation in many Latin American countries, although here it is necessary to stress the colonial origin of the large landholdings (*latifundia*) which exist side by side with small holdings (*minifundia*). In Chile and Peru, these latter are said to be the result of the progressive subdivision of holdings originally granted to lower rank officers, also during the colonial period.[23] Capitalism strengthened the social relations deriving from this colonial and semi-feudal background and has encouraged the large landholding class to take a firm grip on the apparatus of state. As conditions of the small-holder farmers deteriorate and the number of landless increases, emigration to the urban areas becomes the only option open. Unlike Europe in the nineteenth century, when rapid industrialization and employment creation made it possible for the cities to absorb such labour, the accumulation of urban unemployed becomes the complement of a spatial structure based on a capitalist system of social relations.

In the underdeveloped countries of Asia, the colonial economy has also encouraged the emergence and growth of rural classes of landowners, farm labourers, tenants and squatters of various degrees of legality. The position in Africa is nascent but shows the possibility of growing rapidly. While this system of social relations may or may not serve the objective of efficient production, it can hardly be disputed that it does not contribute much to the goals of social equity and national integration. The rural situation remains one of pent-up instability, with the needs of the landless and dispossessed for a meaningful relation to land as a means of production always constituting a threat to the interests of the large landholding class, who currently monopolizes access to this resource.

From this perspective, and from arguments already presented in Chapter 4, it is clear that for most underdeveloped countries, a realistic path to rural development involves a spatial reorganization based on varying degrees of co-operative relations. The exact nature of co-operation, and the range of activities it can undertake, will have to be determined in each country or region on the basis of national experience, cultural values, agricultural crops and practices. Such relations can extend from land preparation to harvesting, credit-supply, marketing and storage, as well as to other areas of social life. On the other hand, it can be phased in such a way that initially co-operativization is limited to only a few aspects of productive activities. Whatever the appropriate intensity, what is important is to strive to be comprehensive both in terms of involving all farmers within the unit of territorial organization and all units in a country. Such comprehensiveness gives the whole exercise a national visibility and ensures maximum impact on the solution to problems which are likely to arise in the process of this transformation.[24]

It is, of course, important to recall that attempts at developing farmers' co-operatives in many underdeveloped countries in the past have met with failure more often than success. In his study of rural co-operatives in Africa, Apthorpe noted that part of the difficulty arises from the conflict of objectives, between economic profitability on the one hand and social equity and participation on the other.[25] He observed that while quick economic benefits may be feasible for certain groups in certain conditions, and necessary to attract new members and retain old ones, social benefits are generally on a different time scale and more difficult to achieve. Moreover, the voluntary nature of many of these co-operatives without full-hearted commitment to them by governments, has also been among the reasons for failure. None the less, experience from various countries such as Kenya and Malaysia, shows that given a clear and consistent objective, a pre-conditional land reform, an adaptive strategy which recognizes the need for local variations from the basic European model, forthright leadership and a mechanism for constant assessment of performance, co-operatives provide the surest means of rapid rural transformation in countries with a preponderance of small-holder farmers.[26]

Consequently, if co-operativization is accepted as the appropriate basis for new social relations, it has wide-ranging implications, not only for other rural institutions and organizations, but clearly for the types of settlement that result. Villages or at least groups of contiguous hamlets become the more likely form of settlement. The acceptance of this close relation between social process and spatial form should enable a country to exercise more deliberation in the determination of the form, composition, layout and network relation of the new rural settlements.

But perhaps the most important implication of a new spatial form based on co-operativization is that concerning the degree of participation of the rural masses in the control of events directly affecting their lives. Clearly, even the very act of getting them to become members of co-operatives involves engaging them in dialogue as to the goals and the means of achieving this purpose. The cadre of change agents required for the purpose must clearly be trained differently from the present generation of extension workers or co-operative officials. As much as possible such individuals must come from the co-operating community and preferably be selected by the community itself. The question of educational qualifications should not be allowed to confuse the issue and illiteracy must not be made synonymous with unintelligence. In other words, for the step-by-step process of rural transformation it should be possible to train bright and talented young people in the local language and to make the training continuous over a number of years during those periods when their presence in their communities is not crucial.

The emphasis on locally based agents is crucial for comprehensiveness and the involvement of rural communities in the choice of such agents is likely to obviate the paternalistic attitude of officials in the rural areas, and the ease with which they may be co-opted to serve the interest of the large landowners. Moreover, familiarity with the agent should make it possible to stimulate real dialogue and mutual exchange of ideas and viewpoints. It should also reduce the likelihood of an agent riding roughshod over local susceptibilities. In this regard, however, the question of integrating women into local development could present special problems in some countries. The influence of local agents in breaking traditional prejudices against women may need to be reinforced by the central authority or may even require special and parallel institutions.

The territorial base of the co-operative should also provide a veritable mechanism for the continuous diffusion of new ideas, attitudes, and value systems among the rural population. Development is nothing if it is not the effective diffusion of innovations. It is from this perspective that development of the type discussed in this book impinges directly on various aspects of social life.

Perhaps of singular interest in this respect is the opportunity for a territorially based co-operative to influence the rate of population growth and the desirable size of family. The interaction between households in the economic sphere can more easily be extended for an open discussion of the relationship between the two critical variables of production and population. It should not be difficult to show that the importance of planning and rationality in the former makes sense only in so far as equal attention is paid to similar demands in the latter.

The spatial structure resulting from co-operativization should also make it possible to extend social services, particularly schools and health centres, rationally to all parts of the rural areas. The fact that these are likely to stretch resources is an important element since countries will be constrained to take innovative and indigenous initiatives of both a technological and organizational nature in meeting this challenge. The Chinese certainly would not have struck on the idea of 'barefoot doctors' without the strain and stresses of their pattern of development. Indeed, it may be rightly argued that an important consequence of development when properly conceived, is the series of new and indigenous answers to pressing social problems which it continuously forces a country and a society to make. The spatial distribution of social services and the need to make the best use of trained and skilled manpower, will serve to reinforce the hierarchical structure of the settlement system in a manner complementary to that imposed by production and marketing organizations in the rural areas.

Finally, the new spatial structure will greatly facilitate a democratic two-way sequence of administration and decision-making in the country. In the first place, the conveying of information and directives from higher authorities to the people will be considerably eased. At the same time, the greater degree of popular mobilization also ensures that the reactions of people and the problems encountered at the lower level are relayed to the authorities more easily. Second, the new structure should facilitate the collection and collation of basic data on production, consumption, distribution and the general quality of life of the population. This in itself is bound to temper the character and orientation of administration in a country.

All in all, comprehensive spatial reorganization forces a country to redefine new social relations not only between a society and its environment but, more importantly, between individuals and classes in the society. In the present situation of the world, in which there are very limited opportunities for mass emigration to relieve the problems of rural areas, a country is constrained in its development process sooner or later to confront the only real option open to it. This is to create a new rural structure which would possess two important characteristics: first, it must be able to ensure meaningful participation of all citizens in productive activities; and second, it must be able to keep the population in the rural areas until industrial development in the urban centres has grown to such a level as to be able to provide alternative employment without compromising the capacity of the rural areas to supply food and industrial raw materials. Such flexibility is easier to incorporate into the social and economic life of a country when its development has been consciously planned and is based on a determined attempt to spatially reorganize the land and the population resources available to it.

Conclusion

Because spatial reorganization and the creation of new settlement structures represent a major aspect of establishing new forms of social relations, an important issue in rural development concerns the strategy for bringing it about. The question here is whether rural development can be undertaken on a gradualist, piecemeal, project-by-project basis or must involve a speedy and dramatic, comprehensive and total transformation of the rural areas of a country, accomplished by the people themselves, with their muscles, resources and organizational talents and without, or at the most a minimum of, foreign financial assistance. The issue of the necessity for a 'big push' strategy for rural development re-echoes ideas that were vigorously canvassed in the early 1950s and then pushed away by most experts who came to dominate the development scene over the last three decades. The issue was that the 'big push' came to be confused with 'big investment' and since underdeveloped countries in general had so little capital to invest, the idea was supplanted by that of 'unbalanced growth'. Yet, the 'big push' implies more than investment. Indeed, its real import is on the scale of conceptualization and implementation of critical development effort – that is, the extent to which the population is involved in this effort. As Myrdal pointed out, the two aspects of the thesis of the big push are the size of the initial effort and the speed with which it is applied. According to him:

> The general case for the big push is based on the interrelationships of all conditions in the social system. . . . The big push must jerk the system out of the grip of the forces of stagnation. . . . At the start, big efforts are needed to set the process in motion. Thereafter, the planners can relax or they can harvest

proportionately ever larger and quicker yields from sustained efforts. It is for this reason that underdeveloped countries cannot rely on a 'gradualist' approach and that a growing number of economists have come to support the 'big plan'. Backwardness and poverty naturally make it difficult for a country to mobilize enough resources for a big plan but they are precisely the reason why the plan has to be big in order to be effective.[27]

On the question of size, Uma Lele, in reviewing the impact of various agricultural and rural development projects in Africa, argues as follows:

Because the bulk of the rural population in Africa is poor and because this poverty is spread over the entire rural sector, target groups in Africa are large relative to the financial resources and, in particular, to the trained manpower and the institutional capability frequently available for development. Therefore, if the emphasis in rural development is to be on mass participation and on the viability of the process of rural development, it would seem necessary that rural development programs be viewed as part of a continuous dynamic process rather than as an extensive versus an intensive or a maximum versus a minimum effort. The emphasis on mass participation also means that a sequential approach may be necessary in planning and implementing a rural development strategy, involving establishment of clear priorities and time-phasing of activities.

Given the low productivity of the subsistence rural sector, in many cases an initial emphasis on broad-based increases in productivity through a certain minimum level of services and institutional development, may well be a more effective way of ensuring visibility of mass participation than the substantial initial concentration of resources in a few regions.[28]

In the next chapter, this issue of a strategy for rural development is taken up. A conceptual basis and empirical examples derived from the experience of present developed countries are offered to show that only a 'big push' strategy offers a realistic means for incubating the processes of cumulative rural development. The idea is not to argue that situations are comparable between those countries and the underdeveloped world. Rather, it is to highlight the similarity in the essential approach of countries with very dissimilar socio-political systems to the problems of rural development. Furthermore, it is to show that only strategies of comparable amplitude and directness can set underdeveloped countries on a course that would guarantee wide-ranging and continuous improvement in the conditions of living of their rural population.

6 Strategy of rural development

The strategy of rural development has both an ideological and a methodological dimension. The former relates to ideas concerning preferred societal organizations and relations; the latter to the means and processes of achieving the former. In real life, both dimensions are closely interrelated. For the purposes of providing a clearer perspective on the development process, however, it is possible to separate them. In this book, the emphasis will be on the methodological aspect of strategy, on the means and processes of transforming rural areas so as to ensure greater effectiveness in the interaction between a given society and its environment.

Part of the methodology of rural development has already been outlined in the preceding chapter. It touches on the questions of land reform, of spatial reorganization and of redefining new social relations. The other part which relates to the importance of speed in carrying out this transformation has also been hinted at. In this chapter, therefore, the concern is to present a conceptual basis for such a swift and comprehensive strategy of change in the rural areas of underdeveloped countries and to show that something similar characterized the transition of developed countries to their present status.

Conceptual basis for the strategy of the 'big push'

The conceptual basis for the strategy of the 'big push' in rural development is rooted in that branch of the social sciences concerned with social perception. Social perception refers to the effects of social and cultural factors on the cognitive structuring of the physical and social environment by individuals. Cognitive structuring involves selectivity, the ability to establish a meaningful relation with selected objects in an environment. At any given moment, the exercise of such a selection process by individuals depends upon a great number of antecedent factors such as their interests, needs, values and goals. Because many of these factors are of social origin, the word 'social' has often come, explicitly or implicitly, to refer to motivational factors in the perception of individuals.

Social perception varies from one cultural group to another. It also varies within the same cultural group at different stages in its evolution. It has been found that this variation depends essentially on three factors:[1]

(a) the functional salience of selected aspects of the physical environment;
(b) familiarity with the material products of the culture;
(c) the communication systems employed in the culture.

Each of these factors has important implications for a strategy of transforming an economy and establishing new social relations. The significance of the functional salience of selected aspects of the physical environment, for instance, lies in its effect on the growth of an appropriate and specialized terminology within a culture. This terminology facilitates efficiency in the perceptual responses of an individual to various aspects of his environment and his appreciation of their resource value. But it also provides him with a wide-ranging motivational reinforcement against change. In the African context, for instance, the persistence of bush fallow has been indicated as one of the obstacles to agricultural development. None the less, this particular agricultural practice has a long history in the continent and has influenced the social perception of many rural dwellers as to what items are functionally salient in the soil, plant, animal life and even the weather conditions of the region. This perception is, for example, reflected in the close relation between the agricultural year and the organization of the social calendar of festivals and community celebrations. To transform such a system of agriculture requires more than tinkering with the type of crops grown or the yield of these crops. It involves a replacement of the functional salience of this type of agricultural system by that of the desired type. Baldwin, for instance, indicated the difficulty of doing this half-heartedly among farmers used to cultivating on a subsistence basis when they were subjected to the greater time discipline of large, mechanized farming.[2]

The second factor, namely, familiarity with the material products of a culture, defines the scope of the prevailing technology. Familiarity with such material products serves to condition perception as to the extent of available resources and the potential for their development. For instance, in societies in which agricultural equipment comprises no more than a hoe and a matchet and whose members have never seen a tractor, the possibilities of large, owner-operated or co-operatively operated farms, or of substantial wealth based on agriculture, are not likely to be appreciated. In other words, these man-made artefacts set a limit to perception and therefore to the motivation required to bring about change and progress.

The comunications system or the language of a culture is the third essential factor in social perception. It is a most important notational system whose structure deeply influences the view of the world as perceived by that culture.[3] While there can be controversy as to how deeply language influences a culture, it cannot be disputed that a society will not invent words for artefacts, ideas, concepts, values and goals which it has never experienced. Since the goal of development and of sustained growth in material well-being are new ideas to the rural majority in most underdeveloped countries, it can be argued that unless the relevant language is made available for translating these objectives into an integral part of the local culture, the goals of development will be difficult to realize.

Appreciation of the factors determining social perception constitutes the first step in the demonstration of the logical necessity of the strategy of the big push in effectively bringing about rural development. The second step is to show how this relates to the need for the push being relatively swift and dramatic. This necessity follows from the relationship between individual and social perception. According to Chefit, one of the most important discoveries of experimental social psychology concerning individuals is that, when in a group, their perceptions tend to converge.[4] In other words, an individual's perception and judgement concerning events or changes, tend to be strongly influenced by the social norms of the group to which he belongs. This aspect of individual behaviour implies that, in a developmental context, unless social norms are seriously altered and forced towards a new level of convergence, it may not be easy to change the perception and behaviour patterns of individuals. However much the few exceptional individuals may see, appreciate and adopt new values and styles of behaviour, the persistence of traditional social norms acts as a major factor inhibiting widespread change and development.

It is from this perspective that Lewin, in his field theory of human behaviour, offers a model strategy for engineering a successful transformation in any community or society.[5] This model emphasizes the importance of three sets of actions:

(a) the 'unfreezing' of the current level;
(b) the 'moving' to the new level;

(c) the 'freezing' of group life at the new level.

The 'unfreezing' process may involve different problems in different situations. Where the situation is an institution or customary usages, 'unfreezing' may require legislation designed to remove the binding force of the usages. It may also necessitate the introduction of new judicial processes to give strong support to the social consequences of the legislation. In some cases, as in rural development, it would not be sufficient simply to legislate. Concrete physical action of the type entailed in land reform and massive spatial reorganization will be required to underline the fact that the current level or norm was no longer to be desired. Very often, the activities of social commentators and writers are vital in this 'unfreezing' stage. Their role is to provide the rationale for the act of unfreezing and to explain it in terms understandable and easily acceptable to the general public. It has been suggested that the goal in all of this should be to produce a 'catharsis'. This is the releasing of emotion, concern and anxiety which is essential if prevailing prejudices are to be removed and the shell of complacency and self-satisfaction broken open.[6]

Theoretically, the transition to the new level must be swift. This strategy does not hold with a process of gradual change since it insists that gradualism enables traditional norms to reassert themselves. Instead, it requires that the critical structural framework of the new dispensation must be installed directly to replace the older system so that there can be no turning back. Once the futility of hankering after the past is patent, the idea is that people will direct their effort physically and mentally to adapting to the new system and adjusting it to their peculiar needs.

The 'freezing' process is, of course, not sequential; it is in fact part of the act of 'unfreezing'. It is necessary, however, to establish organizations and institutions to stabilize this new level, to provide a basis for the appreciation of its emergent functional salience and to help generate an appropriate communication system. All of this is to prevent a reversion to the previous level. In other words, it does not suffice to define the objective of the strategy simply as one of attaining a new societal level of development. It is crucial to ensure stability and a high degree of permanence for the new order.

Empirical evidence

This theoretical exposition of a strategy for rural development is valuable because it provides a basis for evaluating the historical experience of other societies in the process of development. The intention here is not to argue that the situations in different countries are similar or replicable, or that the path of development of present-day underdeveloped countries must follow that of the developed. Rather, it is an attempt to get at the essence of the transformation achieved in the various countries and to show how the crossing of this critical threshold laid the foundation for their subsequent growth and development. In order to emphasize the concern for essence rather than specifics, the countries chosen straddle the ideological spectrum and include Britain, the United States, the Soviet Union and China. For good measure and to emphasize how far the lesson is being learnt in a few other countries, the Tanzanian experiment is also included.

Britain

One of the more critical thresholds which Britain crossed in the process of her development is that generally referred to as the 'Enclosure Movement'.[7] This movement is usually regarded as lasting from 1760 to 1820, although the idea of enclosing fields goes back many centuries before this period and continued for much longer thereafter. But the major thrust was during this period, and its distinctiveness was that it was the result of societal (or parliamentary) decision rather than interpersonal agreements. In the words of Birnie, this movement 'destroyed the last vestiges of communal husbandry in England and put in its place a system of large individual farms cultivated by capitalist tenant farmers'.[8]

The 'unfreezing' process for the transformation of the structure of traditional English rural economy from the open field system to the modern, highly capitalistic one had begun much earlier, as far back as the fifteenth century with various piecemeal enclosures undertaken by voluntary arrangements. It has also been suggested that the increasing urbanization of the country towards the end of the seventeenth century was an equally important factor in the unfreezing process as evidenced by the tendency to find many enclosures in the neighbourhood of towns. In the words of Gonner, 'the very presence of a town with its novel and less stable conditions was subversive to the mere rule of custom. The traditional methods of cultivation went for less, and profit, and increased net profit in particular, went for more'.[9] There were, of course, also the agricultural writers and propagandists such as John Mortimer (1707) and John Lawrence (1726) and more important, Arthur Young (1761–1820) and William Marshall

(1745–1818). All of them set out to show the inappropriateness of the existing farming system and the need to replace it with more appropriate enclosed fields on which more modern agricultural practices could be instituted.

What the Enclosure Movement involved was basically as follows. At the request of the majority of traditional proprietors in a village, an Enclosure Bill was introduced and passed in Parliament. Commissioners were appointed whose job was to survey the scattered strips of each individual farmer, note their cropping capacity, evaluate and redistribute the land (including the commons), set out new roads, reorganize the drainage system, settle conflicting claims and finally proclaim an award whereby each proprietor was given a consolidated holding in place of his scattered strips in the open fields, and his rights of pasture over the village commons and waste.

Indeed, it can be said that what the enclosure commissioners did for England was to lay out anew its rural landscape and to reconstruct its human geography. Before 1770, there were fewer than 260 private acts of enclosure in the whole of England and the total area involved could hardly have exceeded 400,000 acres. From 1760 to 1844, there were more than 2500 Acts dealing with more than 4 million acres of open fields, and another 2800 Acts seeking to enclose nearly 2 million acres of common and waste land.[10]

Although the Enclosure Movement appears to have dragged on for nearly eighty years, the effect at the level of individual villages was swift and dramatic. According to Hoskins:

> It is true that the paper plans, as set out in the award made by the commissioners, did not produce all the physical changes at once, ... but the transformation of the landscape was, all the same, remarkably swift. A villager who had played in the open fields as a boy, or watched the sheep in the common pastures, would have lived to see the modern landscape of his parish completed and matured, the roads all made, the hedgerow trees full grown, and new farmhouses built out in the fields where none had been before. Everything was different: hardly a landmark of the old parish would have remained'.[11]

Figure 3 shows, for instance, the parish of Middleton in the Wolds and reveals the dramatic change it went through within the short space of fifteen years. The attempt to stabilize the new dispensation included the diffusion of a new system of husbandry, notably the Norfolk system of alternating roots and clover with corn crops, the introduction of new agricultural implements and machinery and the increasing

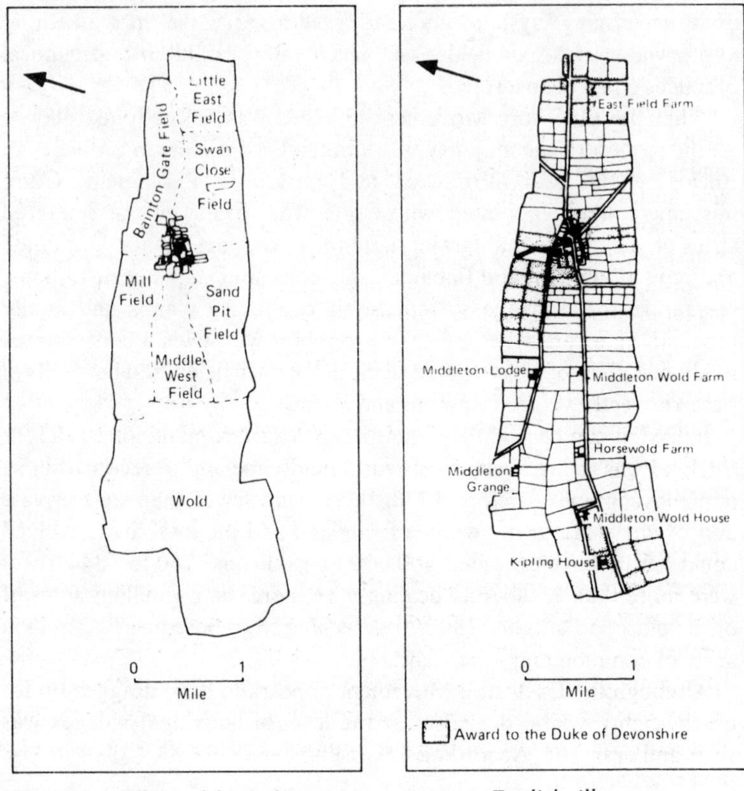

Figure 3 *Effects of the enclosure movement on an English village*

Source: M. B. Gleave, 'Dispersed and nucleated settlement in the Yorkshire Wolds, 1770–1850', *Institute of British Geographers Transactions and Papers*, no. 30 (1962), pp. 108–9.

ease of using land to raise short term loans for agriculture from the emerging number of country banks during this period.

The nature of the 'catharsis' which the rural community had to pass through in this movement is perhaps best appreciated from the description of the economic and social history of the Leicestershire village of Wigston as provided by Hoskins:

By 1795, ... nearly thirty years after enclosures – there were substantial changes. ... The typical owner occupier was farming on a bigger scale as a grazier. In addition to his own land he rented that of his fellow-villagers, and also some from large absentee landlords. But the peasant with no capital could not do this. His holding might still have been large enough to support him under the peasant economy, but that had been destroyed by the enclosure; and he could not fit his one, two or three small closes into the new grazing system. Some let their fields to bigger men, but others, burdened by mortgages, were forced to sell up in the face of the steadily rising price-level, above all of food prices, which had taken a sharp upward turn in the 1760s and had not ceased to rise since then. The old economy would have protected and sustained them in this situation (as it had done during the price-revolution of the sixteenth century) but they had been precipitated from that natural and largely self-sufficing economy into a full-blown money economy in which they had to buy nearly all the things they needed instead of producing for themselves out of their own natural resources.

It is possible to expatiate on the traumatic nature of the transformation that brought rural Britain into the modern period, and to catalogue some of the deleterious effects on a section of the population at the time, as well as the positive consequences on production and productivity which it made possible. What is not usually stressed enough, is how this transformation changed the functional salience of the rural environment, how it weaned the farmers from the familiar objects of a passing culture and taught them a new communication system from which those who mastered it quickly, came to profit immensely. And these dramatic changes in social perception arising from a major effort at spatial reorganization, along with other changes in the industrial and transportational fields, came to be the crucial factor in the subsequent growth and development of British society and economy.

It is, of course, possible to argue that the Enclosure Movement does not fit the conceptual model of change on all fours and has little relevance for underdeveloped countries. It may be pointed out by modern scholars of the event that a significant proportion of Britain was already privately enclosed before the period; that changes that went on in a country for over sixty years cannot be said to be swift and dramatic; and that the movement itself was a response to, rather than the initial cause of, development. All of these reservations no doubt have some truth in them but they are valid only if we choose to ignore the differences that always exist between the trial and error aspects of a first successful attempt, and the more systematic, speedier and confident processes of a second. Of course, there was a significant number of private enclosures before 1760, but until the process became a societal transformation with new infrastructure, laws and institutions established for it, its full potential could not be realized.

Similarly, the sixty year span could be said to imply a gradual process, until one realizes the problem of mass mobilization at the time, the limited facilities for mass communication and the limited understanding and capability of how to increase rapidly the number of people with the practical expertise for the surveying and spatial planning required. When all this has been taken into account, it will be realized that, for its time, the Enclosure Movement was swift and sufficiently dramatic to create the necessary 'catharsis' immortalized in Oliver

Table 10 *Comparisons between levels of agricultural productivity*

Country and stage of development	*Period*	*Index number of agricultural productivity**
Developed countries		
Recent position		
France	1968/72	100.0
United States	1968/72	330.0
Position before or during take off		
France	1810	7.0
Great Britain	1810	14.0
Sweden	1810	6.5
Belgium	1840	10.0
Germany	1840	7.5
Italy	1840	4.0
Russia	1840	7.0
Switzerland	1840	8.0
United States	1840	21.5
Spain	1860	11.0
Less developed countries		
Recent position		
Africa	1960/64–1968/72	4.7
Latin America (excluding Argentina)	1960/64–1968/72	9.8
Asia	1960/64–1968/72	4.8
Middle East	1960/64–1968/72	8.6
Total for all less developed countries	1960/64–1968/72	5.5

Source: Paul Bairoch, *The Economic Development of the Third World Since 1900,* Cynthia Postan (trans.) (London, 1975), p. 40.

* Estimate of the net agricultural production by male labour employed in agriculture expressed in 'direct' calories.

Goldsmith's poem 'The Deserted Village'. The issue of whether it was a response to, or the initial cause of, development is neither here nor there. What is clear is that without it the rural areas of Britain could not have come to play host to the kind of cumulative development processes which have characterized them since then.

Table 10 compares levels of agricultural productivity in a number of developed countries today, and before or during the period of their 'take-off'. Although the index number of agricultural productivity for Britain is not available, comparable indices for France show that productivity today is nearly fifteen times higher than before the 'take-off'! Indeed, Bairoch indicated that the average index of agricultural productivity in the developed countries before their agricultural revolution was about 5, a level which is less than the 5.5 for all under-developed countries but slightly higher than the position in Africa or Asia at the beginning of the present decade.[12]

The United States

The major development process in the United States in the nineteenth century cannot be entirely divorced from events in Europe, particularly Britain which was the source of many social and economic influences. It is worth noting that the American War of Independence (1779–85) took place some two decades after the accepted period of the Enclosure Movement. It is thus not unlikely that the importance and arguments put out to support the need for this massive transformation in England were already well known in the United States. It must, of course, be admitted that the situation in America was in no way comparable to that in Britain. In the United States, there was no traditional society to be transformed. What existed was virtually a whole continent, relatively empty and waiting to have the stamp of a vigorous and developing society put on it. And the remarkable aspect of the American experience is how quickly the lessons of events in Europe were grasped and used effectively to lay the foundation of a dynamic society and economy.

The 'unfreezing' process in this experience must thus be seen as part of the European thrust of redefining all factors of production within a capitalist mode. The economy was not based on the European peasant-landlord relation. The existence of slaves belonging to a different racial group made this unnecessary, at least in the South, but this complicated matters by bringing into the open the contradictions latent in the American colonial economy. This economy was founded on land grants of various types and particularly in the South, on large estates worked by slave labour. In this southern estate economy, Jennings observed that land itself had little value independent of the labour set on it, though the value of a negro was affected almost by a change in the price of cotton.[13] Yet, for efficient functioning of the capitalist system it was necessary for land to have its own market value and to be available to as many people as possible. Hence, as soon as the country became independent, one of the first acts of its government in 1785 was to pass an Ordinance which 'would ensure the easy transfer of public land into private ownership, and provide for each tract unambiguous identification and quantitative areal measurement'. This Ordinance, according to Johnson, was 'the first legislative act of many designed to insure – in the words of Thomas Jefferson – that "as few as possible shall be without a little portion of land" '.[14]

The spatial organizational aspects of this Ordinance need special emphasis. The Ordinance established for the country a base line, known as 'the Geographer's Line', for a new system of surveying the whole country.[15] This line ran from the point where the west boundary of Pennsylvania cuts the Ohio, due west along the parallel of latitude. From this base line, a number of equally spaced north–south lines or 'ranges' were to be surveyed. Each state was to appoint surveyors who were to be directed by a geographer. The surveyors were to divide up the land into square blocks. Each block was again to be divided into square sections, each a mile square, or 640 acres, in area. Each section was further subdivided into square subsections. Each of the sections was defined by parallels of latitude and by locally adjustable meridians.[16] The land division lines always crossed each other at right angles, and the rectilinear grid system of townships, sections and farmland that resulted became the most characteristic feature of the American landscape.

An Ordinance of 1787 also tried to provide for a new and popular tenure system, known as the allodial, that is free of rent or service to anyone.[17] This was to replace a practice of the colonial period whereby some of the land was held under a modified feudal system, that is, by the payment of a fixed labour service or some rent, generally called a quit rent. The new Ordinance made land freely transferable by bargain and sale, and the estates of those who died without a will were to be divided equally among their heirs. The land was held in fee simple, which meant that it was held without condition or limitations with perpetual title, and belonged to the owner, his heirs and assignees forever.

Equally noteworthy is that this Ordinance required the surveyors to take note of geographical features such as salt springs, mineral occurrences and mill sites, to identify trees, outline swamp land, distinguish types of undergrowth, locate bluffs, roads and rapids, and appraise soil conditions, on a simple classificatory system of first-, second- or third-rate soil.[18] In a sense, the surveyors' notes or 'plats' represented a first, if superficial, inventory of the land resources of the country.

When the first Congress under the American Constitution assembled in March 1789, this Ordinance was allowed to lapse and with it land legislation, the imperative of the situation was such that a new Act had to be passed in 1796. This contained a few modifications of the 1785 Ordinance but kept its essential features, including the rectangular system of land division. The debates that led to the passing of this second Land Act reveal not only concern over public revenue but particularly the widespread disquiet over the intense land speculation that had followed from the earlier Ordinance. The problem of land speculation continued to be serious for the first half of the nineteenth century, as long as the government insisted on selling public land to raise revenue. Indeed, according to one report, most of the land sold during that period 'passed through a dozen hands within sixty days' without such transfer resulting in any development.

As the population began to rise, agitation mounted for the government to offer land free to those who would farm it. Workers in the towns and cities were the most vociferous in this demand. Their conventions and papers in the 1830s began to demand that land 'should be treated as an instrument of social reform, to raise the wages of labour, rather than as a source of revenue to relieve tax-payers'.[19] Memorials were prepared and presented to Congress to that effect. One writer in 1835 proposed that 'public lands should no longer be sold, but that any men, unpossessed of land, should be allowed to take possession of a certain portion of the unappropriated domain for the purpose of cultivation, and, to prevent speculation, no one should be allowed to hold more than a certain portion'.[20] Agitation for freedom of the public lands continued with the same arguments as those previously stated, namely improvement of the condition of the working men. *The True Workingman,* a newspaper of 24 January 1846, contained an article, widely circulated as a handbill, which might be called 'vote yourself a farm'. A new political party – the Republican Party – with little to lose and much to gain, took up the demand. On the eve of the American Civil War, this party adopted a plank in favour of the 'free homestead policy' and under Abraham Lincoln was swept into power in 1860.[21]

On 20 May 1862, the United States Congress passed into law the Homestead Act. This Act offered to any citizen or intending citizen who, among other things, was head of a family and over twenty-one years of age, 160 acres of surveyed public domain after five years of continuous residence and payment of a registration fee varying from $26 to $34. The law applied to *all* lands subject to purchase at $1.25 under the Pre-emption Act, and to eighty-acre lots within the lateral limits of railway grants. Under this Act and its successors, the division of the American landscape into quarter sections went on apace, particularly west of the Appalachian Mountains but also in parts of the south and the north-east. As the 100° meridian was passed and humid lands were left behind for the semi-arid and arid regions of the country, the size of the homestead was increased first to 320 acres, then to 640 acres and eventually to more than a square mile, with the terms of residence being made easier still.

Within thirty years, the transition was complete. In 1890, the superintendent of the Census officially announced that the 'frontier of settlement' no longer existed. Much of the land in the public domain had been disposed of. In 1863, over a billion acres were unreserved and unappropriated out of the original 1.44 billion acres in the twenty-nine public land states. By 1904, a mere 0.47 billion remained unclaimed and by 1944, there were none.[22] After 1891, increasing attention came to be paid to adapting the laws to special physiographic conditions, to discouraging land monopoly, to conserving the nation's natural resources and to promoting intelligent use and classification of the land. The period also saw tremendous organizational development in the field of agriculture. This included the rise of the 'grange' movement for concerting the grievances of farmers and organizing them for their own development, the establishment of a government department of agriculture, the expansion of agricultural mechanization, and the development of agricultural education.

The development impact of this transformation which turned the rural areas of the United States almost exclusively into a land of dispersed isolated farmsteads, with their road connections arranged in rectilinear forms (Figure 4) and with the typical European village absent in most parts, is perhaps best appreciated by Frederick Turner. According to him, 'the existence of an area of free land, its continuous recession, and the advance of American settlement westward, explain

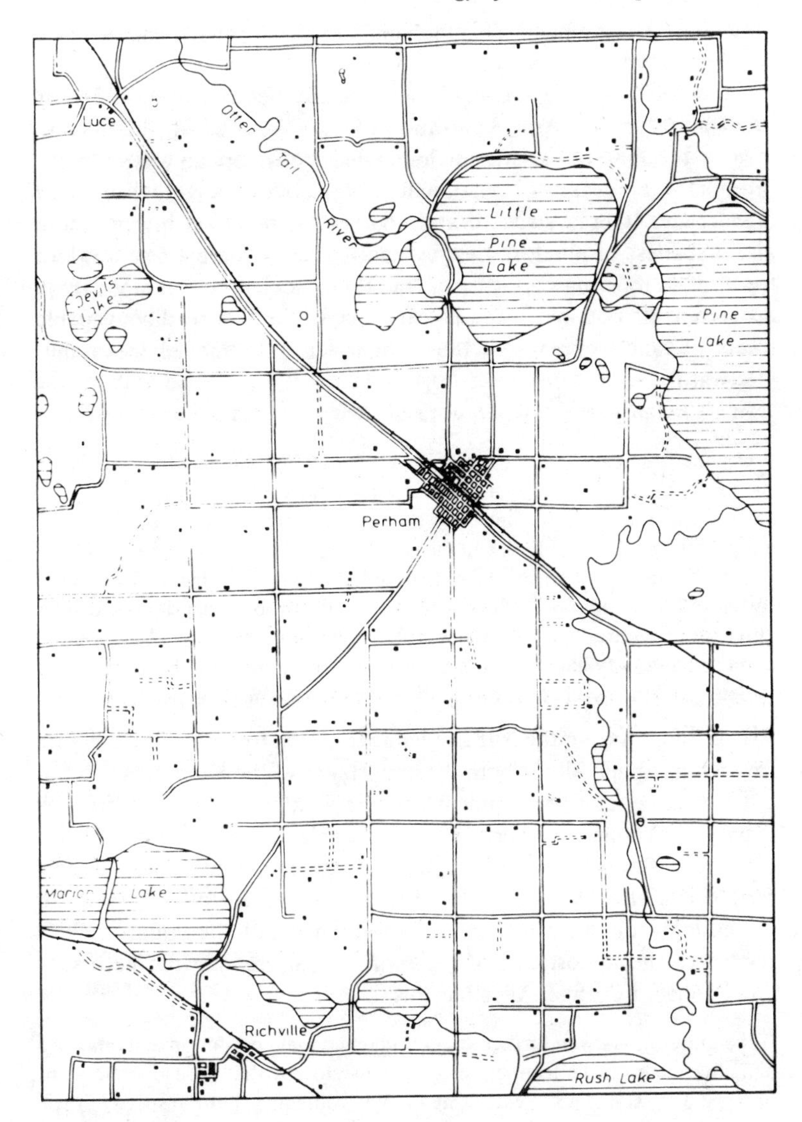

Figure 4 *Organization of rural land in the United States*

American development. This "advance of the frontier" has meant a steady movement away from the influence of Europe, a steady growth of independence on American lines; the frontier is the line of most rapid and effective Americanization'.[23] The West which the moving frontier brought into existence, dominated the American way of life. It promoted a sense of nationality, freedom, individualism and democracy. It bred strength and coarseness, practical inventiveness and quickness of mind, restlessness and nervous energy. The freedom from the hampering garments of traditional society which made these developments possible, owed its existence to the vast readily available tracts of land upon which the frontier fed as it moved inexorably westward.

In a similar vein, Faulkner writes:

It was the frontier, more than anything else, that differentiated the underlying economic conditions in the United States from those of Western Europe; that determined the predominance of the extractive industries, the lines of manufacturing development, and the content and direction of foreign trade. It was the influence of the frontier that moulded our banking and currency systems and determined the course of transportation development. The influence of the frontier upon our economic life is clear enough, but its effect upon social and political development are equally important. It has gone far to determine the psychology and philosophy of the American people.[24]

It is, of course, in the true academic tradition for latter-day scholars to attempt to play down both the importance of the Homestead Act of 1862 in American development and the validity of Turner's 'frontier thesis'. Yet, none of the critics has been able to argue against its psychological impact on the people, and the fact that it was largely responsible for bringing the Federal Government to take an increasingly directive role in the development of the country. With regard to the importance of the swiftness in creating the necessary conditions for change, Allen has this to say:

If the criticism is taken further and it is claimed that very few inventions were actually made on the frontier, let it be remembered that, in economic as in political affairs, it is not the devising of new techniques, but the speed of their adoption – whatever their origin – which is the fundamental condition of great and rapid development.[25]

The relevance of the American experience is certainly not that the situation bears any similarity to any underdeveloped country today. The United States in the nineteenth century was a relatively empty country requiring massive immigration from the Old World to settle and develop it. The particular significance about it is two-fold: the comprehensive nature of the transformation as planned by Congress, and the importance accorded, as in the case of the Enclosure Movement, to the strenuous but necessary surveying and spatial planning. Again it may be objected that all this was possible because it was a response to a perceived need. This is certainly true, although the perception or existence of a need does not necessarily define the form in which a government responds to it. For most underdeveloped countries there is clearly a strongly perceived need for rural development. What is debatable is the efficacy of current response to this need.

The Soviet Union

It may look quite unusual to posit the Soviet pattern of development as essentially no different from that of Britain or the United States. The essence of this position is not in the substance but in the process of development. In both types of societies what was undertaken was moving society, the whole society and its resources, from one level of organization to another and doing it swiftly and comprehensively. In Russia, so much violence and revolutionary struggle accompanied the transformation that it has tended to cloud the issue of what changes were being undertaken at the grass roots. Yet, the experience of China, and, more recently, Tanzania, shows that the process of rapid transformation need not involve such large-scale violence and human suffering.

The task before the communists in Russia in 1917 was perhaps more difficult than that before Britain and the United States in the nineteenth century. The task was to transform a relatively backward country into one of extensive industrialization and modernity and to do this with next to no dependence on foreign assistance. 'Unfreezing' the traditional situation could be said to be entailed in the strong ideological proselytization of the Communist Party and in the series of crises that culminated in the demise of the Tsarist regime. For the communists, the traditional structure of Russian society and landscape was no longer acceptable. New social relations and spatial organization were required, based on the ideology of Marxism-Leninism.

In many ways, the Russian situation in 1917 provides many examples of the dangers of gradualism in the process of socio-economic transformation. In the period just before 1917, an attempt had in fact been made to modernize the country on a capitalist basis in the manner being undertaken in other countries in Europe. Under Stolypin, legislation was passed between 1906 and 1911 for converting communally administered medieval strips in open fields into large, enclosed 'free-hold' farms. As in Britain, when holdings were consolidated into one enclosure, the new farmers sometimes, but not always, left the village and built new isolated farmsteads *(khutor)* on their land. A State Peasants Land Bank created in 1883 was strengthened and authorized to purchase land on its own account for peasant settlement and to offer loans at much reduced interest rates. Not unexpectedly, all this enabled the more prosperous peasants (*kulaks*) and many landlords both to rent and buy land, and so to co-opt the process to serve their own ends. Even then the process was slow. Although it was claimed that by 1917 as much as 24 per cent of communal households had been covered by the changes, more recent reviews estimate that significant results from the reform were in fact confined to peripheral areas where the traditional system had never been well established, and that it made little impact in the central region.[26] There was said to be opposition from the women, who were unwilling to forgo the social life of the village for the loneliness of the farmstead, and also from money-lending merchants who thought the new farmers would be more difficult to exploit than the village peasants.[27]

At any rate, the situation was not resolved until the Revolution of 1917. Again, for the first decade or so of the Revolution, a policy of gradualism was pursued. A Land Reform Law of 1917–18 had led to the immediate breakup of the large estates, the equalization of peasant holdings and a considerable increase in the number of peasant farms. But these changes were not in the direction desired for a socialist society and therefore the necessary new social relations and institutions meant to stabilize them were not forthcoming. What the communists desired was largely co-operative rural communities, a modernized version of the traditional co-operative or *artel,* under which members would keep their own dwellings and gardens, but would have their agricultural land and implements in common, farm collectively and share out the crop.[28] When such *artels* were eventually formed in the late 1920s, it was easier to transform them further into the *kolkhoz* which came to represent the major form of rural spatial and social organization in the Soviet Union. In the period immediately following the Revolution some 14,000 *kolkhoz* were formed. These often included no more than ten to fifteen families, each bringing together no more than 100 to 120 acres and accounting altogether for some 3 to 4 million acres of land. The majority of these *kolkhoz* were not very successful, a considerable proportion of them being dissolved annually and their members dispersing.The same was true of the state farms or *sovkhoz* which declined from 4000–5000 to only about 3000 over the ten-year period. Indeed, on the eve of the First Five-Year Plan in 1928 both the *kolkhoz* and *sovkhoz* farms, together, supplied less than 2 per cent of the total grain crop and covered little more than 1 per cent of the cultivated area of the Soviet Union.[29] The rural area was still overwhelmingly dominated by small-scale, individualist peasant agriculture, in which the richest 10 per cent of the peasantry owned between 35 and 45 per cent of the agricultural means of production and 30 per cent of the draught animals. Government plans for rapid mechanization and increased productivity were thwarted. In 1924–5, some 5000 tractors had been imported and supplied mainly to village soviets for co-operative use. In 1925–6, it was planned to import 17,500 more and to manufacture another 1800 at home. These figures were to rise by the end of the decade to an annual importation of 23,000 and home production of some 10,000 tractors. However, as long as this machinery continued to be employed on small peasant holdings, its impact on agricultural productivity remained inefficiently low.

The crisis came in the period 1925–6 when, despite the recovery both of the cultivated area and the gross harvest, only 17 per cent of the total yield was being sent away to urban markets by the peasants, compared to 26 per cent in the pre-war period.[30] This decline in the marketing surplus of agricultural produce became a fundamental barrier to industrial growth and urbanization as well as to the expansion of foreign trade. It thus became clear that to transform Russia truly and effectively into a modern, socialist state required an ambitious programme of industrialization built with the introduction of large-scale farming on collective lines as its corner-stone.

At the fifteenth Congress of the Communist Party in 1928, the decision was taken for a total and comprehensive organization of the small and dwarf peasant farms 'into large farms based on common, co-operative cultivation of the soil, with the use of agricultural machines and tractors and scientific methods of intensive agriculture' (Stalin). By the summer of 1929, workers were sent from the factories to help in the collectivization campaign and to provide leadership in the collective farms, the rural soviets and the machine and tractor stations. The antagonism and opposition of *kulaka* and landlords, their spontaneous or concerted slaughtering of cattle rather than consigning them to the new collective farms, their acts of arson and violence against the new farms and their personnel, as well as government reprisals in the face of such opposition, all were to give a very brutal twist to the process of transformation in Russia. Yet, within four years, the process was over. According to Dobb:

> The winter and spring of that 'spinal year' of 1929–30 was to witness those tense months of turmoil in the village which are depicted in the novels of Sholokhov and from which the new type of Soviet village was to be born. The birth pangs were sharp; the attendant midwifery was rough. But those few months may well come to be regarded as a turning-point in the economic history both of Europe and of Asia in the twentieth century.[31]

By the end of 1932, the membership of collective farms in terms of the number of peasant households had passed the 14 million mark. This figure represented more than 60 per cent of the peasantry. The 200,000 *kolkhoz* which they represented also embraced nearly two-thirds of the total sown area (Figure 5). Together with the state farms, they supplied 84 per cent of the marketed surplus of grain and 83 per

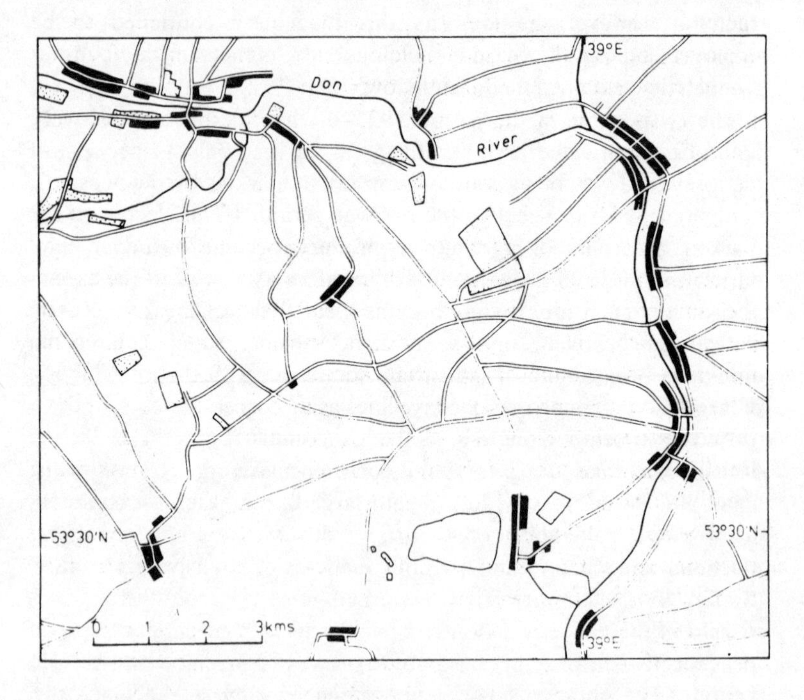

Figure 5 *Socialist reorganization of rural land in the USSR*

cent of the cotton. In the year of 'consolidation' in the middle 1930s which were quickly to succeed the rough and ready methods of the pioneering years, increasing attention came to be paid to improved rotation, the introduction of fodder-crops into the rotation, and the early ploughing of fallow. The number of tractors increased six or seven times and while the density of farm population relative to sown land declined by 25–29 per cent, productivity per head in agriculture probably rose by 60 per cent between 1928 and the end of the 1930s.[32] Clarke in a more recent calculation showed that at 1913 prices, gross agricultural production in the USSR had more than tripled by 1970.[33]

The frequent news of failures of Soviet harvests in the Western press in recent years, and reports suggesting that the USSR may be reverting to an acceptance of a higher proportion of private agricultural production may tend to detract from the significance of Soviet agricultural achievement and its implications for the development process in underdeveloped countries. No one can deny the colossal nature of the task the Soviets set themselves in 1929–30 and the enormity of their accomplishment. Its lessons in the comprehensiveness of the programme and of the mobilization of the population, its speediness of execution and the continuous effort to increase productivity are there for all underdeveloped countries to learn from.

China

Compared to the Soviet Union, the transformation of the rural areas of China was accomplished with little violence although predicated on the same ideological premise of a class struggle. Here again, there was a clear ideological basis for the 'unfreezing' process which was made more necessary by the fluid social situation in China in the period after the First World War. The initial preference for the communist rural transformation of China was for a gradualist process of change. But the problems inherent in a slow pace became so self-evident that there was no alternative but to accept a faster tempo and complete the rural transformation speedily. The result was the emergence of the rural commune within less than a decade of the seizure of power by the Chinese Communist Party.

In 1949, when the party acceded to power, the condition of the peasants and the state of agriculture in China was abject. The holdings of a peasant were fragmented and existed as small, scattered plots. He himself laboured under conditions of heavy rents, increasing absentee landlordism, traditional and often inefficient methods and technology of production, primitive tools, poor seed selection, restricted and controlled markets, lack of credit and indebtedness to moneylenders, and oppressive taxation by warlords or an indifferent government.[34] Added to all this, the peasants suffered incessantly from the vicissitudes of natural hazards of floods and drought, which, without a stable government to minimize their effects, occurred with depressing and distressing regularity.

Land reform and rural development had been a major promise of the Communist Party in the long and bitter years of struggle between 1927 and 1949. However, until the autumn of 1955, every statement made by the Party or the government stressed the importance of introducing socialism over a long period. Such reform, it was argued, must wait until the ideological level of the peasants had been raised. Recognizing the 'deep attachment' of the peasants to their land, these statements affirmed that private land ownership would dominate rural China for many years. Thus, in the period between 1950 and 1954, soon after the Party came to power, the land reform movement redistributed 46 million hectares of land among 300 million poor and landless peasants.[35] Table 11 shows the share of and average size of

Table 11 *Some results of land reform in China, 1949–52*

	Percentage of households	*Share of crop area owned*		*Average crop area owned*	
		Before reform (per cent)	*After reform (per cent)*	*Before reform (mou)**	*After reform (mou)*
Landlords	2.6	28.7	2.1	116.10	11.98
Rich peasants	3.6	17.7	6.4	35.75	26.30
Middle peasants	35.8	30.2	44.8	15.81	18.53
Poor peasants and farm labourers	57.1	23.5	46.8	6.25	12.14
Other	0.9	0.0	0.0	–	–

Source: Peter Schran, *The Development of Chinese Agriculture, 1950–59*, (University of Illinois Press, Chicago, 1969), pp. 21, 22 and 25. See also John G. Gurley, 'Rural Development in China 1949–72 and the lessons to be learned from it', *World Development*, vol. 3, no. 7/8 (July–August 1975), p. 459.

* 1 mou of land = one-sixth of an acre

farmland owned by various categories of the rural population before and after the reform. However, by 1952, a reconcentration of holdings into the hands of a few peasants was becoming apparent while the elimination of the landlords had also removed a major source of rural credit. Moreover, labour productivity was not rising significantly and was still affected by high seasonal fluctuations in the demand for labour.

A solution was therefore sought in the organization of the farmers into mutual aid teams. Each team consisted of seven to eight households who pooled their labour only during peak seasons. At the end of the season, any debts outstanding between members were settled. In every other respect agriculture was managed on an individual private basis, each household controlling the disposal of its own produce. In order to solve the problem of seasonal unemployment and labour shortage, the government attempted to make the teams operate on a permanent, all year basis. They were urged to co-ordinate agricultural and subsidiary activities, developing the latter (especially livestock rearing) in the slack months of winter. In addition, they were to build up a small, collectively owned stock of implements and draught animals, financed by a levy on each member equal to 1–5 per cent of his output each year. The long term goal was to transform such teams into 'land co-operatives', in which the land, though remaining under private ownership, would be pooled and managed as a single unit. By the end of 1954, over 60 per cent of all peasant households in China were grouped into mutual aid teams. Only 45 per cent, however, were members of permanent teams, while 11 per cent were in 'mutual aid co-operatives'.

However, it was becoming clear that unless socialism could be firmly established in the countryside, capitalism would take over. Already, the old inequalities of ownership, income and exploitation which existed before the land reform were reappearing. Poor peasants, lacking capital and managerial experience, had failed to maintain their independence on becoming landowners. They had fallen into debt and sold land to the rich peasants who were their creditors and who combined farming with peddling. Furthermore, the production of cotton, silk, tea, jute, oil crops and livestock failed to meet the planned targets and revealed the growing contradiction between agricultural and industrial growth. In the circumstances, at the National People's Congress of July 1955, it was decided to push vigorously for the complete co-operativization of the Chinese peasants.

The main characteristic of a co-operative was that member

households agreed to pool their land, the use of which would be planned centrally by the co-operative's management committee on the basis of targets handed down by the government. The co-operative paid rent to the owners and allocated to each household a small 'retained plot' for private use, up to but not exceeding 5 per cent of the average arable area per head in the village. Farm implements, draught animals and trees also remained in private hands. If they were hired or entirely managed by the co-operative, a payment for their use was to be made to their owners. The peasant's labour was rewarded on a basis of the number of 'labour days' he earned. In short, the co-operative was only semi-socialist in character. The private sector was still an important source of income. Control over the use of almost all factors of production and the annual output had, however, passed into the hands of the co-operative.

At this stage, there was considerable argument within the Party whether the final thrust towards collectivization should be undertaken. Collectives were bigger aggregates and more socialist in character than the co-operatives. Land became the property of the collective without compensation. The payment of rent to the previous owners ceased. Draught animals, large implements and groups of trees were collectivized at agreed prices 'according to local value'. However, the private plot, allocated as in the co-operative on the 5 per cent criterion, still remained. Similarly, domestic livestock (mainly pigs and poultry), scattered trees and small implements were left in private hands while the piece-work system for labour working in the fields was improved. Those fearing strong opposition to the process had counselled great caution. Mao Tse-Tung, however, insisted that the pace of change should not only be maintained but increased to enable the agricultural sector to serve the requirements of the five-year plan and to prevent a resurgence of capitalism.[36] Thus, although in December 1955 only 4 per cent of households were in co-operatives, by February 1956, the figure was already 51 per cent, in June, 63 per cent and in December, 88 per cent.[37] Thus within a space of two years, the Chinese had managed to set on the path of development and modernization over 120 million households in some 752,000 settlement units of which, by the spring of 1958, 99.7 per cent were organized as collectives.

Since then attempts to resolve the problems of incentives, management and complementary economic activities have further pushed the Chinese to replace the collectives with the people's commune. According to Walker, the commune differs from the collective in three major respects:[38]

(a) As a planning unit, the commune is much bigger both in size and scope. It is concerned with the co-ordination of every type of activity: agriculture, industry, education and defence. Its predecessor, the collective, was merely an agriculture unit.

(b) The *hsien* or district level government was merged with the commune administration.

(c) The private plot, land beneath houses and all trees were communized. Small numbers of domestic animals, however, might still be privately reared.

The transition to communes took place within six months and was completed between June and December, 1958. The communes themselves are now divided into production brigades, which in turn are divided into production teams comprising some 20 to 50 households and often co-terminous with a small natural village.

There have been some reservations as to how much all of this restructuring actually achieved in production terms. In spite of the well-known fact that data on output of agricultural produce are very hard to come by for China, Healey shows that grain output in China from 1950 to 1970 rose by about 60 per cent although the trend was marked by periods of growth and stagnation particularly in the

Table 12 *Some output data for the People's Republic of China, 1952–72 (in million metric tons unless otherwise specified)*

Year	*Grain*	*Cotton*	*Chemical fertilizer*	*Electric power (kwh x 10⁶)*	*Coal*	*Crude oil*	*Steel*	*Cement*
1952	154	1.3	0.2	7.3	67	0.4	1.4	2.9
1957	185	1.6	0.8	19.3	131	1.5	5.4	6.9
1959	170	1.6	2.0	42.0	300	3.7	10.0	11.0
1965	208	1.5	8.0	42.0	220	8.0	11.0	11.0
1970	240	1.7	14.0	60.0	300	20.0	18.0	13.0
1971	250	1.6	18.0	70.0	325	25.0	21.0	16.0
1972	240	1.6	21.3	n.a.	n.a.	29.0*	23.0	20.0

Sources: Joint Economic Committee's publications on the Chinese economy derived from various issues of the *Peking Review, China Reconstructs, Far Eastern Economic Review,* and from the work of US scholars. See also J. G. *Gurley*, 'Rural development in China 1949–72', *World Development,* vol. 3, no. 7/8, p. 457.

**Far Eastern Economic Review* (19 February 1973, p. 5) reports 42 million tons.

drought years 1959–61.[39] None the less, Eckstein argued that the Chinese achieved a more rapid expansion of factor inputs rather than output and Table 12 provides ample evidence of this. The consequence of this according to him is that in purely economic terms total factor productivity must be seen as having declined, indicating a degree of resource misallocation and waste.[40]

Gittings, however, points out that part of the difficulties facing Western assessment of Chinese achievement until recently were attempts to apply performance criteria of a capitalist economy to a very different set of societal situations.[41] With regard to Chinese rural development, he argues, no assessment can be meaningful unless its premises are based on the fundamental problems of socialist development which the Chinese set out to deal with in their countryside. This problem, simply stated, is

> how can one move towards a higher level of social ownership and distribution in such a way that (i) it will cope with existing vested interests and win popular support, (ii) it will reduce existing inequalities without lessening the incentive to production, and (iii) it will advance on an adequate material base and, before too long, stimulate the productive forces to higher output.[42]

Against such a background, aggregate statistics of production tell less than half the story of Chinese developmental effort. The other, and perhaps the more important, half is, as Gray puts it, the emphasis upon investment in the economic improvement of local communities, which would feed back into modern industry, instead of upon investment directly in modern industry in the hope that the effect would spread downwards.[43] Indeed Gray concludes that this choice is always there to be made by 'peasant countries seeking to modernize, whatever their political system or ideology.[44]

Tanzania

The Tanzania experience is of special significance, both because the basic problems are similar to those existing in many underdeveloped African countries today, and because in confronting them squarely, the government was unwillingly forced to move from a policy of gradualism to one of speedy transformation. Up to 1973, the situation in most of the rural areas of Tanzania was similar to that described in the study of the Rungwe District by van Hekken and Thoden van Velzen.[45] The impact of the capitalist mode of production on the rural economy of the country had given rise to emergent class forms which threatened or were superseding the basically egalitarian structure of traditional Tanzanian society. The position of this emerging rural capitalist class was becoming entrenched in two major ways: first, through their providing credit of land or money to poorer cultivators who thus became dependent on them; and second, through their forming alliances with intermediaries who controlled the channels to the outside world, notably government personnel stationed in rural areas.

Attempts to improve rural conditions prior to 1967 had taken the form of resettlement programmes through village settlement schemes. These programmes which were initiated in the 1950s and carried on by the national government soon after independence in the early 1960s, had, however, minimal impact on the rural situation and tended, if anything, to aggravate it. The task of 'unfreezing' the situation could thus be said to begin with President Nyerere's open declaration in 1962 that the country was going to opt for a socialist form of societal organization. The implication of this for the rural areas was later spelt out clearly in a document titled *Socialism and Rural Development*.[46] This indicated that rural development in Tanzania would be prosecuted through the establishment of '*ujamaa* villages'. These are 'co-operative communities in which people lived together and worked together for the good of all'.[47] Ujamaanization, it was hoped, would enable the government to deal with the largely dispersed nature of the country's rural settlement and population. This had made it difficult to achieve greater returns from agriculture on individual holdings, and to provide essential productive and social services to the farming population. Ujamaanization thus entailed:

(a) voluntary collectivization of the population into nucleated *ujamaa* villages;
(b) communal ownership of land resources;
(c) formation of village governments with power to plan and implement local projects;
(d) provision of social services such as schools, dispensaries and water supply.

In the period 1967–69, a selective approach was adopted in which, through exhortation by the President and the Party, a few villages agreed to serve as models for the whole country. An Act of Parliament in 1968 codified the principles of ujamaanization. However, on the eve of the Second Five-Year Plan in 1969, only 180 *ujamaa* villages embracing 60,000 people had been formed. As a consequence, the new Plan spelt out in greater detail total government commitment to the *ujamaa* programme and enjoined all party and government institutions to assist in its vigorous prosecution. The plan, however, still left wide open the issue of a *modus operandi* for achieving nation-wide ujamaanization, of making appropriate resource allocation to the programme and indicating realistic production targets. Consequently, because the discipline involved in this form of societal reorganization was considerable, initial progress was slow. Indeed, in the richer agricultural district of Kilimanjaro, there was open hostility to the programme. None the less, some considerable progress was made, especially in regions with special problems such as flooding, drought or famine. Many *ujamaa* villages were established in Rufiji, Mlwara, Dodoma, Kigoma, Iringa and Chunya. The total number of *ujamaa* villages rose from 650 in 1969, to 5628 in 1973, and the population involved from 300,000 to over 2 million.

As a result of this slow pace, the sixteenth biennial TANU (Tanzania African National Union) Conference of 1973, in reviewing the situation, decided on a policy which recognized a three-stage sequence in the evolution of a *ujamaa* village. The first stage was the physical creation of villages; the second, the introduction of co-operative organizations; and the third, the move towards a collective mode of production. This three-stage process was reflected in the new terminology.[48] Development villages (*vijiji vya maendelo*) are nucleated settlements, or in some cases areas of dispersed but concentrated rural settlement, where individual and family modes of production are to be diversified and supported by the village primary co-operative society. In the third stage, *ujamaa* villages (*vijiji vya ujamaa*) can be formed when the production activities are conducted on a fully fledged collective or communal basis. The conference also decided that the first stage of villagization must be completed throughout the country by 1976.

Regional executive committees were instructed to draw up plans for the immediate implementation of 'Operation *Vijijni*', that is, the mobilization of the entire rural population to take up residence in new village settlements. On 1 March 1974, the first phase of the operation started under the supervision of district development directors. This phase comprised both the measurement and evaluation of old housing and the identification of suitable sites for the new settlement. The latter involved consultation at ward level and with district planning officials. It took into consideration such criteria as land availability, road accessibility, existing services such as co-operative societies, dispensaries, schools, and water supply.[49] At the same time, the pattern of homestead plots of about one and a half acres was laid down. And, although in addition to these plots, farmers were allocated land in the area surrounding the village, during the first year of operation they were allowed to continue cultivation on their old plots so as to avoid the loss of inputs, especially labour already expended.

The movement of people to the new villages started during April 1974, depending on the need to complete surveys and establish sites. By July, over 70 per cent of the population was in the process of shifting to the new villages, although many houses were still in the course of construction. By August, the process had been completed everywhere. Figure 6 shows the change in settlement pattern at the beginning and at the end of that single year.

Inevitably, the speed with which the operation was carried out coupled with the rather rudimentary pre-planning process, has led to a number of faults in the resulting village structure. A process of readjustment is thus continuing in all districts, with a reappraisal of existing villages with a view either to establishing new ones or to absorbing some of the population of large settlements, or to reorganizing existing ones. A further phase in 'Operation *Vijijni*' was the enactment of the Villages and Ujamaa Villages (Registration and Administration) Act of August 1975, where provisions were made for every village to become a multi-purpose co-operative society.

It is still too early to embark on a critical evaluation of the Tanzanian effort at societal transformation. For one thing, the new social relations implied in the two stages subsequent to villagization are still to be firmly established. But it is obvious that the Tanzanian leadership fully grasped the close relationship between spatial form and social process. Hence, the significance of the assertion that, for Tanzania, the village is not only the focus of development itself but the basic unit for the emergence of a new Tanzanian society. An impressionistic view of what that society is already becoming is provided by Blue and Weaver in their critical assessment of the Tanzanian model of development in 1977, soon after the completion of the villagization process.[50] According to them:

> In comparison to other underdeveloped countries in Africa, Asia and Latin America where we become accustomed to seeing masses of poverty-stricken people in the cities and rural areas, the striking thing in Tanzania is the absence of such people. We were able to travel freely and the people we saw looked relatively well-fed and well dressed. We saw few beggars on the streets of the cities, nor did we see little children with bloated bellies and ribs sticking out in either the cities or the villages. We saw no extensive slums, and we also

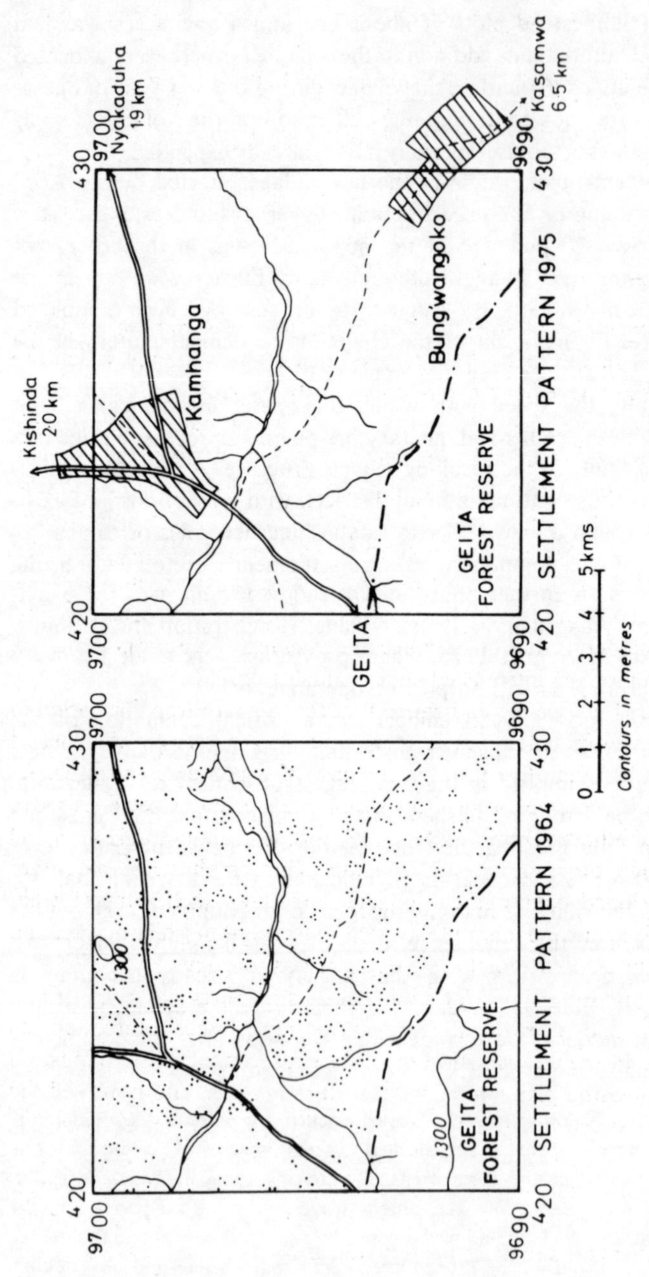

Figure 6 *Tanzania: Operation* Vijijni, *Geita district*

noted that almost all school-age children wore school uniforms. . . . The converse of this was the almost total absence of luxury goods. There are very few cars. There is a great shortage of luxury foodstuffs. Tanzania knows the meaning of income equality. It means that everyone has access to basic necessities, but the rich are denied most of their luxuries. This is in stark contrast to the pattern seen in Kenya, India, the Dominican Republic, Jamaica, etc.[51]

Conclusion

The five examples of societal effort at rural transformation represent only a few of the many that can be cited from countries which are either already developed or are consciously progressing in that direction. Their experience shows that development requires a massive effort of societal change that is comprehensive in its scope and speedy in its execution. The comprehensiveness is entailed in the fact that at any given point in time, production organization accounts for a large part of the social relations in a given society. To change this organization therefore is to change the society. Hence, the failure of rural development projects which are conceived as isolated schemes, and the contradictions that quickly become manifest when programmes of transformation are introduced on a gradualist basis.

It may, of course, be contended that the speed of transformation in the five examples cited, varied from nearly sixty years in the case of Britain to less than five years in the case of China and Tanzania. As has already been mentioned, this is not unexpected as societies move into the twentieth century and the means of mass communication and mobilization are more easily to hand. This emphasizes that other underdeveloped countries can equally undertake such a transformation. It also underscores the fact that indecision and dilatoriness can bring about an abortion of the process just as speedily.

None the less, the relation between such massive rural transformation and growth in production cannot be established as automatic in the very short term. It depends, among other things, on how detailed the planning of the transformation is, and how well the various relevant programmes are co-ordinated. Where the planning has been well done and the transformation process results soon after in increased production and productivity, other problems may emerge, notably a depression in the price of agricultural produce and in rural income. In either situation, what makes a developed or really developing society different from an underdeveloped one is the greater capacity to cope with either type of crisis. For, as indicated in the earlier chapters, development is not simply economic growth but the enhanced capacity of societies and individuals 'to cope with the changing circumstances of their lives'. Rural transformation should thus make it comparatively easier for a country to deal more effectively with the vagaries of agricultural production, whether this arises from weather conditions or from internal and international trade conditions.

Part Three

Urban Development

7 Emergent urbanization

The development process in any country can hardly take off successfully if efforts are concentrated solely on rural and agricultural development. This is because increased productivity within the rural sector can only be sustained through simultaneous development in urban and industrial activities. Apart from the social services provided for the rural population, urban centres constitute the location for industrial enterprises engaged in the final stage of processing agricultural raw materials and serve additionally as markets for food crops from the rural areas. Hence, in the development process, rural and urban transformation must be seen as two sides of the same coin.

None the less, the interrelationships in the transformation process between urban and rural areas are very complex and need to be understood within a specific historical context. For most underdeveloped countries, for instance, present-day urbanization has been one of the more potent means used in achieving the colonial and capitalist penetration of their traditional economies. Initially, its role was to consolidate colonial control over formerly recalcitrant but virtually self-sufficient societies whose technological backwardness made it easy for them to be subjugated and pacified. This was the classic administrative role of colonial cities which was to maintain law and order. Taxation was the mechanism whereby this control was exercised and the penetration initiated. The colonial administration determined the specie in which the tax was to be paid and thus compelled the colonized group to enter into a money-exchange economy in which, at least for that portion of their production which they intended to turn over to pay their tax, they had to seek an equivalent in cash. Colonial taxation was thus a form of accumulation of social surplus value and a factor in the emergence of urban centres in the nineteenth and early twentieth centuries in many underdeveloped countries particularly those in Africa.[1]

Taxation, however, was only the first step in mobilizing surplus value available in traditional economies. Its proceeds were certainly needed to maintain the class of colonial administrators and their local collaborators. More important, however, in mobilizing the surplus was trade which offered foreign goods and services in exchange for local produce, usually on terms dictated by foreign firms. None the less, irrespective of these terms of exchange, there is no doubt that the availability of a wide range of cheap manufactured goods encouraged the rise of production specifically for an overseas market. In this way, urban centres emerged in many underdeveloped countries as the point of trade where local produce was bought and bulked for export and imported materials were sold and distributed in exchange.

Typology of colonial urbanization

It must not be assumed that all parts of the underdeveloped world had no urban centres of their own prior to their colonization by European powers. Indeed, particularly in areas of well-organized social polities as in parts of Asia and West Africa, such traditional urban centres had established mechanisms for articulating and accumulating surplus from surrounding rural areas well before the colonial period. All that the colonial administration did, therefore, was to superimpose itself on the traditional political elite and make them dependent on it for their share of the accumulated surplus.[2]

None the less, the presence of a traditional urban sector had other far-reaching implications. First, it meant there already existed a fairly well-organized class of non-agricultural producers who were to feel more poignantly the impact of exposure to foreign trade and foreign goods. Second, the difference in the basis of social class differentiation between the traditional and the colonial systems was to give a particular twist to conflict in the emerging colonial society. Third, the difference in the spatial confrontation of particular functions in the traditional and colonial society and economy, and the close physical juxtaposition of the two types, was to produce a new and infinitely more complex urban form and organization. The complexity of the colonial city had not, until recently, attracted the attention of students of urbanism. As Wheatley argued:

> The cultural hybrid of the colonial city, which typically subsumes elements of both the traditional and the modern world, and which consequently might have been expected to have excited the curiosity of urbanists, has in fact attracted little more attention than the traditional city proper . . . only recently has there been any attempt to particularise the specifically colonial features of the genre of city and to integrate it into a general theory of urbanism.[3]

It must, however, be obvious that the term 'colonial city' can be applied to three different types of urban formation.[4] First, the city formed *de novo* by the colonial administration, a large proportion of whose population, none the less, is indigenous to the area; second, the traditional city to which has been grafted new suburbs of colonial administrators and merchants and their local agents; third, particularly in those underdeveloped countries with a 'white settler' population, the so-called 'colonial' city is to a large extent no different from cities in the developed countries, except for one notable feature: the neighbouring shanty town of underprivileged and disadvantaged indigenous population who provide cheap labour for the settlers' economy. Where, as in southern Africa, a formal racial policy exists, the contrast between the two areas may be very sharp.[5] In other situations, there may be a graduation through suburbs of relatively more affluent groups of indigenes on to the 'bidonvilles' of their poor compatriots.

Irrespective of the type of colonial city being considered, one common characteristic is that for most of the colonial period their economy was centred mainly on trade.[6] Even when demand had grown large enough to support local manufacturing, there was a clear policy decidedly against such development. The growth of trade, on the other hand, was facilitated by the development of a modern transportation system based initially on the railway but later, particularly after the Second World War, on a road network. Rail development in the colonial period was strongly selective in its regional emphasis. In general, rail construction was concentrated in regions which could produce exportable crops, while regions which were only capable of producing local food surpluses tended to be neglected.

The development of urban centres in the colonial period thus strongly reflected these biases in the transport system. The port was *par excellence* the colonial city. All transport systems focused on it and it was the main centre for the export–import trade. Most of the major foreign commercial houses had their headquarters there. Especially where there was no settler community whose needs may, as in Kenya, administratively outweigh in importance that of trade, the port city was also the capital of the territory. Other major centres were developed at critical nodes on the transport system with a tendency for those within the export-crop producing regions to grow much faster than any other urban centre apart from the port city. Provincial administrative headquarters, where they were not already on the transport network, also attracted more urban functions to themselves and showed, if rather slowly, a significant rate of population growth.

In short, whatever the traditional situation, the establishment of a colonial economy imposed on the landscape a new system of central places. This system provided the means whereby the whole country, now constituting one large colonial market, was flooded with cheap, mass-produced articles from the factories of the metropolitan country. While these trading activities gave the appearance of vigorous economic growth, they had long term deleterious consequences for the overall economic and technological development of the country. The significance of these consequences is better appreciated by examining in some detail the nature of the colonial impact on the traditional non-agricultural activities found in urban centres.

Colonial impact on the traditional urban economy

As in most pre-industrial societies, the urban economy in many under-developed countries prior to their colonization was essentially a response to the needs of the rural population. This economy comprised, apart from administration, trading and various social services, largely of cottage craft production. Many of these craft activities were undertaken here and there in the rural areas but urban centres were the scene of their greatest concentration. Using the situation in West Africa for illustration, one finds that as early as the fourteenth century, Leo Africanus described the urban economy in Gobir, the capital of one of the seven Hausa states in what is now Nigeria, as comprising 'great stores of artificers and linen weavers; here are such shoes made as the ancient Romans were wont to wear, the greatest part whereof is carried to Timbuktu and Gao'.[7] In much the same vein, Hinderer described the economic activity of Ibadan (also in Nigeria) around 1851, a half century before its colonial subjugation, as follows:

> . . . there is a good deal of industry to be seen in and around the town. There are the weavers, the tailors, the tanners, leather dressers and saddlers, the iron smelter and the blacksmith, a kind of country sawyer and carpenter, and last but not least the potters, the palm-oil and nut-oil and the soap manufacturers of the female sex to be seen in all parts of the town, some of them very busily engaged in their respective occupations. There are a great many markets all over the town for vending the product of their farms such as yams, beans, corn, cotton and food prepared, as well as other necessaries of life, but there is one large market especially for home and foreign manufactures, with all kinds of European articles which have yet been imported'.[8]

This picture of active cottage craft industry did not long survive the establishment of the colonial economy. One of the best studies of the nature of the colonial impact on traditional craft industry is that by Nadel on the Nupe Kingdom of Nigeria between 1934 and 1936. On the blacksmith's craft in particular Nadel wrote as follows:

The blacksmiths are called *tswata* in Nupe, or sometimes *tswata ghagha*, 'heavy smiths', to distinguish them from the *tswata muku*, the brass and silversmiths. In Bida, we find three groups of blacksmiths, whole blacksmith 'wards' in fact, each with its cluster of workshops, and each situated in a different *ekpa* (or section) of the town. According to tradition, the three blacksmith groups are of different origins. . . . The largest in the Etsu Umaru quarter traces its descent to the blacksmiths of Raba, the former capital of the kingdom. They came to Bida together with the king and the court when Bida became the capital of Fulani Nupe . . . and . . . established as the supreme guild authority over the blacksmiths of Bida and all Nupe. The craftsmen of this group call themselves *dokodzazi*, after their head, who bears the title *Dokodza*.

The blacksmith wards of Bida are of very different sizes; the smallest contains two workshops, the largest seven. In some cases, the entrance hall of the compound in which the families of the craftsmen are living serves also as workshop or forge. But usually, the forge is in a separate hut which stands by itself a small distance from the compounds.

Iron work involves four subsidiary forms of production: the manufacture of the bellows, of charcoal, of the tools, and finally of the raw material used in blacksmith work, that is, the mining of iron ore. With the exception of the last all are undertaken by the blacksmiths themselves.

Formerly all iron-ore used in blacksmith work was mined and smelted in the country. The production of natural iron-ore has been greatly reduced in recent years. Today, a large proportion of the crude iron is bought from the European stores. . . . The competition of cheap European iron has made the upkeep of the large native foundries unprofitable; many have been abandoned and allowed to fall into disuse, and even where they are still in use only two or three out of perhaps six furnaces are still workable. . . . Nupe iron-smelting can in fact no longer meet heavier or more regular demands, although many Nupe farmers still prefer a hoe made of native-mined iron. It is stronger and lasts better, they say; European iron is fit only for repairs, for patching hoe blades, etc. The tools of the blacksmiths, and, above all the anvil, are invariably made of native-mined iron. . . .

The organization of blacksmith work and iron-ore mining involves two different types of co-operation, the actual working together of the *efako* unit (embracing a number or brothers, their children and grandchildren) and the wider co-operation in the framework of the occupational unit or guild.

The control of the Bida guild system extends in certain respects over the whole country and the various independent local blacksmith groups. This widest professional unit is the result of political rather than economic factors; it is expressed mainly in the allegiance to a common head, the master of the *Dokodza* section in Bida. As this group of Bida blacksmiths is associated historically with Nupe kingship its head also holds rank and position by appointment of the king of Nupe. Within his section, the *Dokodza* is the agency through which important large orders reach the group. On a smaller scale, this task also falls to the heads of other blacksmith groups. But in the case of the guild-head, the *Dokodza*, it gains special significance; for these 'royal blacksmiths' worked for the largest customers of the country, the Nupe court. They could be certain of a constant large demand, for tools for the farm hands working for the king, or various iron implements needed in the royal household, for horse bits and stirrups for the king's horses, or flint-locks for the guns of the bodyguards. These orders are given by the king or court officials to the *Dokodza*, who then distributes the work among the craftsmen of this group according to their capacity and skill, or, if the order proves too large, among the workers in other guild sections as well.

But the blacksmiths in the country are bitterly complaining of the decline of native blacksmith trade. Some of these country blacksmiths have turned farmers, or have at least taken up farming on a larger scale than formerly, in order to balance the losses of their industry. This retrograde movement is due to a variety of causes. One which has affected all blacksmith groups in the country is the competition of European trade. Thus, the Nupe blacksmiths have lost almost their entire trade in bush-knives to the European trading stores; nor does the once very extensive trade in Bida hoes which the Hausa caravans used to take north exist any more. . . .

A more subtle cause of the decline in blacksmith trade lies in the failure to meet successfully the demands of what we have called 'maintenance of labour'. . . . Nupe blacksmith work demands comparatively large group co-operation for a successful maintenance of labour (in all its aspects). In a weakened organization, a markedly reduced co-operative group, maintenance of labour becomes increasingly difficult. In Kutigi, I saw one of the two local blacksmith workshops left dilapidated for a whole year because the workers could not muster sufficient labour. The most conspicuous example of all is that of iron-miners who had let their means of production, their furnaces, fall into disuse.[9]

This lengthy quotation is to make more vivid the impact of the colonial regime on the traditional urban economy. Similar consequences on traditional urban economies are recorded for countries like India, Morocco and Iran.[10] On the face of it, it is easy to compare this situation with what happened, for instance, in Britain in the early decades of the Industrial Revolution. Clapham pointed out, for example, that 'technological unemployment' was very serious among handloom weavers of cotton, flax and worsted, from the 1820s to the 1840s, and to a lesser extent among nail and chain makers, handmade lace workers and framework knitters, as machinery slowly conquered their trades.[11] However, by 1860 the industry of the country as a whole was expansive enough to absorb most of the people so affected since in a sense there was some organic continuity in the transformation taking place. But in the underdeveloped countries no such continuity was allowed or encouraged to develop. Hence, in at least three respects, the colonial impact had a deleterious effect on the traditional urban economy. First, it brought about a certain diminution in the knowledge concerning local resources. With the closedown of various iron mines, for example, Nigeria and other underdeveloped countries in the same circumstances, went through a period when they hardly realized they had significant deposits of these resources. Again, as happened in the case of agriculture, the colonial regime simply selected those local resources of interest to it, such as tin ore, coal, and gold in Nigeria, and concentrated on a vigorous geological survey of areas where the 'natives' already knew these minerals existed. In short, one outcome of the manner in which underdeveloped countries were, as it were, 'de-technologized' was the loss of interest in a detailed inventory of their national resources. This disinterest has not altogether disappeared and is often manifest in the over-reliance on foreign technical assistance to undertake an inventory of such resources, even where there is at least a core of well-trained local personnel.

The second aspect of the impact is in terms of the loss of traditional knowledge and skills of coping with the peculiar problems of the environment in underdeveloped countries. The issue here is not whether traditional knowledge or production processes were more efficient or could turn out more products per unit of man or time. Rather, it relates to insights and accumulated experience which the traditional society had gained through continuous investigation into the special characteristics of particular natural resources or into how to process them in the particular circumstances of climate and ecology. What most underdeveloped countries in particular, and global society in general, have lost in this way will never be fully known.

Third, there is the impact in the incapacity of many underdeveloped countries to develop their own authentic production organization. How the Nupe society, for instance, might have adapted organizational structures like the *efako* or the guild in the context of changing situations, will remain a moot issue. At present, many of these countries are struggling to adjust to foreign production organizations derived mainly from the peculiar European experience. It is instructive to note, for instance, that the Japanese have achieved the same or even better results in efficiency by adapting their own traditional structures to industrial production. Or, as Hazama puts it, 'the virtually diametrical confrontation of behavioural patterns between the West and Japan was manifest not only on the individual level but on the institutional management level as well. The outcome was compromise; traditional Japanese employment relationships were incorporated, emerging as inextricable components of the modern Japanese management system'.[12]

In place of adaptive possibilities, the colonial city offered the population in underdeveloped countries the option of participating in the new capitalist economy through a narrowly conceived elementary liberal education. Such education prepared its recipients to serve as clerical officers in various types of government institution or commercial houses, and as teachers to train still more people for essentially white-collar employment. In terms of social prestige and financial remuneration, white-collar workers, as a group, became the next most important class to the colonial administrators and merchants. Families seeking to ensure the future of their children sent them to schools in the hope that this would eventually guarantee them access to white-collar employment. The colonial city was thus also the place to migrate to, not only for employment but also for the educational opportunities which it offered. None the less, the volume of migration was kept in check by the rather restricted number of schools and of vacancies in employment houses. Thus, although urban centres registered some growth during the colonial period, this was on the whole quite modest.

Much more important, however, was the effect of the type of education and the relatively higher wages of white-collar workers, of stimulating taste for imported, largely non-durable consumption goods. The society in the colonial city came to pride itself on how up to date it was in adopting the latest fashion from the metropolitan country, and of consuming increasingly large amounts of foreign goods. Itself producing very little, the colonial city became a major centre for diffusing new and alien tastes and culture throughout the country. The adoption of such tastes and appetites came to be equated with modernization, and participation in the distribution of goods and services to satisfy this demand came to constitute a major road to modest prosperity.

A colonial city culture gradually emerged. It was a culture which combined aspects of the culture of the metropolitan society and that of the indigenous, pre-colonial society. King refers to it as a 'colonial third culture' and emphasizes that it has emerged 'not simply as a result of interaction with a second culture in a "neutral" diffusion situation, but necessarily, as a result of colonialism, operating in a dominance-dependency power relation'.[13] This culture went beyond patterns of consumption to embrace a whole 'institutional system which comprehends ideational systems, meanings and symbols, social-structure, systems of social relations and patterns of behaviour. They are a form of government, an educational system, a type of family organization, economic institutions, forms of knowledge, language and literature, technology, and a whole range of social beliefs, groupings and cultural artefacts – including built form and urban patterns'.[14] Some aspects of this culture were, from the outset, shared with members of the indigenous society, for example certain items of dress, diet and language. Others were shared and diffused only gradually over the period of colonialism, such as aspects of government, administration, education, ideologies, knowledge systems and cultural artefacts. This cultural colonialism remains a most pernicious link which ties underdeveloped countries into a dependency relation with the metropolitan country. Breaking out of this circle is perhaps one of the most difficult tasks before most underdeveloped countries. How difficult the task is will be further appreciated when it is realized that post-colonial industrialization has added one more strand to that link.

Post-colonial urbanization and import-substituting industrialization

For most underdeveloped countries, the period after the Second World War has been notable for the movement towards and the eventual attainment of political independence. Twenty years after the end of that war, a high proportion of the countries in Africa and Asia were being administered by governments composed largely of indigenes. As indicated in Chapter 4, the leaders of these governments had rallied their people to struggle for independence under the banner of eradicating from the country within a foreseeable future, the three-fold societal disabilities of ignorance, disease and poverty. With independence, therefore, most of these governments directed considerable energy to the achievement of these goals.

The eradication of ignorance was conceived simply in terms of expanding educational opportunities. More schools were built, particularly at the primary level, and located not only in the cities and towns but extensively in villages and hamlets. The number and proportion of children of school age who were actually enrolled rose rapidly and the level of illiteracy, especially in the younger cohorts of the population, diminished correspondingly. In Africa, for instance, the number jumped from 9.5 to 66.4 million between 1950 and 1984. During the same period, enrolment rose in Asia from 52.7 to 202.6 million and in Latin America from 15.1 to 60.9 million.[15]

But, in all this, no major changes were made to the curriculum. The children were still taught in a way to make the securing of white-collar employment the main passport to a successful life career. The consequences of this major social development effort were particularly critical in the rural areas. The education offered to the children alienated them from their environment, introduced them to new tastes and styles of life but failed to equip them adequately to earn the necessary income. The result was the phenomenon of massive rural–urban migration which in most underdeveloped countries today has reached real crisis proportions.

The same contradiction between intent and result became obvious in the attempts by various governments to eradicate disease. In most years, these governments make impressive financial outlays on epidemiological control, preventive and social medicine and particularly on hospitals, antenatal and maternity clinics. Equally important allocations are directed to the provision of modern water supplies to an increasing proportion of the population. The effect of all this has been to bring about a marked drop in mortality rates. This was noticeable in both urban and rural areas, but particularly in the former where most of the health agencies were concentrated. On the other hand, less attention was paid to the reproductive behaviour of the society, with the result that in most underdeveloped countries the situation today is characterized by high population growth rates of between 2.0 and 3.6 per cent per annum. Such growth rates mean a doubling of population every twenty to thirty-five years. Given the collapse of the traditional rural structures and the mass exodus of youth from the rural areas, the problem of feeding this rapidly increasing population is another major challenge of development.

But perhaps the most challenging task has been that of eradicating poverty. Here again, returns on effort have been the very obverse of what was hoped for. In the early 1950s, deriving from what was going on in Latin America, most underdeveloped countries were counselled to undertake the task of rapid economic development through industrialization based on a strategy of import substitution. The rationale of this strategy was that through starting with 'final touch' industries, a country would be led systematically by backward linkage effects, to developing domestic production of intermediate, and finally basic, industrial materials.[16] This strategy was adopted and has resulted in numerous consumer goods industries being established in many underdeveloped countries in the last two decades. Table 13 emphasizes the remarkable rate of growth in output in the manufacturing industry in the period after 1953. None the less, far from this helping to solve the problem of poverty for the masses in any meaningful way, this strategy has tended to exacerbate it in at least four ways.

(a) *Growth of urban unemployment*

The expectation from import substituting industrialization was that by establishing factories which would produce formerly imported articles, employment opportunities in urban centres would be greatly enhanced. However, most of the machinery was imported from countries which, in response to labour costs, had gone in for capital inten-

Table 13 *Index and rate of growth of output of manufacturing industry in underdeveloped countries (excluding communist countries), 1938–70*

Year	*Index of output 1963 = 100*	*Annual rate of growth (per cent) Period*	*Total*	*Per caput*
1938	29	1938–1950	3.5	1.8
1948	40	1950–1960	6.9	4.5
1950	44	1960–1970	6.3	3.6
1953	49			
1958	73			
1960	86	1938–1960	4.9	3.0
1968	138	1938–1970	5.3	3.1
1970	158	1950–1970	6.6	4.1

Sources: Derived from United Nations, *The Growth of World Industry 1938–1961*, New York, 1963: United Nations. *Monthly Bulletin of Statistics*, New York, May 1973.

sive industrial processes. In consequence, industrialization in most underdeveloped countries tended to result in a high capital–labour ratio and to provide relatively limited employment opportunities. Yet, the establishment of factories had raised hopes of employment and this continues to stimulate large numbers of migrants from the rural areas, many of whom happily join the growing substrata of urban unemployed in the abiding hope of eventually finding a job in this modern sector of the economy.

The problem of urban unemployment is discussed in greater detail in the next chapter. What is important to stress at this juncture is 'the paradox of the import substitution type of development which, while it discriminates severely against the agricultural sector and appears to be partly responsible for the fast growth of large urban centres, has led to excessive patterns of dualism within the urban sector and made the absorption of new urban dwellers more difficult, not easier'.[17]

None the less, evidence from Taiwan, Korea and Colombia suggests that if an underdeveloped country succeeds in going beyond the import substitution to the 'export substitution' phase of its development, expansion in industrial output may go hand in hand with growth in employment.[18]

(b) *Inflation and heightened rural – urban inequality*

In spite of the high capital cost, the productivity of these local factories is never such that they can face price competition from continued imports of their products. As such, 'infant industry' arguments are usually advanced to make governments in underdeveloped countries erect a substantial tariff wall of protection around them. Such protection not unexpectedly encourages shoddiness and the production of relatively inferior goods, for which the price remains as for previously imported ones or even rises higher. Since, under the aegis of organized trade unions, urban wages are made to keep up with such price variations, the inflationary effect is less problematic for the urban dwellers, except perhaps for the unemployed and the low income groups not employed in the formal sector. For the mass of rural dwellers, the major crop production is not only taxed internally but, where exported, suffers constantly from price fluctuations on the world market. Yet, they cannot derive advantage from cheaper and better imported goods but must buy the highly priced and poorly finished local production. A situation of unequal exchange is established which progressively aggravates the disparity in real income between the rural and urban areas of the country, and in turn, stimulates the growing exodus from the former.

In short, import substitution industrialization, because of the high degree of protection, penalizes agriculture and the rural sector in two ways. First, it raises the price of industrial as against agricultural goods in the domestic market. Second, the artificially high exchange rate, which is part of the protection package, reduces the receipts in terms of domestic currency from a given volume of agricultural exports thus discouraging agricultural expansion. Indeed, Little, Scitovsky and Scott noted that during the post-war period 'the price of manufactured goods in relation to farm prices, over much of the period, has been twice as high on the average as the world market prices would be'.[19] They concluded that 'the general pattern though not unambiguous suggests that domestic terms of trade have moved too greatly against agriculture'.

(c) *Foreign exchange crisis and development strangulation*

The import substitution strategy failed to consider the likely behaviour of the actors involved, particularly the multinational corporations who dominated the establishment of the industries. For most of these, the strategy, if logically pursued, especially in terms of its backward linkage relation to the rest of the economy of underdeveloped countries, would disrupt large areas of corporate organization and create a crisis of confidence with long term suppliers. As a result, far from reducing dependence, import substituting industrialization has meant greater importation of raw materials, fuel, spare parts, equipment, technology and capital. Because most of these represent 'lumpy' investment, most underdeveloped countries soon find themselves in a real foreign exchange crisis. The nature of this crisis is perhaps best exemplified in the recent publications by the United States Senate of the implications of multinational firms for world trade and investment and for trade and labour in the United States.[20] According to this report, the United States had obtained significant overall benefits in the late 1960s from the world-wide operations of US-based multinational enterprises. In terms of overall basic balance of payments such firms had improved their position by 2.8 billion in 1966-70. The report concludes:

> The most consistent of the conclusions is that the US-based transnational enterprises in their transactions with the United States, exert a uniformly

large, negative impact on the current accounts of balances of payments of the host countries. Conversely, of course, they have a favourable impact on the corresponding account of the US balance of payments.[21]

Not unexpectedly, therefore, import substitution industrialization has been accompanied in many underdeveloped countries by a progressive trend into international debt and insolvency. Indebtedness not only slows down the rate of industrialization but greatly compromises overall development. For many countries, the recourse often advised is to create conditions still more favourable to the importation of foreign capital and thereby to lock them more securely into the dependency relation with advanced capitalist countries.

Hughes noted that the debt of underdeveloped countries grew from $8 billion in 1955 to $16 billion in 1960 and some $35 billion in 1965. By 1977, it totalled $210 billion.[22] It jumped to over $424 billion by 1983. Furthermore, although in 1977, about 40 per cent of the outstanding debt of underdeveloped countries was from official sources, mostly with a substantial concessional element, this had dropped by 1983 to less than 18 per cent, the remaining 82 per cent coming largely from private sources. Much of the latter was incurred in the process of attracting foreign investors to undertake the industrialization of most of these countries. The generous investment environment created for multinational enterprises often makes it such that much of the profits from their operations are repatriated surreptitiously through various forms of transfer pricing or charges such as management fees, royalties or interest which accrue to the parent firm leaving the burden of the capital debt unliquidated. Indeed, except for low income countries (with a per caput income of less than $265, that is largely those in Africa south of the Sahara and South Asia), debt has grown at about the same rate as exports, on average remaining fairly stable at about 90 per cent of annual exports from the mid-1960s. For low income countries debt was about 200 per cent of annual exports, for oil exporting countries less than 50 per cent of annual exports, and for the other 'middle income' developing countries, it rose from about 80 per cent to just over 100 per cent between 1967 and 1976.[23]

(d) *Low technology transfer*

Because of the highly capital-intensive nature of most of the machinery and the overriding motive of profit maximization on the part of the multinational corporations, import substitution industrialization has provided very minimal opportunity for the transfer of technology from developed to underdeveloped countries. Streten suggests that two major obstacles have made such transfer difficult: a communications gap and a suitability gap.[24] The communications gap arises as a result of the high costs of transfer, international restrictions, or the exercise of monopoly power. Even if this difficulty is overcome and transfer could be effected, the suitability gap will emerge. This reflects the fact that most techniques which are available were designed to suit developed country factor endowments – including the types of labour available and not merely the aggregate. On the other hand, few underdeveloped countries have set out to reduce the suitability gap by investing seriously in technological research and development (R & D). Until recently, it was estimated that the share of underdeveloped countries in world R & D expenditure is only 2 per cent.[25]

The difficulty of technology transfer through the current strategy of import substitution industrialization can undermine any future prospect of independent and autonomous industrial development in underdeveloped countries. As Frank puts it:

> The import substitution process, instead of reducing import requirements, thus increases them. Moreover, it tends to raise the cost of imports as it becomes necessary to import ever more technically complicated, advanced, *monopolized,* and thus costly, equipment from the metropolis. Yet this import substitution cannot reach the point at which the satellite ceases to be dependent on the metropolis for equipment, technology and critical raw materials, since this industrial blind alley leads the satellite country away from rather than toward the production of these necessities . . . Technology intercedes in the metropolis–satellite relationship and serves to generate still deeper satellite underdevelopment. It is technology which is rapidly and increasingly becoming the new basis of metropolitan monopoly over the satellites.[26]

It is, of course, arguable that the issue here is one of a time scale and that many underdeveloped countries, particularly the medium sized or large ones, can, over two or three generations, hope to break out of this dependent relation. This is no doubt a possibility but for it to be realized a heightened consciousness as to the complex implications of the current strategy of industrialization is required and a determination to pursue policies which increasingly enhance an ability genuinely to acquire technological know-how. Concern on this matter is already being voiced in a number of underdeveloped countries. India, for instance, provides a good example of a country in which indigenous scientific effort is being geared up to complement, and in time, to displace imported technology.[27]

Emergent urban systems in underdeveloped countries

In spatial terms, import substituting industrialization has left a marked imprint on the emergent system of urban centres in most underdeveloped countries. It has, more than any other factor, encouraged a tendency towards a primate city configuration in the urban system. Primacy, in this context, implies a situation where the first city in a country is so much larger than the second as almost not to belong to the same genre. In most underdeveloped countries, particularly in Africa, this first city is not only the most important industrial centre but is often also the capital city and the premier port.

The basis for this primate city development was laid quite early in the colonial period when for ease of communication with the metropolitan country, a port city was usually chosen as the administrative capital of a colony. This rule varied only where a country was landlocked or where, because of better physiographic and climatic conditions, the interior situation encouraged settlement of colonists from the metropolitan country.[28] The port city also attracted the headquarters of most commercial houses involved in the import–export trade and was the terminus for the network of transportation and communication linking the far interior to the outside world. It also had the best developed infrastructural facilities in the country.

In the era of industrialization, all of these factors were to be critical in attracting industries to the port-cum-capital-cum-commercial headquarters. However, three other considerations proved equally decisive. First, since the strategy of industrialization simply involved the replacement of importation of consumer goods with heavy machinery and equipment, a port location was of prime importance. Second, since most of these industries were simply producing articles which were formerly imported and whose retail outlets were already articulated by commercial houses with headquarters at the port, the importance of effective distribution also constrained the industries to locate at the port. The third decisive factor was that of the market provided by the port capital city. Most of the imported articles tended to be luxury goods which only people with a relatively high and steady income could afford to purchase in large quantities. The port capital city had a disproportionate share of this class of people in the colonial period and this was to be reinforced in the era of industrialization.

Table 14 *Percentage share of manufacturing in capital cities in Africa*

City	*Percentage*	*City*	*Percentage*
Dakar (Senegal)	81.48	Brazzaville (Republic of Congo)	33.30
Banjul (Gambia)	100.00	Kinshasa (Zaïre)	30.28
Conakry (Guinea)	50.00	Addis Ababa (Ethiopia)	47.09
Freetown (Sierra Leone)	75.00	Khartoum (Sudan)	60.00
Monrovia (Liberia)	100.00	Kampala (Uganda)	27.78
Abidjan (Ivory Coast)	62.50	Nairobi (Kenya)	41.67
Accra (Ghana)	30.43	Dar-es-Salaam (Tanzania)	62.50
Cotonou (Benin)	66.67	Bukavu (Rwanda)	100.00
Lagos (Nigeria)	35.00	Bujumbura (Burundi)	80.00
Douala (Cameroon)	50.00	Blantyre (Malawi)	72.73
Bangui (CAR)	100.00	Lusaka (Zambia)	35.00
Libreville (Gabon)	100.00		

Source: UN Economic Commission for Africa. Also A. L. Mabogunje, 'Manufacturing and the Geography of Development in Tropical Africa', *Economic Geography*, vol. 49, no. 1 (January 1973), p. 11.

The market factor encouraged the primate city structure of urban centres in many underdeveloped countries. Particularly in Africa, most of these countries are relatively small in population, the median size being less than five million. Given the poverty of the vast majority of the population, the market available for the product of even a modern medium-sized factory is very limited and much of this is concentrated in the port capital. It is thus not possible for many countries to build another production plant apart from the one in the capital city. Thus the port capital commercial city becomes the only, or at least the most, important industrial centre in the country and its primacy status is established. Table 14 (p. 167) shows the percentage share of manufacturing in the capital city of various African countries.

In a study of seventy-five developed and underdeveloped countries throughout the world, El-Shakhs found that while there is a negative correlation between primacy values and level of development among developed countries, there is a significant positive correlation in the case of underdeveloped countries.[29] Furthermore, he noted that primacy values tended to change at a higher rate during the early stages of development and become more erratic and less responsive to changes in development in the middle or transitional stages of development. The index of development used in this analysis is not specifically indicated as industrialization but is an adaptation of a composite measure utilized in the *Atlas of Economic Development* by Ginsburg.[30] None the less, Sovani demonstrated convincingly that, particularly in the early stages of economic development, there is a strong correlation between industrialization and urbanization.[31]

In terms of the development process, however, the tendency to primacy in the system of urban centres in underdeveloped countries has two important implications. First, there are the implications for the rest of the country, the other cities and towns and particularly the rural areas. Because of the 'enclave nature' of the industrial processes within the primate city, very little backward linkage is developed. This means that the apparent prosperity of the primate city is not reflected by vigorous economic activities in other settlements within the country. This is why it is generally argued that capital cities in underdeveloped countries are 'parasitic' on the rest of the economy. Indeed, it was estimated in 1963 that in Pakistan the annual subsidy to large-scale manufacturing based in urban centres represented 6.6 per cent of total implicit tax on agriculture and the rural areas.[32] The conventional contribution of industry was measured as 7.0 per cent; its actual contribution, after allowing for protection, was a dismal 0.4 per cent of domestic value added. Clearly, a primate city structure reflects a certain incoherence in the socio-spatial articulation in underdeveloped countries. Johnson suggests that partly to compensate for such poor articulation of activities at the local and provincial level, governments in these countries tend to centralize authority rather than delegate or diffuse it.[33]

In the second place, the primate city of underdeveloped countries operates quite effectively as an element in a global system of metropolitan centres. Indeed, both in its spatial forms and its economic activities, it bears closer resemblance to these other cities than to the rest of the country. The tall skyscrapers, the wide roads with their elaborate flyovers, the five star hotels and the sprawling suburbia: these are features of the primate cities of underdeveloped countries which clearly show the stamp of their origin. It is from this perspective that these primate cities are seen as 'channels for the extraction of quantities of surplus from a rural and resource hinterland for purposes of shipment to the major metropolitan centres'.[34]

Patterns of urbanization

Because of the difference in the current level of industrial development among underdeveloped countries, there are distinct variations in their urbanization experience which are best taken into account through a typology. The basic criteria of differentiation are consequent on the current level of industrialization and urbanization. This typology, however, does not indicate that the character or problems of urbanization in the various categories or types of countries are different in kind. Rather, it provides a convenient way of grouping the growth paths likely to be followed by most underdeveloped countries. On the basis indicated, four categories or types of urbanization patterns can be identified among underdeveloped countries.[35]

The first category of countries comprises those in Latin America such as Argentina, Brazil, Chile, Colombia, Mexico and Venezuela. These countries have attained a relatively high degree of industrialization and the process of urbanization has been going on for over a century. Table 15 shows that at least 60 per cent of their population is now urban and only a diminishing proportion remains in the rural areas. Production activities in the cities have grown steadily since about 1900 with relatively high incomes and employment accompanying the growth of industry and commerce. High urban incomes have encouraged rural–urban migration and the growth of many large

Table 15 *Urbanization patterns in a sample of underdeveloped countries*

	Country	*Size of urban population (millions) 1984*	*Percentage of total population 1984*	*Urban growth rate (1973–84)*	*Rural growth rate (1973–84)*	*Per caput GNP level in 1984 (US $)*
Type I	Argentina	25.3	84	2.1	−0.3	2230
	Brazil	95.5	72	4.0	−0.1	1720
	Mexico	53.0	69	4.0	0.0	2040
Type II	Algeria	10.0	47	5.4	1.5	2410
	Egypt	10.6	33	3.0	2.5	720
	Korea	25.7	64	4.6	2.1	2110
	Malaysia	4.7	31	3.6	2.0	1980
	Philippines	20.8	39	3.7	2.1	660
Type III	Ivory Coast	4.6	46	8.3	2.2	610
	Kenya	3.5	18	7.9	3.4	310
	Nigeria	29.0	30	5.2	2.0	730
	Senegal	2.2	35	3.8	2.3	380
Type IV	China (mainland)	226.4	22	2.9	1.0	310
	India	187.3	25	4.2	2.4	260
	Indonesia	39.7	25	4.5	1.7	540
	Pakistan	26.8	29	4.4	2.4	380

Source: *World Development Report, 1986* (Washington D.C., 1986)

cities. In some cases, however, urban growth has been concentrated in a single primate city as in Argentina, Chile and Peru. Per caput gross national product is not necessarily low (see Table 15) but their distribution leaves a sizeable proportion of the rural population in conditions of abject poverty and fuels the ever increasing exodus to the cities.

Even within the cities, differential access to employment and urban services, highly protected modern sector wages and governmental policies excluding large portions of the urban population from the benefits of urban productivity, have resulted in a large number of poor households. High population growth, particularly in the last quarter of the present century, will exacerbate this situation in a number of countries until a stable low level of population is reached in the rural areas. This is likely to bring about a slowing down in urban growth, most likely by the turn of the century.

The second category is represented by semi-industrialized countries whose urbanization experience is more recent. Most of these countries are to be found in Asia and North Africa and include such countries as Algeria, Egypt, Korea, Malaysia and the Philippines. These countries have become increasingly urban with the growth of manufacturing activities and already between 30 and 50 per cent of their population live in urban places. The rural areas, however, suffer from some degree of population pressure which has tended to keep rural per caput income relatively low. By contrast, the productivity of the cities has sharpened rural–urban income differential. The rapid urban growth has meant both open unemployment and some underemployment in some of these countries, in spite of their relatively successful effort in generating employment and raising incomes. The result is the inevitable lag in satisfying the urgent demand for services by the urban population. Careful policy measures are needed to alleviate the magnitude of these problems. Given the limited natural resource base in most of these countries, control of the population factor could be a critical element in continued successful development. None the less, if rural population pressure can be eased and resource constraints overcome, this group of countries, by the turn of the century, can be expected to attain levels of urbanization similar to those found in the countries of the first category.

The third group of countries is found in Africa south of the Sahara where urban growth is a relatively recent phenomenon. The countries remain predominantly rural even though high rural–urban income differentials have stimulated massive rural–urban migration. Manufacturing and other productive activities in the cities are in an early stage of development. Consequently, in spite of the relative importance of rural–urban migration, the urban sector remains small, generally below 35 per cent of the population. The majority of the population still live in the rural areas and, given the relative abundance of land in most of these countries, agriculture, if properly developed, can be expected to absorb growing numbers of people with increases in productivity.

However, the continued importance of the rural sector has serious implications for the rate of population growth and makes the eventual outcome of the urbanization experience in many of these countries most unpredictable. Given the low level of national economic resources, the adoption in many of these countries of a high standard of infrastructure and protected modern sector activity, is likely to lead to the eventual impoverishment of the urban population.[36] More realistic, low cost solutions could avoid the extremes of inefficiency and inequity. The relatively small size of the urban sector in these countries suggests that, at least for the moment, these problems are more manageable than those found in other underdeveloped countries. None the less, if unrealistic policies are adopted, cities in most of these countries will be unable to absorb future rural–urban migration which is likely to result when the limits to increasing rural incomes are approached.

The fourth category of urbanization pattern is found in the large countries in Asia such as Bangladesh, China, India, Indonesia and Pakistan. These countries are predominantly rural and have low incomes, yet the absolute size of their urban populations is very large. Cities in these countries have grown steadily with industrialization and with almost unbelievably large numbers of people crowding into already difficult circumstances. None the less, the level of urbanization remains low, being generally below 30 per cent of total population. Indeed, in the case of India and Indonesia, it is estimated that it will still be low even by the turn of the century.

As with countries in the third category, the potential for rapid urban growth remains substantial. Unlike them, the intense pressure of the rural population on the land means that even a small percentage increase in migration from the rural areas will lead to massive growth in the cities. For example, India's urban population increase between 1975 and 2000 is estimated at 223 million as against an increase of 260 million in the rural areas, and this without allowing for any substantial migration out of the latter. The experience of China contrasts sharply with that of other countries in this category. By achieving noticeably lower rates of population growth within a generation, increasing incomes and improving their distribution, the country has succeeded in becoming increasingly urban but at a pace which permits the generation of employment and the provision of services for the growing urban population, without the resource constraints generated by a large and growing rural population living in conditions of almost absolute poverty[37]

Conclusion

The typology of the urbanization experience in underdeveloped countries helps to emphasize that, with few exceptions, the process has involved tremendous and vigorous redistribution of population. Whether it has been as productive as possible in terms of the political, social and economic well-being of the present and future inhabitants of the city and the overall development of the country, is a question which cannot be answered unambiguously. It depends very much on the specific circumstances of each country and the manner in which its government has tried to contain some of the many negative aspects of the current strategy of industrialization. For many of these countries, industrialization, it was hoped, would help the cities in attacking the national problem of poverty at its roots. Instead, the effect to date has been for these problems to be further aggravated. Particularly for the countries in the third category, there is a real danger today that, after the giddy years of early industrialization, some of them may be facing a backward-sloping curve of industrial development. This possibility is indeed very real, particularly for the smaller countries, since it is becoming clear that most of them cannot successfully adopt and manage the intricate backward linkage relation implicit in import substitution industrialization. The experience of Ghana in this regard is perhaps the most instructive to date.[38]

None the less, it must not be assumed that this danger is real only for the underdeveloped countries of Africa. Indeed, there is a growing reaction, particularly in countries in the first category, as to how far problems of their current style of development are due to this strategy of industrialization. This disaffection prompted Hirschman, one of the chief protagonists of import substitution industrialization as a viable strategy of development to observe as follows:

> Among the characteristics of import substituting industrialization, the possibility of proceeding sequentially in tightly separated stages, because of the availability of imported inputs and machinery, plays a particularly com-

manding and complex role, direct and indirect, positive and negative. The sequential or staged character of the process is responsible not only for the ease with which it can be brought underway but also for the lack of training in technological innovation and for the resistance to both backward linkage investments and to exporting that are being encountered. The most important consequence of sequentiality, however, is the fact that it has become possible for industrialization to penetrate into Latin America and elsewhere among the latecomers without requiring the fundamental social and political changes which it wrought among the pioneer industrial countries and also among the earlier group of latecomers.... This fact that import substituting industrialization can be accommodated relatively easily in the existing social and political environment is probably responsible for the widespread disappointment with the process. Industrialization was expected to change the social order; all it did was to supply manufactures.[39]

If the strategy of industrialization as currently conceived is thus not helping to bring about critical changes in the social order, then clearly a reconsideration is called for. Such a reappraisal must not only concentrate on alternative strategies of industrial development but must pay special attention to a choice of technology. For implicit in the question of the choice is that of the locational matrix of cities and towns within which industrialization takes place. Most underdeveloped countries have still to evolve a system of cities appropriate to the task of stimulating self-centred productive specialization of economic activities throughout their national space. A discussion of the basis for such a development must await further detailed analysis of the crisis situation which seems today to characterize urbanization in all the various categories of underdeveloped countries.

8 Urban crisis of underdevelopment

In order to appreciate the nature of the urban crisis currently confronting all categories of underdeveloped countries, it is useful to indicate what implications should stem from the urbanization of a country. In this regard, the statement by Senator Ribicoff at the opening of congressional hearings on urban problems in the United States in 1966, provides the most succinct expression of what urbanization should entail for the people of a country. To quote him:

> The city is not just housing and stores. It is not just education and employment, parks and theatres, banks and shops. It is a place where men should be able to *live in dignity and security and harmony*, where the great achievements of modern civilization and the ageless pleasures afforded by natural beauty should be available to all.[1]

Anyone who has visited cities and towns in most underdeveloped countries must come away with the overwhelming impression that whatever validity this observation has in the developed countries, it does not describe the reality of conditions in the former. A significant proportion of urban residents cannot be said to 'live in dignity and security and harmony' and the national resources represented by urban services and opportunities for employment are far from being available to all.

The sharp social and economic inequalities that characterize life in the urban centres of underdeveloped countries would not necessarily have given rise to a crisis situation if their magnitude were determinable and the tendencies kept within manageable proportions. But it is precisely the intense dynamism of these conditions which is fuelling the crisis and making it difficult to achieve a realistic resolution. That dynamism is, however, not the consequence of healthy vigour but the rosy flush of sickness, what McGee has referred to as 'pseudo-urbanization'.[2] It is rather the product of urban economic activities growing at a rate not fast enough to provide employment opportunity for the rapidly increasing population of these cities.

Growth of the urban population

Two facts are generally conceded in discussing urbanization in underdeveloped countries. The first is the rapid rate of growth in the urban population, particularly in the period just after the Second World War; the second is the concentration of much of this urban increase in large cities of at least 100,000 people. According to Table 16, between 1920 and 1970, the proportion of the population of underdeveloped countries living in cities of at least 20,000 people rose from under 7 per cent to nearly 20 per cent. This varied from 4.8 to 16.0 per cent for Africa, and 14.4 to 37.8 per cent for Latin America. The present prognosis is for a still faster rate of growth by the turn of the century. This remarkable upsurge in the size of the urban population has been brought about by a combination of a relatively high rate of natural increase and an increasing rate of rural–urban migration. The high rate of natural increase has followed in the wake of a significant reduction in death rates due to improved medical and sanitation conditions, at a time when the birth rate remained quite high. Indeed, Arriaga's study of urban population growth in Chile, Mexico and Venezuela shows that natural increase accounted for 58–70 per cent of the urban population growth, with the lowest contribution from natural increase being found among cities with populations over 500,000 and the highest in intermediate-size cities of 100,000 to 500,000.[3] This situation contrasts sharply with that in the developed countries in the early

Table 16 *Levels of urbanization in underdeveloped regions, 1920–2000 (percentage of total population in cities of at least 20,000 people)*

	Africa	*Latin America*	*Asia*	*Total*
1920	4.8	14.4	5.6	6.7
1930	5.9	16.8	6.6	7.8
1940	7.2	19.3	8.6	9.7
1950	9.7	25.1	11.6	12.9
1960	13.4	32.8	14.5	16.7
1970	16.0	37.8	17.1	19.7
1980	20.0	43.1	20.6	24.2
2000	28.3	53.6	28.1	32.6

Source: United Nations, *Growth of the World's Urban and Rural Population, 1920–2000* (New York, 1969).

phase of their industrialization, when due to a higher risk of epidemic and degenerative diseases urban mortality rates were higher than those in the rural areas. Davis, indeed, noted that 'as late as the period 1901–10, the death rate of the urban counties in England and Wales, as modified to make the age structure comparable, was 33 per cent higher than the death rate of the rural counties.'[4] Equally significant is the fact that in underdeveloped countries no fundamental change in reproductive behaviour accompanied the reduction in mortality. The result was that with death rates around sixteen per thousand and birth rates still above thirty-five per thousand in many of these countries, the excess kept the population growing at a steadily increasing rate.

There has, however, been some disagreement as to the relative contribution of natural increases and rural–urban migration to total urban growth rates in underdeveloped countries. On the one hand, a World Bank study shows that with the exceptions of Bogota and Taipei, migration has accounted for 50–76 per cent of growth in the metropolitan centres of most underdeveloped countries.[5] However, figures from the United Nations Population Division indicate that while this might have been true for many of these countries particularly in the 1950s, the general trend is for a decreasing share of migration since then. Table 17, for example, shows that the contribution of migration to urban increases in Africa declined from 56.3 to 45.2 per cent between 1950–60 and 1970–75, while for Latin America the decline was from 38.4 to 29.5 per cent during the same period.

Table 17 *Rural-urban transfer as percentage of urban growth in underdeveloped countries, 1950–1990*

	1950–60	*1970–75*	*1980–90*
World total	48.7	32.5	33.0
Developed countries	48.8	46.2	49.7
Underdeveloped countries	59.3	42.0	42.2
Africa	56.3	45.2	38.7
Latin America	38.4	29.5	24.1
East Asia	71.7	46.6	52.1
South Asia	43.1	40.0	41.0

Source: United Nations, Department of Economics and Social Affairs, Population Division, *Selected World Demographic Indicators by Countries 1950–2000, Medium Variant* (New York, May 1972) (ESA/P/W.P. 55).

Findley suggests that much of this disagreement arises from different estimates of the urban rate of natural increase used by the Population Division of the United Nations for different periods.[6] Irrespective of the issue of relative value, there is no doubt that with the rapid rate of population growth in most underdeveloped countries the absolute size of the contribution of migration to urban growth is enormous. The reasons for this recent spate of migration have been indicated in the previous chapter. They reflect the dominance of push factors out of the rural area as well as significant pull factors from the urban areas themselves.

Up to the 1950s, the average annual urban growth rate in most of these countries was of the order of 4.5 per cent. Since then, this has

Table 18 *Urban population in selected African countries, 1955–84 (in millions)*

	Population in towns with over 20,000 people			*Percentage of total population*			*Rate of growth*	
	1955	*1970*	*1984*	*1955*	*1970*	*1984*	*1955–70*	*1970–84*
Angola	0.31	0.72	2.38	7	13	24	5.8	8.9
Cameroon	0.28	0.77	4.06	7	13	41	7.0	12.6
Congo	0.15	0.35	1.01	20	29	56	5.8	7.7
Ethiopia	0.78	1.47	6.36	5	6	15	4.3	11.0
Ghana	0.68	1.73	4.80	12	20	39	6.4	7.6
Ivory Coast	0.23	0.86	4.55	8	20	46	9.1	12.6
Kenya	0.40	0.89	3.53	5	8	18	5.5	10.3
Mozambique	0.22	0.52	2.14	4	6	16	5.9	10.6
Nigeria	4.50	9.50	28.95	11	15	30	5.1	8.3
Senegal	0.50	1.06	2.24	18	27	35	5.1	5.5
Sudan	0.58	1.20	4.47	6	8	21	5.0	9.9
Tanzania	0.34	0.76	3.01	4	6	14	5.5	10.3
Uganda	0.15	0.44	1.05	3	5	7	7.4	6.4
Zaire	1.50	3.60	11.58	11	17	39	6.0	8.7
Zambia	0.44	1.12	3.07	15	27	48	6.4	7.5
Others	1.74	4.50	18.40	–	6	20	6.6	10.6
Total tropical Africa	12.80	29.50	101.60	8	12	25	5.7	9.2

Source: A. M. O'Connor, *The Geography of Tropical African Development*, 2nd ed. (London, 1978), p. 173 for 1955 and 1970; World Bank, *World Development Report 1986* (Washington D.C., 1986) for 1984 data.

risen sharply to between 5.5 and 8.0 per cent. Such high rates would mean that some cities would more than double their population in the short time span of between ten to fifteen years. Table 18 gives a general idea of the remarkable increase in the urban population of a select group of tropical African countries between 1955 and 1984. Amazingly, in the first fifteen years, the urban population more than doubled in nearly all cases. The last fifteen years have witnessed even more phenomenal growth, of three- to fourfold increases. However, as the table also reveals, this impressive rate of growth is partly due to the small initial size of the urban population. None the less, as the table on urbanization patterns in the preceding chapter shows, even the countries of Latin America still indicate a high growth rate such that their urban population can be expected to more than double before the end of the century.

A very high proportion of these urban gains took place in the larger urban centres and in particular in the primate cities. Here, according to Table 19, the rates of growth are even more impressive for African countries with many of these cities achieving three- to four-fold

Table 19 *Population growth of major tropical African cities, 1955–75 (in thousands)*

	c.1955	*c.1965*	*c.1975*	*Rate of growth 1955–1975*
Abidjan (Ivory Coast)	128	340	900	10.2
Accra (Ghana	290	530	1,000	6.4
Addis Ababa (Ethiopia)	430	580	1,000	4.8
Dakar (Senegal)	231	460	800	6.4
Dar-es-Salaam (Tanzania)	120	240	500	7.4
Douala (Cameroon)	110	200	—	(6.2)*
Kampala (Uganda)	110	240	—	(8.1)*
Khartoum (Sudan)	246	500	900	6.7
Kinshasa (Zäire)	349	800	2,000	9.1
Lagos (Nigeria)	360	1,000	2,500	10.2
Luanda (Angola)	180	300	—	(5.2)*
Lusaka (Zambia)	70	150	430	9.5
Maputo (Mozambique)	130	250	—	(6.8)*
Mombasa (Kenya)	130	210	340	4.9
Nairobi (Kenya)	200	380	720	6.6

Source: O'Connor, p. 177.

* Statistics not available for the full twenty-year period.

increases during the fifteen-year period 1955–1970. Moreover, the population of these centres came to represent between 55 and 75 per cent of the total urban population in many of the countries. The position for the underdeveloped regions as a whole between 1950 and 1970 confirms this same tendency irrespective of their category. Table 20, for instance, shows that while the total urban population just more than doubled, the proportion living in millionaire cities increased nearly four times and came to constitute 29.4 per cent of the total urban population in 1970, as against 19.5 per cent in 1950. The position with regard to the developed regions is also quite instructive. It shows during the period a remarkable closing of the gap between the two regions in terms of both size of total urban population and its distribution between different levels of the urban hierarchy. What all this means in effect is that whatever problems arose as a result of the remarkable urbanization rates, will be found in their most acute forms in the large metropolitan centres of underdeveloped countries.

A major factor in the configuration of the problems arising from rapid urbanization is the resultant age–sex distribution in the urban

Table 20 *Distribution of world urban population by city size*

	Underdeveloped regions (millions)		*Developed regions (millions)*		*World total (millions)*	
	1950	*1970*	*1950*	*1970*	*1950*	*1970*
Below 100,000	133.3	281.0	183.5	245.1	316.7	526.0
Percentage	(49.0)	(42.5)	(42.7)	(36.6)	(45.2)	(39.5)
100–199,999	24.2	50.2	37.0	59.5	61.2	109.7
Percentage	(8.9)	(7.6)	(8.6)	(8.9)	(8.7)	(8.2)
200–499,999	33.2	78.2	50.8	85.4	84.0	163.6
Percentage	(12.2)	(11.8)	(11.8)	(12.7)	(12.0)	(12.3)
500–99,999	28.5	57.5	39.7	65.9	68.2	123.4
Percentage	(10.5)	(8.7)	(9.3)	(9.8)	(9.7)	(9.3)
Over 1 million	53.1	194.1	118.3	214.5	171.4	408.5
Percentage	(19.5)	(29.4)	(27.6)	(32.0)	(24.4)	(30.7)
Total	272.2	660.9	429.3	670.4	701.5	1,331.3
Percentage	(100.0)	(100.0)	(100.0)	(100.0)	(100.0)	(100.0)
Level of urbanization	16.5	25.7	51.6	64.0	28.2	36.8

Source: United Nations, *City Projections 1950–2000, Medium Tempo with Medium Variant,* New York, 10 December 1970. See also G. Beier *et al,* 'The task ahead for the cities of the developing countries, *World Development,* vol. 4, no. 5 (May 1976).

population. The recent upsurge in the growth of these cities means that a high proportion of their population is young. This is further emphasized by the selectivity of the rural–urban migration process in favour of youth. Hence, one of the most striking characteristics of urban centres in underdeveloped countries is the high proportion of the population in the age-class fifteen to thirty-five. In many of the countries, this age selectivity goes hand-in-hand with a sex imbalance, in which there is a disproportionate number of males in the population. This is more so in the case of African and Asian cities than in Latin America.[7] However, as a town gets older, there is often a move towards sex equalization.

The most important implication of the youthfulness of the urban population is the relatively high proportion of economically active people. Although for the countries as a whole the enhanced infant survival rates would imply that the proportion of children below fifteen years of age is constantly rising and the relative size of the economically active population consequently decreasing, the position in urban centres can be expected to be the exact opposite. With continued in-migration, this proportion has continued to rise and the task of finding them gainful employment is perhaps one of the greatest challenges in underdeveloped countries today.

Urban unemployment and the growth of the informal sector

It has already been indicated that the current strategy of industrialization has had the two-fold effect of attracting to the urban centres a significant proportion of the economically active population while being able to employ only a small fraction of them. The situation is perhaps best illustrated by a consideration of Table 21 which shows comparative changes in urbanization rates and the proportions of the labour force employed in manufacturing industry, both for Europe (excluding England) from 1850 to 1930 and for underdeveloped countries from 1920 to 1970. In continental Europe, according to the table, the level of urbanization remained lower than that of employment in manufacturing industry until about 1830 when the percentage of persons employed in manufacturing reached 18 per cent. From that point onwards, the gap between the two widened progressively with urbanization growing faster than the level of employment in manufacturing. By contrast, in the underdeveloped countries, the stage at which the two levels were the same was reached between 1930 and 1940, at a time when most of the countries concerned had hardly begun to be industrialized. Henceforth, the gap, measured by the

Table 21 *Comparison between levels of urbanization and percentages of active population employed in manufacturing industry*

	Rate of growth of urban population (A)	*Percentage of active population in manufacturing industry (B)*	*Relationship of A to B*
Europe (excl. England)			
1850	11	16	−30
1880	16	18	−11
1900	24	20	+20
1920	29	21	+38
1930	32	22	+45
Underdeveloped countries (non-communist)			
1920	6.7	8.5	−21
1930	7.8	8.5	− 8
1940	9.7	8.0	+21
1950	12.9	7.5	+72
1960	16.7	9.0	+86
1970	21.0	10.0	+110
Africa			
1960	13.4	7.0	+91
Latin America			
1960	32.8	14.5	+126
Asia			
1960	13.7	9.0	+52

Source: Paul Bairoch, *The Economic Development of the Third World since 1900,* Cynthia Postan (trans.) (London, 1975) p. 149.

relationship between the two indices, widened rapidly and within less than thirty years, that is, by 1970, it had exceeded 100 per cent. Change of a similar magnitude in Europe took over eighty years as against less than twenty years in underdeveloped countries. What makes this gap even more significant is that as of 1970 when the rate of urbanization was 21 per cent for underdeveloped countries, their per caput gross national product (at 1970 US prices) was only $340 as against the figure of $650 attained in Europe in 1890 with a comparable level of urbanization.[8]

The failure of manufacturing to activate the overall economy and provide employment opportunities in different sectors has been a major cause of urban unemployment in underdeveloped countries. None the less, it is increasingly being realized that this situation does not manifest itself wholly as one of open unemployment. Although such unemployment is recorded particularly among young and 'inexperienced' workers, by far the majority of the unemployed find inadequate outlets in various low productivity enterprises, especially in the tertiary sector of the urban economy. Table 22 shows how over the period 1900 to 1970, this sector has been the fastest growing sector of the urban economy, particularly in the period since 1950.

One of the more interesting research developments in recent years has been the attempt to understand more closely the characteristics of

Table 22 *Changes in the structure of the active population of underdeveloped countries by type of employment, 1900–1970*

	1900	*1920*	*1930*	*1950*	*1960*	*1970*
Percentages						
Primary						
Agriculture	77.9	77.6	76.6	73.3	70.7	66.0
Secondary						
Extractive, manufacturing and construction industries	9.8	9.9	10.0	10.0	11.5	13.0
Tertiary						
Trade, banking, transport and communication services	12.3	12.5	13.4	16.7	17.8	21.0
Total	100.0	100.0	100.0	100.0	100.0	100.0
Absolute figures (millions)						
Primary						
Agriculture	213.0	238.0	249.0	304.0	366.0	435.0
Secondary						
Extractive, manufacturing and construction industries	26.5	30.0	32.6	41.2	59.8	85.0
Tertiary						
Trade, banking, transport and communication services	33.5	38.2	43.5	69.5	92.1	140.0
Total	273.0	306.2	325.1	414.7	517.9	660.0

Source: Bairoch, p.160.

this sector of the urban economy. Called variously the 'informal', the 'bazaar', the 'lower circuit' economy, this category of urban activity is gradually being appreciated as representing more than just employment for the majority of urban residents in underdeveloped countries.[9] The situation in the informal sector is in sharp contradistinction to that in the 'formal' sector of the urban economy. The latter embraces activities which are identifiable by legal definition and by the protection offered to workers by such legal recognition. It comprises public services and private enterprises of a certain size which, through the availability of capital and formal management control are exposed to less fluctuations in revenue and relatively smaller labour turnover in the process of production. In consequence, they are able to obey the labour laws, to offer higher wages and even provide certain social welfare services to their workers. By contrast the 'informal' sector comprises numerous owner-operated activities, and job opportunities offered by individuals who purchase merchandise or services generally on a short term basis.[10] These include such activities as traditional crafts, petty trading, small scale repair services, construction works and domestic services of various kinds. Operating outside the ambit of legal definition, this sector is characterized by low wages, occupational instability and the absence of a social welfare system.

Depending on the particular countries and cities, employment opportunity within this sector seems to be influenced by such factors as sex, age, length of urban residence, and education. In some countries of Latin America and Asia, for instance, unemployment and less favourable occupations were encountered particularly among the women and the youth just entering the market.[11] A good example is provided by the study by Merrill of the informal sector in Belo Horizonte, Brazil, part of the results being shown in Table 23.[12] Using actual deviations of percentages in the informal sector of the different classes of workers concerned, from the overall average of 31.3 per cent of total labour force in the sector, he found that female workers are quite disproportionately represented, as well as workers outside the primary age group who have not completed primary education and who are not heads of household. Similarly, in many African countries, especially in West Africa, women dominate major subsectors of the informal economy, particularly petty trading, in conditions which are far from unfavourable. In general, however, and with respect to the age factor, there appears to be a lower incidence of unemployment among adults due, in many cases, to them having been able to secure a number of irregular employments at which they can

Table 23 *Deviations of percentage of workers in informal sector from the overall average, Belo Horizonte, Brazil*

Sex		*Age*	
Males	−12.7	15–24, and over 50	+9.1
Females	+22.9	25–49	−7.5
Position		*Education*	
Head of household	−10.1	Incomplete primary	+19.4
Non-head	+ 9.0	Completed primary	−9.5
Overall percentage of all workers in informal sector			31.3

Source: T. W. Merrick, 'The informal sector in Belo Horizonte: A case study' (ILO Geneva, 1973), Table 6.

work severally on a part time basis. The length of urban residence factor tends to work generally in much the same manner, with recently arrived migrants tending to concentrate in the lower strata of the occupational structure, although their greater willingness to accept such jobs means that they are not too affected by open unemployment. None the less, it needs to be stressed that the urban informal sector is not dominated only by migrants. Indeed, in the example of Belo Horizonte just cited, it was found that the expected association between migration and the informal sector was rather weak and the proportions of native-born and migrants were more or less the same. This notwithstanding, there is evidence that in some subsectors, such as domestic servants, migrants, especially recent arrivals, tend to dominate.[13]

The educational factor, however, adds a certain constraint to this fact. To the extent that migrants with lower educational qualifications are predisposed to accept more precarious employment, underemployment rather than open unemployment tends to be of greater incidence among them. The more educated migrants tend to seek more stable and better paid employment. In this connection, the work of Callaway in Nigeria shows that as hope of direct employment in the formal sectors recedes for this category of migrants and urban residents, they become more willing to enter the informal sector as 'apprentices', to learn a trade which could enhance their prospects of eventually securing a job in the formal sector, or establishing on their own in the 'informal' sector but with a better income generating prospect.[14]

The relation between the 'informal' and 'formal' sectors of the urban economy shows some interesting differences between the larger

Table 24 *Characteristics of the two sectors of the urban economy in underdeveloped countries*

Characteristic	*Upper sector (formal)*	*Lower sector (informal)*
Technology	Capital-intensive	Labour-intensive
Organization	Bureaucratic	Primitive
Capital	Abundant	Scarce
Work	Limited	Abundant
Regular wages	Normal	Not required
Inventories	Large quantities, and/ or high quality	Small quantities, poor quality
Prices	Generally fixed	Generally negotiable between buyer and seller
Credit	From banks and other institutions	Personal, non-institutional
Benefits	Reduced to unity, but important due to the volume of business (except luxury items)	Raised to unity, but small in relation to the volume of business
Relations with clientele	Impersonal and/or through documents	Direct, personal
Fixed cost	Important	Negligible
Publicity	Necessary	None
Re-use of goods	None; wasted	Frequent
Overhead capital	Indispensable	Not indispensable
Government aid	Important	None or almost none
Direct dependence on foreign countries	Great; outward-oriented activity	Small or none

Source: Milton Santos, *L'Espace Partagé* (Paris, 1975).

and the smaller urban centres.[15] The basis for this is set out in Table 24, which contrasts the two systems. This table emphasizes that the informal sector provides services and goods for the poorer class in the population and low capitalized activities, neither of which can have regular access to identical services or goods produced in the formal sector. As such, the relative importance of informal sector activities increases with decreasing urban size. Eventually, in the small town, it encompasses the whole of the urban economy and constitutes a major means of maintaining constant relations with the countryside from where most of the poor come. However, in absolute terms and with respect to specialization and diversity, the importance of the informal sector increases directly with the size of cities. This is because in the big cities, not only is the size of the poorer population more considerable but subsistence consumption includes a larger number of goods and services, which tend to be expensive when sold in the formal sector but which can be offered in the informal sector in very small quantities and often on personal credit. Thus, the larger the city, the greater the proportion of those in the low income group and the greater the need to elaborate and refine the retail outlet of goods and services or engage in various forms of technical hair-splitting activities in catering to their needs. Such elaboration and refinement in turn means the creation of marginal employment for more hands so that a process is created which feeds on itself and gives the informal sector the appearance of having an infinite capacity to absorb increasing labour. This process has been referred to by McGee as one of 'urban involution'.[16]

The critical issue, however, is how far the growth of the informal sector represents a developmental response in the cities of underdeveloped countries. On the one hand it has been suggested that although earnings per worker are low in the informal sector as a whole, it has been increasing over time, and doing so much faster than average incomes in agriculture.[17] It is thus possible to see the sector as playing a benign role in the process of development in present-day underdeveloped countries and thus deserving of policy recognition and programmatic support. This, for instance, is the stance of the International Labour Office especially as revealed in the policy recommendations of most of the country case studies undertaken under its World Employment Programme. As against this is a school of thought which sees the growth of the informal sector in the urban economy of underdeveloped countries as part of the response to the exploitative relation imposed by largely foreign-owned activities dominating the formal sector.[18] Representing this point of view is Frank who argues with regard to Chile that:

> far from being a mark of development . . . this structure and distribution are a reflection of Chile's structural underdevelopment: 60 per cent of the employed, not to speak of the unemployed and under-employed, work in activities that do not produce goods in a society that obviously in a high degree lacks goods.[19]

Moreover, the gross inflation of the urban population which it encourages is regarded as inimical to the social and political stability of these cities since it leads to the creation of an 'impoverished and explosive lumpen-proletariat'.[20] Yet, in spite of the frequent predictions of the explosive and revolutionary potential of this group in the population, no such widespread and consistent reaction has been noted in their behaviour. Perhaps this is due to the high degree of heterogeneity within the sector, both with regard to status and to earnings. Among proprietors and self-employed, for instance, it is not unusual for the more enterprising to spawn new businesses that increasingly show features of activities in the formal sector. Their scale of operation grows and with time shows a greater concentration of capital and a decline in labour input. Such enterprises increasingly come to exhibit a reluctance to spread business widely among fellow traders or to encourage the flow of goods in hundreds of little trickles.

This ostensibly positive function has, however, been seen as the very seed of destruction of the system. It is argued that this is precisely the element which could trigger off the revolutionary potential in the situation.[21] The basis for such an eventuality lies in the employment-destroying implication of the emergence of local capitalists out of the informal sector. Sir Arthur Lewis, for instance, points out that although Karl Marx's belief 'that in the capitalist system employment-destroying innovations would always be excessive relative to employment-creating innovations, and therefore create an ever growing army of unemployed', may not be true for the advanced countries, the statement may prove to be correct in the context of present-day underdeveloped countries.[22] What the emergence of modern type enterprises out of the informal sector represents is thus the greater capitalist penetration of the urban economy. Such penetration means a reduction in the capacity of this sector to serve as a sponge to absorb large proportions of rural–urban migrants and indigent native-born, and as a cushion against urban and rural discontent. When the urban economy has been completely penetrated, this cushion is destroyed. Similar comprehensive capitalist penetration of the rural economy under these conditions would mean throwing dispossessed peasants into cities where there are little or no outlets for informal subsistence economic activities. With its back to the wall, the lumpen-proletariat is ripe for revolutionary struggle. According to McGee, this is one way of looking at the Cuba revolution and he suggests that similar symptoms are already evident in many other countries of Latin America and indeed of the underdeveloped world as a whole.[23]

Environmental problems of cities in underdeveloped countries

Given the irregular and low wage characteristics of employment for a large majority of the urban population, it is easy to appreciate that their disadvantaged position will be reflected and reinforced by a way of life characterized by substandard housing, precarious conditions of nutrition and health, low levels of education and consumption, and a generally degraded environment.

The housing problem is perhaps the most critical of the environmental conditions. In pre-colonial urbanization, the right to urban housing was predicated on having a social referent in a town, which in turn meant some right to urban land. Lineage membership *ipso facto* determines and confirms an individual's right to housing. The capitalist penetration of the urban economy which turned urban land into a commodity whose economic value was determined by the self-regulating market forces of supply and demand, also came to affect the importance of urban housing. Events both in the urban land and urban housing markets, today conjoin to influence the access of a large majority of urban dwellers to housing.

A major implication of this fact is the importance of land speculation in determining the amount of land available for housing construction and the price at which this is offered to the market. High land values combined with other aspects of economic life, notably the high cost of imported building materials, have placed most conventional types of dwelling units beyond the means of the majority of urban residents. The situation is aggravated by the rigid adherence of urban authorities to standards of materials and construction which derive from experience in the metropolitan countries and bear little relation to local resource availability.[24] Usually, it is against such unrealistic standards that governments in many of these countries proceed to argue the existence of a housing deficit, a deficit which, were it to be met, would still leave the housing problem of a large section of urban population unresolved because they are in no position to pay for or meet the maintenance obligations of such houses. It is this type of unrealistic estimate which is reflected in Table 25 according to which there was a deficit of nearly 110 million dwelling units in underdeveloped countries between 1960 and 1975.

It is, however, undeniable that various housing efforts of governments in underdeveloped countries do meet some needs. Often described as 'low-cost' housing and offered to the public at generous interest rates, these houses are bought, lived in or let out by the middle

Table 25 *Estimated urban housing needs in underdeveloped countries*

Housing required to provide for:	*Average annual requirements (in million dwelling units)* 1960–65	1965–70	1970–75	*Total requirement* 1960–75
Population increase				
Africa	0.4	0.5	0.7	7.8
Asia	2.2	2.7	3.2	41.0
Latin America	0.9	1.3	1.5	18.7
Sub-total	3.5	4.5	5.4	67.5
Replacement of obsolescent stock				
Africa	0.1	0.1	0.1	1.8
Asia	1.1	1.1	1.1	16.5
Latin America	0.3	0.3	0.3	4.1
Sub-total	1.5	1.5	1.5	22.4
Elimination of existing shortages				
Africa	0.1	0.1	0.1	1.8
Asia	0.7	0.7	0.7	14.6
Latin America	0.2	0.2	0.2	3.4
Sub-total	1.0	1.0	1.0	19.8
Total	6.0	7.0	7.9	109.7

Source: United Nations Department of Economic and Social Affairs (1965), p. 4.

class of public officers, traders and businessmen. Such activities, none the less, represent an unwarranted and inequitable transfer of social resources from the less to the more affluent class in society. This transfer, in detail, entails more than the provision of greatly subsidized housing. It includes easy access to credit for people building their own houses, direct government intervention in the land market to make cheap land available, the provision of adequate urban infrastructural facilities and tax concessions of various types. Some of the houses built under these conditions are then let to the lower strata of wage-earning groups at rents which force many of them to crowd into limited space. On the other hand, if the character of the neighbourhood permits, some of these houses are developed for the large foreign population to whom they are also let at exorbitant rent.[25]

In short, the cities in underdeveloped countries show a strong investment bias in favour of the relatively well-off.[26] Much of the emphasis on road construction is to improve facilities for the cars of the rich. In many Latin American countries, only 59 per cent of the urban population is served by piped water and another 17 per cent by public standpipes. The position is much worse in cities of Asia and Africa. Studies in Bogota, Cartagena and Mexico City show that variations in neighbourhood income explain most of the variations in the distribution of piped water. Similarly in Libreville, Gabon, only 25 per cent of all households have piped water but 5 per cent of the total water supply is available to the rest of the urban population. Similar results were obtained for sewerage and electricity.

For many migrants into the city, however, and particularly those in the informal sector, not even the overcrowded accommodation and limited services available to the low wage earning group are within their means, given the highly irregular nature of their employment and income. For such individuals, the only solution is self-construction of a shelter, using materials that are to hand, and squatting on any vacant land, usually on the periphery of the city or in enclaves suffering from specific disabilities like swampiness. The result is the phenomenon of squalid squatter settlements which have become a distinctive feature of cities in underdeveloped countries especially in the last thirty years.

As with the informal sector activities, the last decade or so has witnessed a better understanding of conditions in these squatter settlements. In Figure 7, Turner provides a typology for these settlements based largely on the Latin American situation but with aspects which are similar to those found in the other underdeveloped regions.[27] According to him, most squatter settlements go through levels of development reflecting the degree of security of land tenure which the squatters believe they have. In general, as they feel more secure they tend to improve on the quality of their self-constructed houses and to add to it modern amenities of various types. On the face of it, therefore, Turner argues that the existence of these squatter settlements does not constitute a problem (except that they are uncontrolled and that their forms are so often distorted) and that they are both the product of and the vehicle for activities which are essential in the process of modernization.

Turner's argument is based largely on the inevitability of these settlements and the virtual impossibility of doing anything with them short of 'eradicating their inhabitants with them'. Yet, the fact is that governments in most underdeveloped countries do not accept this fact and are constantly engaged in projects to evict the squatters and to clear their slums. Such periodic misapplication of effort is undertaken to improve the international public image of the particular city and to

	A Legal occupancy	B Semi-squatter or semi-legal	C Established squatter	D Tentative squatter	E Itinerant
A Complete structure and utilities to modern standards	complete legal	complete semi-squatter			
B Incomplete structure or utilities but built to modern standards	incomplete legal	incomplete semi-squatter	incomplete squatter		
C Incipient construction of potentially modern standard		incipient semi-squatter	incipient squatter	incipient tentative squatter	
D Provisional construction of low standard or impermanent materials		provisional semi-squatter	provisional squatter	provisional tentative squatter	
E Transient temporary and easily removed shelter			transient squatter	transient tentative squatter	nomad

Degrees of security of tenure

A Legal occupancy: institutionally recognized forms of tenure (e.g. freehold, lease, rental)

B Semi-squatter or semi-legal: without full recognition but with some rights

C Established squatter: *de facto* and secure possession but without legal status

D Tentative squatter: occupancy without any legal status or guarantee of continued tenure

E Itinerant: transient occupancy with no intention of permanent tenure

Figure 7 *Typology of squatter settlements*

ensure that housing standards are maintained. Sometimes, of course, this is done because in market terms the land on which the squatter settlement is established is now prime development land. Yet, as Engels had observed,

No matter how different the reasons may be, the result is everywhere the same: the scandalous alleys disappear to the accompaniment of lavish self-praise from the bourgeoisie on account of this tremendous success but they appear again immediately somewhere else and often in the immediate neighbourhood! . . . The breeding places of disease, the infamous holes and cellars in which the capitalist mode of production confines our workers night after night, are not abolished; they are merely shifted elsewhere.[28]

It is this refusal to accept the legitimacy of these settlements more than anything else that precipitates them into an environmental hazard. In the first place, once government agencies move in to destroy them, confidence and expectations are also shattered in the process. Instead of striving to improve the constructions and upgrade the environment, subsequent constructions are done in the most expendable materials and little investment of time and energy is devoted to the area. In the second place, the government continues to treat the area as virtually non-existent. Hardly any services are provided except perhaps transportation and the people are forced to make their own arrangements particularly for water supply and refuse and sewage disposal. In the third place, given the rather spontaneous manner in which these settlements developed and the consequent absence of an orderly layout, the disposal of refuse and sewage poses peculiar problems and their inefficiency constitutes a major factor in the degraded condition of urban development in underdeveloped countries.[29]

Of course, it is often the case that many of these settlements are outside of the official city limits. It is also sometimes true that most of their population do not pay rates or taxes. Yet, as Charles Abrams noted in writing about the squatters (the *rancheros*) of Ciudad, Guayana:

Rancheros will settle where they can if they are not told where they may. They will build what they can afford if they are not helped to build what they should. I am less worried, however about what they build than where they will build it and less concerned about initial standards than about initial layout. Rancho houses will improve with time and with better economic conditions if the rancheros are given a stake. [30]

In short, one of the most significant environmental problems of urbanization in underdeveloped countries is how to give the majority of the population a sense of belonging, a feeling of having a stake in the future of the city. This problem is bound to assume even greater importance in future unless the current strategy of urbanization in these countries is drastically reviewed.

Social alienation and urban management

The more immediate problems that the poor and the low income group in urban centres of underdeveloped countries are confronted with are those of security, discipline and justice within their own ranks. Not being generally welcome to the city and, in some cases, because of the illegal position of the shanty towns, they cannot look to the urban administration to provide these services. The result is the emergence of an innovative form of social organization that borrows from the traditional system but is uniquely a response to the new urban situation with which the migrants are confronted. In many African countries, for instance, ethnic origins and other forms of traditional association provide the basis for these organizations. Writing on the situation in West African cities, especially during the colonial period, Little notes:

> Allowing for local differences in custom the rules of most 'traditional' associations follow a common pattern. They prescribe a specific code of personal and moral conduct which is designed to regulate the public behaviour of members as well as their relations with each other. For example, a member who is reported for quarrelling in the town, for abusing elderly people, or for putting curses on others, may be suspended, fined or expelled. Similarly, in addition to adultery, where members are known to steal, or cause disorders at gatherings, they are warned to correct their behaviour. . . . Ordinarily, breaches of these regulations are dealt with by the Committee. . . . In addition, of course, to these formal tribunals, there are well-tried and salutary methods of a traditional kind for checking anti-social or improper conduct.[31]

On the other hand, in Latin America, in spite of the absence of an ethnic basis of organization, it is noteworthy that in the cities the large body of poor immigrants feel the necessity to establish similar voluntary associations. Writing on the situation in the squatter settlements around Santiago in Chile, Castells remarked as follows:

> . . . the illegal position of the camps had the effect, especially in the early

> days, of bringing about the establishment of a defence system against police repression. This situation also led to the organization of an independent system for the prevention and repression of delinquency and a judiciary system of communal life. 'Popular militia' and 'surveillance committees' were created at the start, and gradually disbanded after the arrival of the Popular Unity, since the left-wing parties maintained that it was better to rely on the 'police of the popular government'.... These experiments with popular justice were characterized not only by the creation of new 'institutions' but also, in some cases, by the new content given to justice; protection of communal values and consideration of matters ignored by bourgeois law. For example, absence from meetings or disorderly behaviour during the assemblies was considered a fault and behaviour within the family was closely watched. Drunkenness was severely reproved: alcoholic drinks were banned in several camps and shelters were set up at the entrance where residents who returned a little too merry could sleep it off. These measures were completed by a re-education programme and an attempt to attack the social roots of alcoholism.[32]

Such organizations among the poorer members of the urban community could, however, protect their interests only in respect of certain issues of communal life. There were equally important problems of education for the young, health for all members of the community, water supply, refuse disposal, cultural and leisure activities and social security for the aged in particular, which were not so easy to organize because of the financial resources required. In some cities, however, such resources were mobilized and, with or without some assistance from the urban administration, attempts were undertaken to resolve these problems. Such resolutions were possible where the administration maintained a policy of 'benign neglect' rather than one of 'malevolent aggression' against such settlements.

In terms of overall management, the existence of the squatter settlements and their pervading air of poverty and dreariness constitute perhaps the most intractable problem of current urbanization. Yet, it has been shown that the poor are not themselves without initiative and resourcefulness in these matters. Indeed, it is a common experience in many of these cities that these areas of acute environmental degradation are not necessarily areas of serious social delinquency. None the less, with authority in the squatter settlements themselves based on consensus rather than on legality, the social organizations developed in these areas are fragile institutions which can collapse under certain conditions. These include, for instance, an increase in differential incomes and circumstances within the settlement and co-optation into the political processes in the city as a whole. Such development weakens the cohesiveness of the people without necessarily integrating them effectively into the urban society. This fractionalization of communal consciousness, to the extent that it weakens the effectiveness of actions by the poor can be the most serious impediment to the improvement of living conditions in cities of underdeveloped countries.

Thus, lacking the means of involving popular participation, urban management tends towards authoritarian rule by technocrats and political bosses serving the interests of the middle and upper classes in the community. The result of their limited concern is most manifest in the sharp contrast in living and environmental conditions between the rich and the poor neighbourhoods in these cities. It is also reflected in the pattern of distribution of social amenities such as schools, health clinics and hospitals and recreational and cultural centres in the city. But the poor are not easily wished away and the neglect with which their problems are dealt is reflected in their alienation from the rest of the urban environment and their indifference, except under compulsion, to helping to maintain a quality of cleanliness and orderliness in public places outside of their own settlements.

The problems of the urban poor clearly highlight the difficulties of urban management in underdeveloped countries without necessarily indicating the basic causes and their consequences. These include the multiplicity of agencies dealing with urban growth problems, weak planning and management, scarcity of municipal finance and inadequate policies on pricing and taxes. One of the striking aspects of urban management problems in most underdeveloped countries is the general lack of experience in the organization, institution-building and implementation of metropolitan-wide programmes. This lack of experience has meant the persistence of a multiplicity of agencies dealing with problems within the metropolitan areas. As many as five levels of government – national, state, regional, metropolitan and municipal – may be involved in programmes within the area with little communication between them. For example, more than 30 municipal authorities are said to be in charge of the metropolitan area of Calcutta.[33] The situation in Latin American cities is no different and Cornelius noted that many of their metropolitan areas are no longer governed by a simple entity but are fragmented into numerous politico-administrative subdivisions.[34] Moreover, many policies and programmes affecting these cities are determined by national government agencies and decentralized public authorities operating within their boundaries. The results are frequent cases of overlapping responsibilities, budgetary conflicts and bureaucratic competition, which tend on the whole to hinder public service delivery, control of land use and planning for future growth.

Of the other problems, the financing of urban growth deserves special mention. Most urban centres in underdeveloped countries, but particularly the metropolitan areas, have generally been unable to generate the tax revenues necessary to finance their development. In general, they have all typically depended on the national or state government for a large portion of their resources. The result is inefficient and ineffective delivery of urban services, urban infrastructure left in a poor state of disrepair, or, where allowed, extensive deficit financing. Part of the poor environmental quality of large parts of urban centres in underdeveloped countries is certainly due to this fact. Yet, as Gilbert noted in his study of Bogota, the inadequate revenues for financing urban development 'are due mainly to the reluctance of politicians to raise taxes or to introduce an effective tax on land'.[35] Indeed, studies from other areas show that the problem is not so much one of an inadequate base of taxable income and property, as fear that the institution of the necessary fiscal measures would set the authorities on a collision course with many of the most powerful interests in the cities whose wealth is derived largely from extensive urban landholdings.

In the African context, the financial problems of rapid urbanization are further aggravated by lack of qualified and experienced staff. It is claimed in the case of Nigeria, for instance, that this is due in part to the inability of municipal or metropolitan authorities to pay commensurably this category of staff and to the fact that the career prospects for such staff appear much more limited compared to the situation in other areas of government. The situation is indeed the classic one of the chicken and the egg. Qualified staff are needed for more effective management and revenue raising yet, without the latter it is not possible to employ and retain such staff.

Conclusion

The rapid growth of urban centres in underdeveloped countries has brought to the fore various problems which are manifested not only in the physical forms of the city but also in the ways they function. These problems have been grouped in this chapter under three broad headings: urban unemployment, environmental degradation and social alienation. Each of these problems is linked to a basic weakness in the current conceptualization of development, which does not see it as a mobilization of the total population with a view to enhancing its capacity to cope with the daily challenges of existence. As with other aspects of current development, the cities in underdeveloped countries reflect a dual structure, of a small minority enjoying many of the resources of urbanization and the poorer majority having to manage or eke out a living as best they can.

Indeed, in respect of each of the broad categories of problems, one of the notable features has been the innovative approaches which the masses of the population on their own have brought to their resolution. An 'informal' economic sector has developed to provide a means of creating employment for the majority of migrants and a training ground for their accession into local entrepreneurship or into formal sector employment as workers. A settlement form has emerged to cater for their accommodation needs, while forms of social organization have been evolved to pave the way for their entry into urban society.

Yet, in all these instances, the most notable fact is the difficulty of integrating these innovative responses into the economic and management structures of the modern cities in these countries. The increased capitalist penetration of informal sector activities constitutes a threat to their continued ability to absorb more and more migrants; the constant destruction of the squatter settlements, while not resolving the problem of accommodation, inhibits their evolution into improved residential districts; and the co-optation of the social organizations of migrants by so-called modern political processes encourages fractionalization and weakens their ability to deal with urgent social problems. The urban crisis of underdevelopment is thus not so much unemployment, poor housing and social delinquency but how to integrate the indigenous solutions to these problems within an overall strategy of development.

9 Urban system and national development

The two preceding chapters have attempted to provide a view of current patterns of urban development in most underdeveloped countries, and the problems which have been attendant on the evolution of this pattern. A tendency towards primate city development has been shown to be consequent on the strategy not only of colonial exploitation, but particularly of import substitution industrialization which has been adopted in many underdeveloped countries. This tendency, if anything, emphasizes the outward orientation of the economies of these countries and their continued dependency on erstwhile metropolitan countries. This orientation has encouraged an unrealistic attitude to problems arising from the prevailing situation. Certainly, in hardly any underdeveloped countries does one feel that the tripartite problems of urban unemployment, environmental degradation and social alienation are being confronted with imagination and the necessary initiative.

It has, however, been emphasized that an economic system gives rise to its own corresponding spatial order. The thrust of the argument in this particular volume is that an appreciation of the nature of a desirable spatial order could be conducive to a more realistic set of policies for the overall development of a country. In this chapter, therefore, the main question is: What type or pattern of urban development would ensure that the capacity of individuals in a country is enhanced by realizing their inherent potential and effectively coping with the changing circumstances of their lives? Or, to put it differently, what type of urban development would bring about the full and total mobilization of the population of a country?

The conceptual basis of urban development

In his concern about the problems of underdevelopment and some of its spatial determinants, Johnson argues that the really astonishing difference between developed and underdeveloped countries is in their relative number of central places and in the dispersion of these towns, small cities, medium-sized urban centres, and larger cities. Using examples from Europe and the Middle East (see Table 26), he emphasizes that in the developed countries, the varied hierarchy of central places has not only made possible an almost complete commercialization of agriculture, but has facilitated a wide spatial diffusion of light manufacturing, processing and service industries. By contrast, underdeveloped countries, lacking such a central place infrastructure, suffer the serious handicap of countrysides inadequately provided with 'accessible market centres where farm produce can readily be sold and where shops filled with consumer and producer goods can exert their tempting "demonstration effects", the incentives to produce more for the market and the inducement to invest in better tools, fertilizers, or better livestock in order to generate a larger marketable surplus'.[1]

Johnson was at pains to emphasize that what is of interest here is not the scattered periodic produce and pedlars' markets which underdeveloped countries have in large numbers, but the lack of towns large enough to maintain daily, competitive produce markets, a certain number and variety of shops and stores and a few light manufacturing

Table 26 *Number of villages per central place of over 2500 inhabitants*

Country	*Census*	*Villages*	*Central places*	*Ratio*
Developed countries (Europe)				
Switzerland	1963	1209	233	5
France	1964	5075	489	10
Sweden	1963	2053	165	12
United Kingdom	1964	4337	277	16
Netherlands	1964	2378	147	16
Underdeveloped countries (Middle East)				
Saudi Arabia	1963	11193	71	157
Turkey	1964	44175	219	201
Iraq	1963	9186	45	204
Syria	1963	7550	25	301
Yemen	1962	9532	15	635

Source: E. A. Johnson, *The Organization of Space in Developing Countries* (Cambridge, Mass., 1970), p. 175.

and processing industries. It is, for instance, instructive that in the communist spatial reorganization of the Chinese countryside, the move from the rural collectivized co-operatives to the commune, was to ensure the co-ordination of agriculture with other activities such as industry, education and defence, and to integrate the administration of the rural district with that of the commune. Similarly, Smith noted, in writing about accelerating modernization in Japan, that many young men before leaving the rural area for the large cities and the more demanding tasks in modern urban industry, had had the benefit of acquiring 'a certain quickness of hand and eye, a respect for tools and materials, [and] an adaptability to the cadences and confusion of moving parts',[2] from light manufacturing establishments in the numerous castle towns that dot the countryside.

In considering the pattern of urban centres that would ensure a more self-centred development in most underdeveloped countries, it is tempting to consider the relevance of Christaller's Theory of Central Places along with its later modification by Lösch.[3] The real relevance of the central place scheme is its apparently comprehensive coverage of a hypothetical national or regional space. Given a country where population and purchasing power can be assumed to be uniformly distributed in space and where transportation is easily available in all directions, the central place theory attempts to postulate the number, size and spacing of urban centres that are likely to be found. Put briefly, these two deductive models of urban development hypothesize that the number of central places in a country is closely related to the classes or order of goods and services produced or available for exchange within the economy. These classes of goods and services have different frequencies and scale of demand and, in consequence, different ranges or distances from which prospective consumers will find it economic to travel to purchase them. The range in turn defines the market area of the goods or service. The convergence of different ranges of market areas defines different sizes of central places, as well as the spacing between them. In general, small central places selling less expensive classes of goods with a high frequency of demand are numerous and spaced closer together. Larger central places offering increasingly more expensive goods and services with a more restricted demand are fewer and more widely spaced.

As against these theoretical formulations, empirical analysis of urban systems in many developed countries shows a tendency to what has been referred to as a rank-size or a log-normal distribution, or some modification of this distribution.[4] The distribution defines the relationship of the largest city in a country to all other cities in the system. In the ideal case, the size of any city in the system can be determined by regarding the inverse of its rank as its proportion of the capital city population. Moreover, when these cities are ranked in decreasing order of size and plotted on a double logarithmic graph with population on one axis and rank on the other, the values form a straight line. The manifestation of a rank-size distribution among cities in a country implies a fairly dispersed urbanization process. It is also suggested that this reflects the achievement of national unity in both political and economic terms.

With regard to underdeveloped countries neither the deductive models nor the empirical findings on city size distribution can, at their face value, be regarded as providing guidelines for urban development. For one thing, the deductive models relate to only a segment of urban economic activity, namely that of marketing goods and services, and even in this case its main concentration is on the demand side. They say very little concerning the significance of industrialization in the evolution of a national urban system. For another, the models are based on a set of limiting assumptions which make it difficult to achieve a direct translation into real world situations. Yet, to the extent that they try to go below the surface of manifest reality to uncover at least one aspect of what towns and cities represent within a spatial economy, their contribution is not without importance. Moreover, although they make no such claims overtly, the models do have an underlying value premise of economic rationality, and some measure of distributive justice in their concern for the provision of goods and services to all individuals in a given society.

The empirical findings, on the other hand, do not admit of an unambiguous interpretation. Indeed, it has even been argued that the primate city system (otherwise referred to as the colonial model), rather than the rank-size distribution, is the normal state for most countries including the developed ones.[5] Most of the cases where this generalization does not apply, the argument continues, can be accounted for by either or both of two situations: first, the amalgamation of several primate systems into one 'national' system, the nature of whose distribution of urban size reflects the relative sizes of the hinterlands so amalgamated; second, the development of subsidiary primate systems based on subgateways in very large areas, usually of continental dimensions. Attempts to justify this contention are provided by the historical experience of Australia, the United States and various countries of Europe.

Irrespective of the inadequacies of both the deductive and the empirical models, what is clear is that they both recognize the need for a dispersed urban pattern without denying the importance of a certain degree of concentration. Urban dispersal and concentration are seen as the products of processes which are not mutually exclusive and whose outcome has been, in the experience of present-day developed countries, not only to ensure that the demand for goods and services across the country is effectively met, but also that opportunities for continued growth and change in the economy are not compromised. For underdeveloped countries, therefore, it is possible to examine factors both on the demand and the supply side which should govern a strategy for a desirable pattern of urbanization.

Demand factors in the strategy of urbanization

The demand factors in a preferred strategy of urbanization refer to those issues of a cultural and institutional nature which determine the form in which demands on urban resources are articulated. The best illustrative example of what is involved here is provided by the history of urban development in Israel in the period after the creation of the state in 1948. This period was marked by a vast immigration of Jews from all over the world and there was considerable concern as to how to distribute them spatially.[6] Up to then, the urban system had shown a strong primate city tendency with over 43 per cent of the national population living in the single city of Tel Aviv with another 25 per cent in the cities of Jerusalem and Haifa. In order to absorb the mass of immigrants and prevent the growth of squatter settlements, the government of Israel, through its National Planning Department, deliberately formulated a national urbanization policy with the following five objectives:[7]

(a) the settling of sparsely populated regions to avoid the growth of regional 'imbalances';
(b) the occupation of frontier regions for strategic purposes and with a view to establishing a national presence;
(c) the opening up of 'resource frontiers', particularly in the southern deserts;
(d) changing the primacy structure of the urban system by limiting the growth of urban concentration in and around Tel Aviv, and creating the 'missing' level of middle-sized towns;
(e) building an integrated regional system of settlement by promoting complete urban hierarchies in each region.

In achieving the last two objectives, the Planning Department based its proposals on Christaller's central place theory. It proposed to develop a hierarchical system of settlements comprising five main levels:

(a) the basic agricultural cell or village unit of 500 persons;
(b) the rural centres serving four to six surrounding villages with agricultural services;
(c) urban-rural centres with a population of 6000–12,000, serving tens of villages;
(d) medium-sized towns of 15,000–60,000 people serving as headquarters for each of the twenty-four districts into which the country is divided;
(e) national centres of over 100,000 people.

It was hoped that the middle three categories, (b), (c) and (d), would supply the 'missing links' in the national system of cities. In the period between 1948 and 1968, new rural settlements and thirty-four development towns were created. By 1970 these towns accounted for 21.3 per cent of Israel's urban population and have helped to reduce Tel Aviv's share of the national population to one-third. The tendency to primate city size distribution had also been replaced by one approximating to a rank-size distribution.

However, notwithstanding the effective implementation of these proposals, it is now generally agreed that the new development towns, particularly at the lower levels, have not been an unqualified success. A major factor which appears not to have been taken adequately into consideration in designing the national system of cities, was, in fact, the preferred social relations of production. The Israeli nation, even before it came into formal existence, showed a strong preference for co-operative organization in its rural areas. Various forms of such organization constitute the basis of the four types of rural settlement found in the country – namely the *kibbutz*, the *moshav*, the *moshav shitufi* and the *moshava*.[8] In the *kibbutz*, everything is collectively owned by the members except the land, which is national property. The basic principle is 'every member gives to the community to the best of his abilities and receives from it according to his needs'. Thus, all needs of the members – food, clothing, education, entertainment, recreation – are fully met by the community. All meals are served in a communal dining hall while communal laundries, clothing stores and shoemakers provide for members' requirements. In general, the largest *kibbutzim* have attained populations of between 1500 and 2000 people. The *moshav* is a smallholders' settlement in which, although each settler works his separate plot of land, lives in his own household and draws income from his farm's produce, he is obligated to participate in joint marketing of produce, joint purchasing of farming and even household implements and mutual aid and responsibility in case of illness, conscription and so on. *Moshav* villages range in population from 150 to 900. The *moshav shitufi* is a co-operative smallholders' village where production is carried out communally as in the *kibbutz* but where family life is self-contained as in the *moshav*. Most *moshav shitufi* have remained much smaller in population than the average *kibbutz*, none of them exceeding 400 inhabitants. The *moshava* is a village of the regular European type where land, buildings, farming installation and so on are all privately owned and the co-operative organizations are relatively weak. Many *moshavot* (plural of *moshava*) pre-date the Israeli nation and because of their *laissez-faire* basis of organization could absorb large numbers of immigrants and transform into towns and cities of 20,000 to 70,000 people.

The significance of the dominant co-operative basis of social relations of rural production for urban development in Israel is best appreciated when viewed against the assumptions of Christaller's central place system. The system was developed on the basis of consumers relating individually to the central place and thus having to absorb the cost of transportation for each shopping trip. Hence, for goods needed often and regularly, there was the need to have small central places close by the consumers. Co-operative organizations with their enhanced capacity for bulk purchases, scale economies and transport cost minimization, would vitiate the local dependence on small towns and relate the rural settlements directly to higher order centres. Viewed in this perspective, it is no wonder that many of the development towns were not as successful as anticipated and, as Berry noted, have generated 'less intense local city-region relations than was expected'.[9]

Thus, in designing a programme of urbanization to serve the need for comprehensive national development, it is important to pay attention to factors on the demand side, notably the institutional arrangements defining the social relations of consumption. Decentralized urbanization is certainly a basic spatial structural element in the strategy of development. The determination of its textural component, however, cannot be divorced from the preferred social relations of consumption of urban services by the rural population since this, to a considerable extent, defines the spatial competitive effectiveness between centres and therefore the viability of different levels of a national hierarchy of towns and cities. The implication of this is that a strategy of urbanization cannot be divorced from that of rural development. They are both intricately interlocked for reasons to be elaborated later.

Technological choices and industrial development

Of equal significance is an approach to the strategy of urban development from the supply side. This approach considers those factors which determine the ease and effectiveness of supplying central goods and services to the population of a country, in particular those relating to the production and transportation processes. The deductive models of central places paid scant attention to production processes and dismissed the question of transportation under the assumption of uniform accessibility. Yet, the available or preferred technology of transportation can also be critical for the texture of urbanization, since the more efficient and cheap transportation is, the less the need for the lower levels of an urban hierarchy. In the example of Israel which has just been discussed, improved transportation facilities were a factor in the limited success of many of the lower order development towns.

But it is the technological and organizational choices in the production process that are perhaps most crucial for the configuration of the pattern of urban centres in the course of national development. It has already been emphasized that the adoption of an import substituting industrial strategy with its dependence on capital-intensive, automated and mass production technology has been an important factor in the tendency towards a primate city structure in most underdeveloped countries. Clearly, in moving towards a new system of cities it is vital to review the issue of technological choices in the overall strategy of development of a country.

An important point of departure in this regard is an appreciation of what is meant by technologies. According to Merrill, 'technologies are bodies of skills, knowledge and procedures for making, using and doing useful things ... technologies are cultural traditions developed in human communities for dealing with the physical and biological environment, including human and biological organisms'.[10] From this definition, it is clear that virtually all human communities have developed technologies of one sort or the other to be able to cope with their physical and biological environments. Such technologies not only determine how effectively they cope but also what types of organization and settlement patterns they evolve in the process. Some technologies, however, are of a more advanced and complex nature than others and therefore require complex organizations and large metropolitan centres for their blossoming. None the less, an important aspect of development is for communities with less sophisticated technologies to borrow and incorporate superior ones into their cultural traditions and in the process to transform both their organizational forms and settlement pattern. Where opportunities for such borrowing and incorporation are restricted, the technological basis of development becomes stultified.

There is considerable doubt as to whether the technology embodied in the current strategy of import substitution industrialization is in a form which provides opportunities for borrowing and incorporating into the cultural traditions of underdeveloped countries. Since for most of the foreign firms who own the enterprises or supply the machinery, the ability to expand output and capacity rapidly, to provide a full range of brands or price lines or to have an assured uninterrupted supply are the major considerations, it is easy to appreciate their preference for the most automated production techniques available. The easy satisfaction of mass consumer demand obscures the important issue of how well the particular society is borrowing and incorporating into its economic body the production technologies and organization. Yet, from the point of view of development, it is this consideration of the actual transfer of technology that is crucial. In this connection, it is worth emphasizing that even in present day developed countries, production processes depend on a broad mix of technologies ranging from the complex, highly sophisticated equipment for large-scale production to smaller powered machinery for small-scale production. For underdeveloped countries, therefore, a deliberate programme of technological choices and mixes is becoming imperative. This will not only help to correct current distortions in their urbanization process but ensure that they strike a realistic balance between their needs for mass produced goods and improved technological knowledge. A judicious emphasis on small-scale industries is considered of great importance for this latter purpose. Writing on this issue with respect to Nigeria, Kilby observed:

> The technology-diffusing potential of foreign enterprises varies inversely with its scale and the complexity of the techniques it employs. The large, capital-intensive soap factory, brewery, flour mill and so on have had no visible 'carry-over' in Nigeria – they have remained technological enclaves. On the other hand, small, modestly financed, individual or family-owned concerns (predominantly Levantine) employing the most rudimentary and unadorned

production processes have spawned hundreds of Nigerian firms in soap-making, metal-working, sawmilling, rubber creping, baking, umbrella assembly, singlet manufacture and construction. The reason for this is simply that learning is a continuous process, comparable to an individual climbing a ladder in the sense that if too many rungs are missing, if the technological gap is too great, upward progress ceases.[11]

In the development context, therefore, an industrialization policy in a country can only be meaningful in so far as it openly and squarely confronts the problem of technological choices. The issue here is not just one of what type and scale of machinery to import into a country, or whether to encourage foreign or indigenous ownership of production organizations, although these decisions are, no doubt, of some importance especially in ensuring the actual resolution of the real problem. The issue is how to bridge the chasm which separates the technological traditions of developed and underdeveloped countries, a chasm which makes the difference between the two traditions one of kind rather than of degree. This generic difference, according to Hetzler, revolves around the lack of a tradition of using power technology in underdeveloped countries. To bridge the gap, he suggests the dissemination of general mechanization as the first and most basic stage of general planning for industrial and technological development. To quote him:

> General mechanization should have specific foci, however. It should have the two-fold objective of indoctrinating the society in the principles of power technology and simultaneously building up the society's social overhead facilities. By indoctrination in power technology is meant the steeping of the society in the operation of energy subsystems. Gasoline and diesel engines, electric motors and chemical-powered units bulk large in every technology, and it is only after a certain proportion of the society's population becomes intimately familiar with the mechanics of these subsystems that it begins to think and behave as systems analysts and designers. As this stage is reached, the society will begin experimenting by putting these power units to various uses, to building various types of applied mechanisms around them. Without a knowledge of power technology, society stands in isolation from the machine. Its members may have learned to operate the machine, but it will remain an enigma if they are unable to maintain or repair it. For a population which has only a limited operator's knowledge of power units, there cannot be sustained technological growth. But the technology does become sustainable, indeed internally self-fuelling, once it has acquired a knowledge of power systems.[12]

Small-scale, power-operated industrial plants have other

significance in developmental terms apart from their greater effectiveness in ensuring real transfer of technology. They provide a better opportunity for the emergence of an indigenous industrial entrepreneurial class, they ensure the development and growth of management capabilities and managerial talent in the country and they offer tremendous scope for widespread inculcation of the requisite industrial discipline among the population at large. More than this, small-scale industrial plants are better adapted to the small size of markets in many underdeveloped countries but have the potential of growing with the market. Such small-scale plants are, in fact, bound to be vital in the initial task of valorizing, through rounds of processing, many of the agricultural and primary raw materials which in many underdeveloped countries are outside the scope of present industrial activities.

The experience of Japanese industrialization is perhaps most instructive with regard to a deliberate policy both for the acquisition, absorption and incorporation of modern technological know-how and for encouraging the development of both large-scale and small-scale industries. Writing on the strategy of technological transfer employed by Japan at the beginning of her industrialization, Hirschmeier noted as follows:

> For Japan things were different. She had to absorb a technology coming from a completely alien culture, it had to be transmitted in Western languages that of themselves posed a formidable barrier; finally, it came at a very advanced stage to an unprepared people. The Japanese government officials who were so much awed by the marvels of European and American factories were not expert technicians. But they realized that there were few others in Japan who knew more than they. They understood then, that their ambitious program of industrialization hinged decisively on effective technical training and on the formation of a native elite of technicians and scientists ... the Meiji government did everything possible to enlist the help of Western technicians and teachers. It did not intend to rely on book knowledge alone or waste time and resources on experiments of the Saga type. In almost all major governmental projects, foreign experts were given the double task of technical supervision and on-the-spot training of Japanese engineers ... the most important function of the foreigners, therefore, was to make themselves superfluous as quickly as possible. The government expended all efforts to build up native technical personnel and to promote technical education in order to dispense with direct foreign technical assistance.[13]

The industrial development process showed the Japanese government concentrating on heavy industries and transport. In the decade after 1868, it built and operated railways and telegraph lines, opened new coal mines and established iron foundries, shipyards and machine shops, set up model factories to manufacture cement, paper and glass and financed the establishment of many new Western-style industries. This phase of direct entrepreneurship passed rather quickly. After 1882, the government relinquished its lead but gave tremendous support for the nascent big capitalists or *zaibatsu* to take over and consolidate Japanese interest in the field of large-scale industrial development.

With regard to the small-scale industries, the government's interest was more circumspect though no less crucial. It established model spinning mills and bought machinery which it sold at discount rates to prospective buyers.[14] In order to stimulate interest in mechanical cotton spinning among the cotton merchants and putting-out masters, for instance, it launched a massive propaganda drive. A series of exhibitions on cotton spinning was staged, and on these and similar occasions speakers were made to hammer away at the point that many more mills must be established to save Japan from the menace of imported yarns. The result of all this was a great flowering of small-scale industrial establishments throughout Japan, especially through numerous medium-sized rural landlords 'with one foot in the agricultural and one foot in the industrial sector'.[15] It was this class of citizen in particular that was responsible for the prodigious growth of the raw silk and silk reeling industries, for the spread of a multitude of medium and small-scale establishments engaged in weaving, footwear, brush, pottery and lacquerware manufacture as well as bicycle assembly, toy making and in recent times electric lamps and machine parts. Many of them were linked together via the putting-out and sub-contracting system with large-scale urban operations at the purchasing and marketing levels.

This mushrooming of medium and small-scale industry permitted maximum play for the adoption of indigenous innovations of an organizational nature. This was particularly important since the early export trade was built largely on these small-scale industries producing such commodities as silk products, tea and straw mats. However, their larger number, their almost total reliance on irregular family labour and their remoteness from the market, made for a haphazard and unsystematic scheme of organization in which there was a chronic tendency to spoil the trade by turning out inferior goods. If exports were to prosper, it was essential to improve and standardize the quality of such products. Between 1884 and 1902 a series of laws was enacted which enabled the government to encourage the formation of local chambers of commerce and guilds of industrialists and merchants for co-operative action.[16] These trade associations were given legal status and tax immunities in order that they might engage in joint investigation, representation and services like inspection. A large number of such chambers and guilds came to be formed and managed to impose some degree of self-regulation on the quality of production in small-scale industries.

However, these organizations did not substantially alter the atomistic structure of most small-scale businesses. This was modified only subsequently as a network of financial and marketing relationships developed, whereby small-scale producers were linked through merchant employers and wholesale traders to big banks, export-import firms and large-scale factories.

This example of the Japanese experience is not meant to imply that conditions in that country at the beginning of its industrialization were in any way similar to those in underdeveloped countries today. For one thing, Japan had reached by then a high level of cottage industry development hardly matched in any of the latter even today. The real point of the illustration is the distinction which the country made between the related, though different, policies of industrialization and the acquisition of technological know-how. This distinction made it possible for Japan to greatly quicken the pace of its industrial development and to achieve within one generation the tremendous transformation of its society as a whole.

This distinction was also crucial for the emergence of indigenous organizational and management structures around the new technological capabilities and for determining the resulting pattern of urbanization. In other words, concern with problems of a desirable system of urban centres in an underdeveloped country can hardly be divorced from a programme not simply of industrialization but of a calculated transfer of technology.

Programme of urban development

The emphasis on a deliberate policy of encouraging small-scale industries in the industrialization programme is not only to ensure a more systematic transfer of technology but also to correct the existing primate city distortion in the urban system of underdeveloped countries. Small-scale industries in terms of their capital and management requirements are more flexible in their locational decision. A huge number would end up concentrating in the larger cities and metropolitan centres close to the big industrial plants to which they are vertically linked or to the large market of low-income workers whom they serve. But numerous others can locate in small rural towns, district and provincial headquarters to service the rural population and provide a critical complement to their productive effort.

Indeed, one of the main missing links in raising rural productivity in most underdeveloped countries is the complementary set of small-scale processing industries which should locate in nearby small and medium-sized towns. The survival and growth of this level of towns in the urban hierarchy will, however, depend on how far agricultural production has been rationalized to ensure an adequate and continuous supply of raw material to the processing factories. It is at this critical level that the close interrelation between rural and urban development is to be found and where planning for both has to be integrated. It is no doubt for this reason, among others, that as indicated in Chapter 6, the Chinese in their organizational strategy were forced to move rapidly from the rural and purely agricultural collective to the people's commune which combined industrial activities and agricultural production. In the case of Japan, the existence of numerous small towns scattered throughout the rural area, especially the castle towns of the *daimyos* or feudal lords with their population of craftsmen, merchants and professional soldiers, provided the basis around which small-scale industrial development serving the rural population developed. Indeed, Ranis noted that as late as 1883, 80 per cent of all Japanese factories were located in rural districts, with 30 per cent of the agricultural labour force engaged in rural industrial 'side jobs'.[17]

Industrial locational considerations, particularly of raw materials will thus guide the number and distribution of this lowest category of urban centres. These considerations from, as it were, the supply side, must be reconciled with those from the demand side which we have already considered. Thus, a programme oriented towards the development of small and medium-sized towns serving not only as marketing, shopping, servicing and entertainment centres but also as industrial centres, is critical for the self-centred development of most underdeveloped countries. For such a programme to stand a chance of success, however, not only industrial processes but other investment programmes need to be co-ordinated both spatially and temporally. It has been suggested that this may require the establishment of a single agency capable of co-ordinating, synchronizing and supervising the various ministry-level programmes to fully utilize complementarities in infrastructure and services.[18] It certainly also requires a greater effort at mobilizing and involving local communities in a manner that will ensure their active collaboration in the development process.

The strategy of dispersed urbanization will appear at first sight to contradict that of 'polarized' development which emphasizes the role of growth in stimulating development within a given region. In a sense this is true, at least to the extent that primary importance is given to small town development as part of a programme package for stimulating rural productivity. But beyond that, a sustained level of activities in these smaller centres will require that their output is absorbed as input in factories and plants in centres of a higher order which can also supply them with goods for the rural population. The developmental significance of a 'growth centre' in an underdeveloped country is thus predicated on a prior restructuring of the spatial economy. Without this, as Penouil emphasized, the 'growth centre' will become no more than 'a creator of dual economies because in the rural milieu its influence soon becomes too weak to bring about rapid changes in the field of occult and magic practices and to change prevailing customs so that they may be adapted to the requirements of modern life'.[19] A growth pole strategy with its technical implications thus depends for success on a very different society from that found in many underdeveloped countries today, where the problem is not just to reactivate productive structures but to create new ones more appropriate to modern needs.

If a growth centre strategy appears inappropriate for many underdeveloped countries especially in the early stages of their transformation, this does not vitiate the necessity for a hierarchy of urban settlements. The implications inherent in the existence of different orders of goods and services, and their need for centres commanding market or service areas of different magnitude, make this imperative. Whether this hierarchy can be predicated on the numerical sequence of the central place theory, is, however, another matter. The size, number and spacing of urban centres at different levels of the hierarchy will have to be determined by the realities of population and activity distribution in a given country. What is important is that all over the rural areas there must be created, within reasonable distance, a diversified economy offering considerable incentives and varied employment opportunities to masses of the population, and in consequence removing the pressure from the few metropolitan centres. In short, for most underdeveloped countries the fundamental strategy of urban development revolves around how to transform a primate city structure to one closely approximating a rank-size distribution with all that this involves in technological choices, industrial location programmes and enhanced rural productivity.

None the less, it is necessary to recognize the wide variation among underdeveloped countries and the difference this makes to the likelihood of success of a growth centre strategy. Many countries of Latin America, for example, with their relatively high rate of urbanization, see the development of such alternative attraction poles as a major means of redirecting the flow of migrants away from a single or few metropolitan centres. In spite of various attempts at implementation, however, the strategy has met nowhere with significant success. Various problems have been encountered and considerable knowledge accumulated as to the causes of failure. The most important relates to the difficulty of ensuring that the multiplier effects of investment in the growth centre do not leak back to the primate city or to developed countries. The major factor here is the conflict between the large, vertically organized, multilocal enterprises usually attracted for this purpose, and the horizontally disposed and strongly local communities whose growth they are meant to induce. Given the well established intra-organizational, national and international linkages along which resources flow within large enterprises, it is easy to appreciate that in the absence of requisite regional structures, to which such enterprises can relate, no amount of legislation can ensure the achievement of the cumulative local growth on which the whole strategy is predicated. The example of Brazil is perhaps instructive in this regard. While there is no doubt that the growth centre strategy of the regional agency (SUDENE), set up in 1959 to develop the north-east region, has brought increased industrial employment to the cities of Salvador, Fortaleza and Recife, it is now generally agreed that this development has largely benefited the elites within and outside the region, rather than the region as a whole. According to Gilbert,

> By offering fiscal incentives to manufacturing in backward areas, the neglect of agriculture has been encouraged. Had strong employment multiplier effects been generated by these economic changes, then the overall social effect might have been beneficial. In general, however, these modifications have tended only to provide profit opportunities for a privileged minority. Worse still, the few modifications that have been made to the economic structure have covered agencies with a veneer of action that has hidden their inability to institute fundamental changes in fields such as land reform, taxation and market structures.[20]

The role of large, vertically organized, indigenous or foreign-owned enterprises in the urbanization process of countries is only gradually being grasped. Even when such enterprises have been fully integrated into the national economy, their growth-inducing capabilities are known to be asymmetrical in relation to expectations from the urban hierarchy.[21] This occurs in part because although these enterprises are themselves hierarchical, they do not at all conform to the hierarchy of the urban system. Not all of them have their headquarters in the largest metropolitan centre, nor do they all share the same locations for their regional peaking since no semblance of similarity can be expected in their geographical spread. Given then that new and sizeable employment opportunities are likely to be created in the wake of growth-inducing innovations that are diffused down this intra-organizational hierarchy, it is easy to appreciate the critical role of these enterprises for the variable pattern and relative stability of urban growth processes.

Urban planning and management

Within a given urban centre, however, governments in underdeveloped countries still have to confront the problem of environmental deterioration and social alienation. Urban planning and management must therefore be directed at mobilizing the urban population to a more wholesome relationship both to the environment and to the society. It is of considerable interest that the basic means by which most communities have tried to achieve this objective is through the division of an urban area into a hierarchy of spatial units, which are meant to embody some type of social relation. The most fundamental of these units are the neighbourhoods which group into quarters or wards, which in turn group into districts and so on until the whole city is encompassed.

In traditional or pre-industrial urban centres in African and Asian countries, the importance of these units was recognized not only for purposes of social mobilization but also for ensuring a much better relation with the environment, especially with regard to sanitation and refuse disposal. In traditional Yoruba towns, for instance, Ojo observes that 'a simple, though hierarchical, system of government was evolved to suit this arrangement of society. A compound head was responsible for governing the inmates of the compound. The quarter head (chief) was responsible for regulating the inter-compound affairs of all compounds under his care. All the chiefs in the council under the leadership of the Oba, who are more often than not kept in the background, regulated inter-quarter matters and those of general interest to the town as a unit.'[22]

One of the tragedies of colonial administration was a certain indifference to the spatial organization of social life beyond a simple maintenance of law and order. In these circumstances, this form of organization in urban areas fell into disuse, later to be replaced by wards (for elected councillors) whose size and significance had little relevance for the problems operative at the neighbourhood level. This lack of social and historical continuity is no doubt partly responsible for the parlous conditions of most pre-colonial urban centres in under-developed countries. It may, of course, be argued that the changes inherent in colonialism and the capitalist penetration of traditional economies would in any case have made the preservation of this traditional arrangement difficult, if not impossible. These arrangements were based on some ascriptive principles of social relations whereas the impact of capitalist productive organization emphasized more universalistic ones. Besides, the massive immigration into these urban areas could be expected to create new and substantial difficulties, particularly with regard to their incorporation into this system of social relations. Yet, it is instructive to note that this is something which Japan has managed to do. Most cities still preserve their neighbourhood organizational units *(tonari gumi)* of five to ten houses as well as the *chonaikai* or ward which comprises a number of *tonari gumi*. The *chonaikai* oversees its own local affairs, arranges for the collection of rubbish, fire watching and the training of the fire brigade, poor relief, the organization and upkeep of the shrine and the arrangements for festivals and the employment of watchmen.[23]

Clearly, some form of internal territorial partitioning and organization of urban areas is necessary to ensure effective social identification and improved participation of all urban residents in the political processes that influence urban planning and management in under-developed countries.

At present, conventional wisdom is to attempt to create the necessary territorial units through some form of zoning of land uses which, with respect to residential districts, tries to bring together households with relatively homogeneous incomes, value systems, utility functions and behaviour patterns. It has been argued that zoning also provides a basis whereby different groups within the urban community share out the negative and positive externalities of urban activities between them on some equitable basis.[24] However, bringing households together through residential location within a zone is a necessary though not sufficient basis for mobilizing them for purposes of bargaining or acting in defence of their own rights. Indeed, as Makielski observed, the history of zoning indicates that such a condition is unlikely to hold in situations prevalent in cities in underdeveloped countries where there is considerable imbalance in the distribution of economic and political power.[25]

If zoning or internal territorial organization of urban centres must be used to reduce social alienation, improve environmental conditions and enhance the effectiveness of urban management, then it is necessary to appreciate the circumstances under which large groups can engage in coherent collective behaviour in their own interest. Clearly, what is involved here is not only a spatial structuring of urban areas but a distribution of political power such that even the slum dweller has a say in the manner in which the resources of the city are distributed. At present, in most underdeveloped countries, the shanty town, the slum, the low income residential areas, are all areas which have lost out in the political competition for the goods which the city can offer such as schools, jobs, garbage collection, drainage, street lighting, libraries and social services. As Sherrard puts it:

> The slum is an area where the population lacks resources to compete successfully and where collectively it lacks control over the channels through which such resources are distributed or maintained. This may suggest some new approaches to metropolitan planning – recognising the necessity for redistribution of power, broader access to resources and expansion of individual choice to those who have been consistently denied.[26]

Internal territorial units, when created with the requisite measure of political power, thus provide even the slum dwellers with the opportunity to react to the dynamics of the urban system. The relative smallness of such a unit is to ensure that the group is not so large as to be unable to act together voluntarily to achieve some collective aim.[27] The participatory effectiveness of small group action has been noted by various writers. Organizing an urban centre territorially on such a basis will allow all groups to have a voice in the decision-making process and have a tremendous impact on urban planning and management in underdeveloped countries. Indeed Kotler observed that 'the poor need neighbourhood government to secure the liberty to achieve prosperity'.[28]

The problem of urban planning and management in the context of development involves the creation of structures through which the people can relate better to their environment and have broader access to urban resources. Two critical resources are jobs and housing. Much of the discussion concerning technological choices and patterns of industrial development indicates ways in which job opportunities can be vastly increased and widely distributed. Housing, on the other hand, has not been adequately treated. It involves access both to land and shelter. A government would have to set some control on land speculation where it cannot take over the land in public ownership. However the government succeeds in making land available to the masses, zoning regulations must be used to give people real choices in the provision of their buildings. There is hardly a human community which has not developed its own technology of construction. Some result in structures which are not sophisticated or do not require heavy stress-bearing materials. But most communities can produce a shelter which will protect a household from inclement weather and ensure for them a high degree of privacy. Environmental quality is not dependent on the material characteristic of individual buildings, but on how these are organized in spatial units to take care of as it were 'community metabolism': the layout of the buildings and adequate roads to take movement between them; adequate drainage to take away slop water and rain; adequate provision for easy evacuation of refuse and garbage. These are the problems which a government is best placed to resolve.

This resolution, according to Turner, should involve governments in underdeveloped countries in a number of specific actions.[29] These include: modification of existing legislation on minimum standards and building procedures; the introduction of legislation and planning practices that set limits, rather than procedural lines, for housing activity; legislative control limiting the concentration of resources in the hands of the wealthier classes and facilitating the supply of land, technology and credit to low income groups; legalization of tenure of land and dwellings illegally occupied by squatters; clear separation of various levels of authority in housing activities, and the restriction of central government and municipal influence to certain well-defined and basic functions of the type indicated above; and finally, the encouragement, if possible, of informal sector activities through proscriptive legislation that gives decentralized technologies and local systems of labour, finance and materials, greater access to resources.

Such a programme of action is certainly unexceptionable and, if implemented, should certainly help in improving housing and environmental conditions in most cities of underdeveloped countries.

However, in a most incisive critique, Burgess argues that Turner's programme of action is likely to be ineffective because he failed to see that the housing problem is a structural condition of the capitalist mode of production.[30] Within such a context, the role of government in the housing process must be viewed from two perspectives: structurally in terms of the limits to its effectiveness set by the social relations of the specific capitalist mode of production; and politically in terms of the forces of the social classes which the government itself represents and whose interest it must be expected to defend. Given these conditions, Burgess argues that if recommendations of the Turner type were implemented through the state, they would simply facilitate the further penetration of industrialized building materials into markets now covered by petty commodity sources. The granting of legality of tenure to existing and future squatters would more effectively integrate those areas at present excluded by their own illegality into the process of capitalist valorization of urban land and it is not inconceivable that more profit could be found in the provision of housing goods and services than in the full provision of the finished housing object. He, therefore, concludes:

> There would seem to be little hope that Turner's policies could be carried out on the scale and in the manner that he considers to be critical to their success. There is a greater likelihood that they will be used on a limited scale to further petty commodity interests in ways that are not detrimental to the maintenance of capitalist relations of production in general. Indeed, if Turner's policies were implemented on the scale and within those conditions of production, consumption and exchange that he leaves unaltered, there would be the most drastic and deleterious consequences for low income groups.[31]

Conclusion

What is clear then is that whether in terms of the specific micro issue of urban housing or the macro problem of a national system of urban centres, no serious resolution can be achieved except it be part of the overall socio-economic transformation of the country. In the urban context, such a transformation will seek to break the stranglehold of the primate city on the urbanizing population and the creation of a viable internal spatial organization within each urban centre, which can ensure not only effective urban planning but also efficient urban management. In both situations, emphasis is on the population caught in the urbanization process. As Brutzkus observed, the dispersed urbanization involved in the notion of a national system of cities:

> . . . makes in social and cultural terms the unavoidable transformation of traditional societies less abrupt and painful . . . than under conditions of a uniform 'modern' and cosmopolitan civilization of a metropolitan complex. . . . The social and personal strains and sufferings linked to metropolitan concentration in most developing countries cannot be considered as minor inconveniences versus the allegedly overriding necessities of more rapid increases in national income.[32]

None the less, however much a national system of cities makes socio-economic transformation less abrupt and painful for a country, it is how far it guarantees for the individual wider employment opportunities and improves his chances for better living conditions that is crucial. It is here that the internal spatial organization of cities has real significance. That significance relates to how far the spatial structure entails real redistribution of local political power and how far the individual's competitive capacity in the struggle for access to resources is thereby enhanced. Effectiveness in this cannot be regarded as automatic or as being likely to be a spontaneous response of the urban underprivileged. Governments have to take the matter openly in hand, to provide external inducements for stimulating efforts at mobilization and apply sanctions to discourage indifference. But whatever the means adopted, the first step is to create the appropriate spatial structures to facilitate the redistribution of local political power. For only such structures can help enhance the capacity of individuals in an urbanizing situation to realize their inherent potential and cope effectively with the changing circumstances of their lives.

Part Four

National Integration

10 Integrating the national population

It has been emphasized in preceding chapters that development is essentially concerned with the prospects of individuals in the process of social change. Those chapters have thus concentrated on how to improve those prospects for the individual, irrespective of whether he lives in the rural or the urban areas of a country. In spatial terms, it was shown that this involves the creation of structures which improve the competitiveness of individuals in terms of their access to and command over societal resources. A necessary corollary of competition and its resultant tendency towards specialization is, however, coordination and integration. In this section, therefore, the concern is how a country may achieve effective integration of the various new structures that are being created in the development process.

A primary area of integrative activity relates to the population of the country as groups rather than as individuals. Integration in this regard is concerned with interactions between the population and the natural environment and between groups in different functional areas of a country. The first type of interaction involves a consideration of the problems of population growth and densities and the relation of these to what might be called the 'carrying capacity' of different areas of a country. The second refers to interactive adjustment which inevitably must take place between the regions of a country, particularly between rural and urban areas, which is mirrored through the movements and migrations of the population. Both population growth and population movements are essentially integrative mechanisms in a country. But as has been emphasized elsewhere, they are not neutral forces. Whether they act to generate and sustain development depends on the degree of deliberation and planning that goes into determining the manner of their operation. Population planning is thus a critical aspect of spatial integrative activities within a country.

Population growth

For most underdeveloped countries, the period after the Second World War has been marked by a remarkable spurt of population growth and significant gains in the expectation of life. Table 27 shows that while between 1930 and 1950, the increase in population among this category of countries was well under 400 million, in the succeeding twenty years the gains were more than double that figure, and the prospect is that the next twenty years will witness a near doubling again. Indeed, on 11 July 1987, the United Nations celebrated the arrival of a world of 5 billion people. In other words, up to 1950, it took the population of underdeveloped countries nearly sixty years to double, whereas the time required to do this again, has been drastically reduced to just about thirty years.

The growth of any population is the joint product of the excess of

Table 27 *Growth of population of underdeveloped countries, 1930–2000*

Population in millions			
Period	*World*	*Underdeveloped countries*	*Developed countries*
1930	2,044	1,285	759
1950	2,486	1,628	858
1970	3,621	2,537	1,084
1990	5,346	4,064	1,282
2000	6,407	5,039	1,368

Years required to double population at average annual growth rate during the period			
Period	*World*	*Underdeveloped countries*	*Developed countries*
1930	–	–	–
1950	71	59	114
1970	32	32	61
1990	35	30	83
2000	39	32	108

Source: United Nations, *Demographic Trends in the World and its Major Regions 1950–1970*, E/Conf. 60/8P/1, Table 1, p. 4 (New York, 3 May 1973). See also World Bank Staff Report, *Population Policies and Economic Development* (Baltimore, 1974), p. 8.

births over deaths and the migration of people to or from other areas. An important feature of the organization of national life in the twentieth century is the heavy restriction of in-migration during peaceful times. During periods of strife and instability, neighbouring countries have found themselves receiving thousands and sometimes millions of refugees fleeing from their beleaguered countries. Substantial additions to the population of countries such as India, Syria, Lebanon, Sudan, Zambia, Tanzania, Mozambique and Senegal have been due to this form of forced in-migration.

More commonly, however, population growth has been due largely to natural increase. The prime cause of the acceleration of the rate of population increase has been unequivocally the sharp fall in mortality levels. Today, underdeveloped countries have achieved average death rates of as low as 11 per thousand while average birth rates remain over 30 per thousand. Life expectancy has risen to an average of fifty-seven years. The situation varies significantly in the major areas of the underdeveloped world. For Africa, the average death rate for the period 1980–5 is still as high as 17 per thousand while life expectancy is only about forty-nine years. By contrast, in Latin America, the figures are 8 per thousand for average death rate and sixty-four years for life expectancy. Differential decline in infant mortality rates accounts for much of the variation found among the underdeveloped countries. In Table 28, for instance, whilst more than 65 per cent of African countries still had an infant mortality rate of more than 100 per thousand in 1985, only two countries (mainly Caribbean islands) fell into this category in Latin America and the Caribbean. Many of the latter, in fact, have rates comparable with those of countries in Europe and North America.

The principal causes of the sharp decline in mortality are two-fold: disease control through massive public health measures and relative improvement in nutritional standards. In many underdeveloped countries, significant investment in public health, especially in epidemiological control measures and mass vaccination and inoculation, have sharply reduced the potency of fatal disease organisms in their environment. The demographic consequence of measures such as the draining of swamps, the spraying of insecticide on standing bodies of water, environmental sanitation, medical care systems, clean water and vaccination of infants, have been literally dramatic. The example of Sri Lanka is often quoted in the literature as one country where the effect of malaria control measures has been spectacular. Here, the first major anti-malarial campaign with DDT coincided with a drop in the

Table 28 *Infant mortality rates in major underdeveloped regions, 1980–5*

Infant mortality rate (deaths per thousand)	*Number of countries*				
	All regions	*Africa*	*Asia*	*Latin America and the Caribbean*	*Europe, N. America, Oceania*
Below 25	62	2	7	15	38
25–49	30	1	11	10	8
50–74	15	2	6	6	1
75–99	23	13	5	4	1
100–124	20	14	5	1	–
125–149	17	11	5	1	–
150 and above	12	9	3	–	–
Total	179	52	42	37	48

Source: *United Nations World Population Chart, 1985*, (New York, 1986)

* The 111 countries include only about 64 per cent of world population, 69 per cent in Africa, 47 per cent in Asia, 56 per cent in Latin America and 100 per cent in Europe, N. America and Oceania.

crude death rate from 20 to 14 per thousand within a year, that is, between 1946 and 1947, and a further fall to 12 by 1960.[1] It was also noted that this sharp decline was accompanied by an improvement in economic conditions. By contrast, a similar remarkable drop in the mortality rate due to malaria eradication occurred in Mauritius but coincided with a drastic deterioration in the economic situation.[2]

Apart from disease control and general public health measures, an equally important factor making for a significant decline in mortality rates is the relative improvement in nutritional standards. In some countries this has been due to increased rural productivity. Underdeveloped countries in this fortunate situation, are, however, few and far between. An FAO survey of seventy-three underdeveloped countries in 1971 showed that only twenty-five, representing a mere 14 per cent of the population, had a growth rate of food grain production adequate to match increases in demand resulting from both income gains and population growth.[3] Indeed, between 1980 and 1985, the growth in food production at 2.7 per cent in the underdeveloped countries had dropped below the 1971–80 average to a level just barely exceeding the population growth rate of 2.4 per cent. At the other end of the scale, 16 per cent lived in countries with wholly inadequate calorie supplies and food production was growing more slowly than the population, while the majority (61 per cent) lived in countries with inadequate calorie supplies but where production was growing faster than the population – but not as fast as estimated demand. Notwithstanding, many underdeveloped countries have been able to maintain nutritional standards at a precariously adequate level as a result of improved distribution following a relatively stable political situation and modern transportation development.

Decline in mortality, however, has not been compensated for by any strong tendency towards a fall in the number of births. Indeed, in most underdeveloped countries, fertility rates have remained remarkably stable at traditionally high levels or have declined only very slowly. For the period 1935-39, the birth rate stood between 40 and 45 per thousand for underdeveloped countries as a whole; it was still 43.9 for the period 1950–5, 43.6 for 1955–60, 42.0 for 1960–5, 40.6 for 1965–70 and 31.2 for 1980–5. These figures mean that in the period since 1950, birth rates in underdeveloped countries have been roughly twice as high as in the developed countries. The current birth rates in the underdeveloped countries, however, vary by region with Asia recording the lowest rates (27.4) and Africa the highest (46.4) for the period 1980–5.

The factors which have maintained birth rates at this level are not hard to find. The first is, in a sense, the result of the declining mortality rate with its effect on survival rates and the age composition of the population. A country with a large proportion of its population in the youthful age classes is likely to record a relatively high rate of fertility. Indeed, for most underdeveloped countries, the percentage of the population below the age of fifteen is usually between 40 and 45 compared to between 20 and 30 in the developed countries.

Age composition of the population, however, need not automatically imply high fertility, if this was not encouraged by various social practices, attitudes and institutions. In most underdeveloped countries, the women tend to marry early and the marital status is regarded with strong favour in the society. In general, less than 4 per cent of women aged fifty in African and Asian countries have never been married compared to between 11 and 29 per cent for European countries, around 1900. At the same time, most of the women were first married between the ages of fifteen and twenty-two, compared to the age class twenty-two to thirty-nine for women in European countries around 1900. What this means, of course, is that women in underdeveloped countries have a longer reproductive career than their European counterparts and so even with a normal spacing of three years between children can end up with quite a large family. This is particularly significant when it is remembered that in most underdeveloped countries there remains a strong social preference for large families.

Two recent factors which are expected to start depressing the high birth rates of underdeveloped countries are education and urbanization. Particularly in African countries, the effect of these factors to date has been ambiguous. Reporting for instance on their survey of family life and structure in southern Cameroon in late 1973, Vagliani noted as follows:

> Reviewing the results of this analysis, our first surprise was the seeming absence of any rural/urban fertility differential. High fertility is, in many African societies, a function of prevailing social customs and beliefs. In traditional Cameroonian culture, wealth and prestige are measured in terms of the number of people one is responsible for; in addition, children are often seen as an economic asset. If these values still persist, we could expect to find continued high fertility even in the urban areas. The city may, in fact, actually encourage a high birth rate and high infant survival by providing superior health and sanitation service . . . Certainly, if the extended family continues to exist even in the city and offers a solution to child care, the number of children will be less decisive in a woman's ability to undertake employment. The propensity we have noted for women engaged in office work to have fewer children than those who work in the fields or in marginal jobs (which are more compatible with children) was, indeed, slight . . . Education levels seem to explain the only major variance we found in individual fertility for women. But Primary education alone did not appear to have a great effect on age of mother at first birth; the threshold seems to be at the Junior High School level.[4]

Similar conclusions have been reported from surveys undertaken in

Table 29 *Urban and rural vital rates (per 1000 population)*

	Underdeveloped countries		*Developed countries*	
Rates in 1960	*Urban*	*Rural*	*Urban*	*Rural*
Birth rate	37.9	44.1	20.1	23.3
Death rate	15.4	21.7	8.9	9.3
Rate of natural increase	22.5	22.4	11.2	14.0

Source: United Nations, *Demographic Trends in the World and its Major Regions,* E/Conf. 60/8P/1, Table 12 (New York, May 1973); see also World Bank Staff Report, p.18.

many other countries of tropical Africa. None the less, for underdeveloped countries as a whole there is some noticeable, if not significant, difference in urban compared with rural birth rates which when put against equally lower death rates, gives urban areas a slight edge over rural areas in terms of natural increase. Table 29 reflects this situation and underscores how much difference exists between underdeveloped and developed countries in their vital statistics.

What is clear then is that, in underdeveloped countries, particularly those in Africa, the rest of the present century is going to witness continued remarkable growth of population whether in the urban or the rural areas. Growth of population is in itself not a problem except as it relates to the availability and use of resources. In its simplest form this relationship is examined in terms of population density which provides a crude index of the pressure being exerted by a given population, particularly on its land and environmental resources.

Population density

Population density is, in a sense, a measure of the interaction between a given population and the environmental resources available to it. In general, this interaction is mediated through the productive organization of the given population and the technology available to it. In most underdeveloped countries, one of the obvious consequences of the rapid increase in population is the equally rapid increase of population density in many areas. It is, of course, true that in a number of instances, increase in population density has resulted in the internal colonization of neighbouring, but relatively sparsely populated, areas. Such colonization results in the expansion of the cultivated areas and in a more even distribution of population in a country.

However, for reasons which are equally obvious, redistribution is not always a viable option open to groups caught up in the process of rapid population growth. It is in this situation that the density situation becomes a problem. At the point where the number of people deriving sustenance from a given area of land starts to have adverse ecological effects on the area, the phenomenon of overpopulation brings to mind the related concept of an optimum population, a finely tuned condition in which a population operating with its technology and production organization is just large enough to be supported from a particular piece of area. None the less, while it is possible to recognize areas undergoing a maladjustment between population and resources, it is not easy to recognize an optimum population density.

Recognition of the situation of overpopulation is, however, not as easy as it may appear at first sight. This is because of the infinite capacity for adjustments in different human communities, besides out-migration. Such adjustments have enabled a larger population to be accommodated without a visible breakdown in ecological conditions. Browning recognizes two types of accommodation: ameliorative or transformative.[5] In the former, the social structure of the community is essentially unchanged whereas in the latter fundamental structural changes take place. The source of the accommodation, whether ameliorative or transformative, may be further classified as internal or external to the community. For example, external ameliorative accommodation would include out-migration, as well as forms of assistance by central or state governments, foreign aid by relief agencies and by ex-natives of the community. These various activities are ameliorative to the extent that they provide subsidy, direct or indirect, which maintains the existing social structure without seriously altering it.

Of greater significance, however, is the internal ameliorative accommodation through which some communities have been able to absorb an increasing population. Especially for the densely populated countries of South-East Asia, Geertz characterized this process as one of 'agricultural involution'. Describing how in rural Java, the population explosion that began early in the nineteenth century and continued right into the present century was contained, he observed that:

> the productive system . . . developed . . . into a dense web of finely spun work rights and work responsibilities spread, like the reticulate veins of the hand, through the whole body of the village land. A man will let out a part of his one hectare to a tenant – or to two or three – while at the same time seeking tenancies on the lands of other men, thus balancing his obligation to give work (to his relatives, to his dependents, or even to his close friends and neighbours) against his own subsistence requirements . . . In share tenancy, the ever-driven wet-rice village found the means by which to divide its growing economic pie into a greater number of traditionally fixed pieces and so to hold its enormous population at a comparatively very homogeneous, if grim, level of living.[6]

In short, the process kept the social system from being disrupted or torn apart. Rather, it was steadily driven in upon itself. This, according to Brookfield, was possible because the foundation of the system was the terraced pondfield, an ecosystem of great stability with a remarkably elastic response to additional inputs of labour.[7] It was this elasticity of the ecological system which facilitated the 'flexibility within fixity' in the allocation of rights to resources. So, although average productivity of a unit of labour declined, marginal productivity continued to remain above zero, and elaboration of interpersonal relationships of mutual aid and indebtedness continued to keep the whole system viable. Brookfield further observed that 'integration into a wider economy was a part of this process, indeed, the principal cause, but in Java integration led to differentiation only between sectors; differentiation within the agricultural sector was squeezed down by the equalization of poverty.'[8]

Whatever its advantages, internal ameliorative accommodation does not raise per caput productivity. Rather, its effect is, as it were, to share out poverty. Yet, its robustness is of singular importance, although its widespread adoption is limited to communities with common values, well-developed networks of reciprocal obligations, effective forms of social control as well as a special type of agricultural and ecological system. Where these conditions do not exist, as in much of tropical Africa, the tendency has been towards some form of internal transformative accommodation, which affects both social organization and technology. Some indications of changes in social organization consequent on a significant rise in population density have already been referred to in Chapter 4. They relate in particular to the breakdown of the land tenure system due to increasing fragmentation and individualization of land holdings, the growth of rural indebtedness and the rising concentration of holdings in the hands of fewer and fewer people.

But perhaps more critical are the technological changes made in response to rising population density. Such changes include, for instance, a reduction in the length of fallow and a greater intensity in the use of land, the increasing susceptibility of farmland to soil erosion due to greater frequency of exposure for purposes of cultivation, the progressive decline in soil fertility with a sharp drop in output of the standard staples, and the adoption of new, less nutritive crops which make less demand on soil fertility. This sequence has been noticeable in many African countries and, in West Africa for instance, has been responsible for the phenomenal spread of cassava cultivation in recent years in place of yam production. Cassava is a hardier crop with lower protein content but able to give an average yield of 11,350 kilogrammes per hectare even in degraded or depleted soils, as against 6600 kilogrammes per hectare of yam on moderately fertile soils.[9]

This type of transformation, it should be stressed, is accommodative rather than developmental. It is a survival mechanism which allows a community to distribute stresses resulting from rapid population growth. Unlike ameliorative accommodation, the breakdown in the social system may move the community into a situation of new strains and conflicts resulting from the relative social displacement of groups. It may even create a situation where a section of the community is placed in a position where they are more likely to be induced to adopt productive innovations. But unless society creates the necessary conditions which will enable them to derive the critical benefits from these changes, all that would happen is a differential in the total output available without significant improvement in per caput productivity.

From these various considerations, Boserup suggests that, particularly for the relatively underpopulated underdeveloped countries such as those in Africa, a certain amount of population pressure may be necessary to induce changes in agricultural productivity.[10] She distinguishes between land-using (extensive) and land-saving (intensive) agricultural systems and suggests that while the former (comprising of food gathering and forest–fallow cultivation) needs no major labour investment in food production, the latter usually cannot be applied without some preliminary labour-intensive land improvements. These labour investments range from clearing roots and stones from land before it is ploughed for the first time, to building wells, ponds, dams, terraces, bonds and so on for irrigation. Once such labour investments have been made, it may be possible to obtain a higher output per man-hour from the cultivation of permanent fields than from long-fallow agriculture on similar land. But for a primitive population that must choose between undertaking the labour investment and starting long-fallow cultivation in a new place, it is a labour-saving operation to split up the group and continue with long-fallow in different places. Only when, as a result of population increases, this second alternative is no longer tenable is the community forced to adopt more intensive systems of agriculture. Boserup therefore concludes:

> Sustained growth of total population and of total output in a given territory has secondary effects which – at least in some cases – can set off a genuine process of economic growth with rising output per man-hour first in non-agricultural activities and later in agriculture. Such secondary effects come about through two different mechanisms. On the one hand, the intensification of agriculture may compel cultivators and agricultural labourers to work harder and more regularly. This can produce changes in work habits which help to raise overall productivity. On the other hand, the increasing population density facilitates the division of labour and the spread of communica-

tions and education. The important corollary of this is that primitive communities with sustained population growth have a better chance to get into a process of genuine economic development than primitive communities with stagnant or declining population, provided of course, that the necessary agricultural investments are undertaken. This condition may not be fulfilled in densely peopled communities if rates of population growth are high.[11]

The concept of a human carrying capacity

The variation in the conditions under which growth in population density can induce beneficial changes in agricultural productivity, leads to a consideration of the concept of the human carrying capacity of a given area. According to Allan, this concept relates to the maximum population density an agricultural system is capable of supporting permanently in a given environment without damage to the land.[12] This maximum density in turn is referred to as 'the critical population density'. For any given area, this density is determined on the basis of five major factors: the vegetation-soil type which defines the cultivability of the area; the total extent of the area; the cultivable percentage or the proportion of the land presently cultivable by the standard of the staple food crop; the period of fallow; and finally, an estimate of the area required per head of population.

The result of this exercise is to give a density figure much higher than the crude population–land ratio and closer to the reality of the ecological life-support capability. One important implication of the concept is to underline the fact that population density is not uniquely a response to natural resources, but to the system which a given people have developed to exploit these resources. In other words, it is not particularly meaningful to talk of the poor agricultural productivity of tropical soils in the manner once so fashionable, represented, for instance, in the work of Pierre Gourou.[13] Indeed, this view is now greatly challenged and it is becoming generally recognized that the unrealized agricultural capabilities of most underdeveloped countries are still substantial.[14] As Vries, Ferweda and Flach demonstrated, in Africa and Latin America especially, great expanses of potentially productive land are but superficially exploited, if at all, and yields (see Figure 8) are everywhere far below what is obtained experimentally or in Western Europe.[15] Even Bangladesh, which is often considered to have very serious overpopulation, still has tremendous scope for increased production. Its soils are uncommonly fertile and if rice yields were to be raised simply to the world average, her problems would be ones of storage and disposition rather than survival. In short then, with regard to carrying capacity, one should be talking about the efficiency or otherwise of particular systems of agricultural production under given socio-political conditions.

In most underdeveloped countries, particularly those in Africa and Asia, it is not uncommon to find different systems of agricultural production still continuing side by side and reflecting the responses of different communities to the carrying capacity of their land. One of the implications of conceiving of development as a nationwide spatial process is to ensure that greater efficiency is achieved in the use of land and human resources. This means that in the process of socio-economic transformation, it will be necessary to encourage internal redistribution of population between not only the rural and urban areas but also different regions of a country. The intention here is to achieve a 'preferred' maximum density of population which can be supported in each vegetation-soil region of the country without damage to the land. The significance of introducing the word 'preferred' before maximum density will become clearer in the next section. For now, it needs to be stressed that this means changing the basic parameters which determine the level of the human carrying capacity of different regions of the country. This involves raising both their cultivable percentage and their land use factor. The former is possible either through the introduction of new crops which can use formerly uncultivated land or the adoption of mechanical or chemical means which can extend the area currently under cultivation. The latter requires changes in the direction of perennial cultivation of farmland using various forms of manure or artificial fertilizers.

This fundamental transformation of the system of production, it has been argued in preceding chapters, is only possible with the restructuring of the rural areas and a fundamental spatial reorganization of the country. In other words, the development process within a country has significant effects on the configuration of population densities, not only to the extent that it brings about a better and more even distribution of rural population, but also because it involves a new adjustment in the interaction between population and natural resources. Better and more even distribution should not be construed as implying higher densities of rural population. Indeed, the increase in agricultural productivity which should accompany development requires that fewer and fewer hands need be engaged in the production of food and agricultural raw materials for industrial activities.

However, before such a rise in levels of productivity can take place,

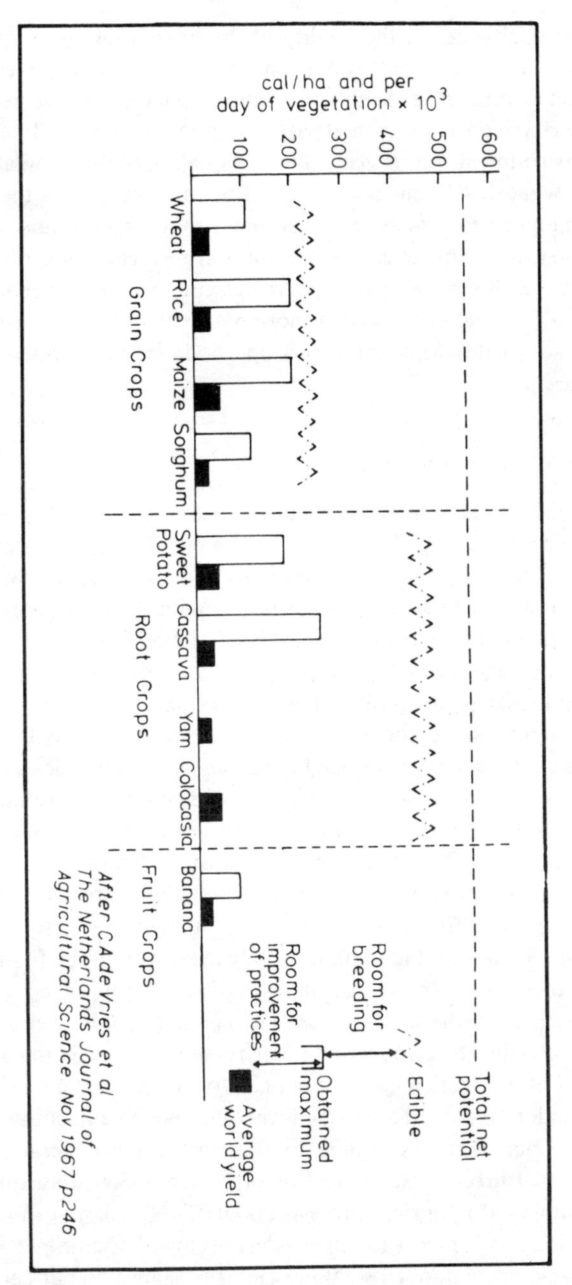

Figure 8 *Potential yields of food crops in underdeveloped countries*

substantial changes in the quality of the population are required. By 'quality' here is meant those characteristics such as education, technical competence, health and culture which make a people able to derive a high level of material and non-material well-being from their environment. Improvement in the 'quality' of the population is, of course, what development is all about but always there is the problem of orientation. Efforts by a government may or may not be closely integrated with the task of structural transformations of the type discussed in this book, and a mis-match may result between intention and actuality. Nowhere is this more patent than in the present situation in social development which one finds in most underdeveloped countries.

The role of social development

By 'social development' is meant those activities of government directed at improving the 'quality' of the population and breaking the vicious circle of ignorance, disease and poverty. Of primary importance among the various activities is education. In the context of most underdeveloped countries today, this education is strongly fashioned after that in the Western industrialized countries and so is basically liberal rather than vocational in its orientation.

The period since the end of the Second World War has been marked in most underdeveloped countries by a tremendous rise in the enrolment of children of school age in educational institutions and in the general level of literacy. In 1950, for instance, enrolment of pupils in primary schools in the non-communist underdeveloped countries was slightly lower than 65 million; by 1960, the figure had risen to 119 million and by 1970 to 202 million.[16] It jumped to nearly 330 million by 1984.[17] Considerable variations, however, were to be found among the major regions. In Africa, the rate of school attendance, although the lowest, grew the fastest, from 18 to 68 per cent between 1950 and 1984. For Asia, the corresponding increase was from 32 to 63 per cent and for Latin America 46 to 89 per cent over the same period. For the whole underdeveloped area, the level of school enrolment in 1970 was about 57 per cent. According to Bairoch, this was about the level reached in Europe (excluding Russia) around 1860 when the labour force employed in agriculture was about 55–60 per cent (as against a little below 70 per cent for the underdeveloped countries today) and in manufacturing about 18–20 per cent (as against 10 per cent for the underdeveloped countries now).

Equally rapid reduction in the level of illiteracy was achieved in the underdeveloped world among the population aged fifteen years and over. According to Bairoch, the level fell from around 80 per cent in 1900 to only about 74 per cent in 1950.[18] But between the latter date and 1970, it dropped sharply to about 56 per cent. These aggregate levels, of course, concealed wide regional and country variations. For instance in 1970, Africa still had a level of illiteracy as high as 76 per cent although this represents a sharp drop from 86 per cent in 1950. For Asia and Latin America during the same period, the drop was even more remarkable being from 78 to 50 per cent for the former and 42 to 24 per cent for the latter.

Secondary and tertiary education showed as phenomenal a rate of increase as primary. Table 30 gives an idea of the growth in enrolment

Table 30 *Enrolment in secondary and tertiary educational institutions in underdeveloped countries (in millions)*

	1950	*1960*	*1970*	*1980*	*1984*
Secondary level[a]					
Underdeveloped countries[c]	7.60	17.74	39.77	101.28	126.47
Africa	0.48	1.66	4.45	13.70	19.39
Asia	5.41	12.19	27.25	77.16	94.88
Latin America	1.71	3.89	8.07	17.50	20.42
China	1.31	8.52[e]	40.23	56.78	48.61
Developed countries[d]	30.50	45.70	77.70	83.35	83.91
World	39.41	71.96	157.70	241.41	258.99
Tertiary level[b]					
Underdeveloped countries	0.94	2.10	7.32	16.82	2[illegible]2
Africa	0.04	0.14	0.40	1.37	1.85
Asia	0.62	1.39	5.28	10.59	13.34
Latin America	0.28	0.57	1.64	4.86	5.93
China	0.14	0.90	n.a.	1.16	1.44
Developed countries[d]	5.38	9.09	20.78	29.24	30.63
World[c]	6.46	12.90	28.10	47.22	53.19

Source: UNESCO, *Statistical Yearbook, 1986* (Paris, 1986)

a General, vocational and teacher training
b Universities and other institutions of higher learning
c Excluding North Korea and North Vietnam
d Including communist countries
e 1958

at these two levels in the period 1950–84. It indicates more than a sixteenfold increase for all underdeveloped countries during the thirty-five-year period. Underdeveloped countries changed from having 25 per cent of the enrolment of developed countries to having over 150 per cent in 1984. However, in terms of attendance rate, the 40 million enrolled in underdeveloped countries in 1970 represented only 26 per cent of the population, a rate which, according to Bairoch, was attained by developed countries in the decades between 1930 and 1950. At that time the per caput standard of living was about six to eight times higher than that in the underdeveloped countries. However by 1985, the attendance rate had risen to 38 per cent of the population. Although expansion in tertiary level education has been equally spectacular, recording over a twentyfold increase over the period 1950–84, the gap between enrolment in developed and underdeveloped countries still remains wide, reflecting a ratio of about two to three in 1984.

None the less, educational progress in underdeveloped countries since 1950 has been by any yardstick, phenomenal. The exposure to new ideas and new modes of living has brought, for at least the younger generations, a real revolution of rising expectations. This has radically altered the desired standard and style of living of the vast majority of the young. This is sure to influence their 'preferred' size of income and therefore their space needs, whether for agricultural or other uses. At present, in the absence of the necessary restructuring of the rural areas, there is very limited opportunity to realize the new expectations in terms of access to bigger farmlands. The result has been a massive migration of the youthful population from the rural areas to the urban centres, particularly the metropolitan centres, and the creation in these centres of serious problems of youthful unemployment.

The origin and consequences of rural–urban migration are, however, more complex than the implication that it is purely a response to poor rural conditions. It is, of course, true that a large proportion of the migration from rural to urban areas in most underdeveloped countries at present, is the result of frustration and limited opportunities to attain a decent standard of living in the rural areas. But even if the necessary structural transformations were to be achieved today, one would still expect a significant shift of population to the cities. The movement of population within a country is thus one of the major factors in the attainment of a high degree of national integration. It is for this reason that it is necessary to put rural–urban migration within the overall framework of population mobility and the increased national integration which such movements induce.

Population mobility and spatial interaction

Within any country, the population is constantly on the move from one place to another. It is possible to classify these various movements on the basis of their motivation and to evaluate how far these movements are integrative in their effect. One of the comprehensive attempts at classifying different types of movements is that by Petersen.[19] Figure 9 presents his typology of population movements in which he identifies four types of migratory interaction, each of which can be further distinguished on the basis of whether its motivation is conservative or innovative. Conservative migrations are those undertaken by people in response to a change in their conditions and in order to retain what they used to have. By contrast, innovative migration is a means of achieving a new experience and standard of living.

Within a development context, it is the last two types of interaction, those involving changing norms and collective behaviour, that are of particular interest both in their conservative and innovative forms. The significance of changing norms in population mobility within a country is that it is the springhead of much internal colonization

Relation	*Migratory force*	*Class of migration*	*Type of migration*	
			Conservative	*Innovating*
Nature and man	Ecological push	Primitive	Wandering Ranging	Flight from the land
State (or equivalent) and man	Migration policy	Forced	Displacement	Slave trade
		Impelled	Flight	Coolie trade
Man and his norms	Higher aspirations	Free	Group	Pioneer
Collective behaviour	Social momentum	Mass	Settlement	Urbanization

Source: Clifford J. Jansen, *Readings in the Sociology of Migration* (Pergamon, 1970), p. 65.

Figure 9 *Typology of population mobility*

movements. Such movements are often undertaken by dissident groups who, in reaction against the reasons for leaving their original area, tend to be both conservative and innovative at the same time. The conservatism is on the whole selective and is particularly strong with regard to those issues which touch on the cause of the original dissidence. Real innovativeness, on the other hand, may be evinced in the effort of such groups to come to terms with their new environment. A good example is provided by the migration of dissident syncretic religious groups into the creek and lagoon region of western Nigeria. This group, known locally as the Aiyetoro Apostles, has been extremely conservative in terms of their social and religious practices but have been quite enterprising in the development of their settlement (on piles within the swamp), their fishing activities and lagoon transportation.[20]

The development of frontier regions also usually calls for a certain type of individual who is prepared to take risks and is determined to make his fortune under the uncertain conditions of uncharted territories. The movement of such individuals often is in response to their perception of a vague but realizable potential in the frontier region. The potential may be the presumed suitability of the region for growing certain crops or yielding certain other resources. The migration of individuals into the forest area of western Nigeria early in this century to grow *gbanja* kola brought from Ghana, can be said to belong to this category of movements.[21] Certainly, these pioneers opened up a whole new region scantily populated and up till then regarded as being of limited agricultural possibilities. Even in countries as densely settled as India, such enterprising individuals continue to migrate across state boundaries in search of unused or under-utilized land which they can make more productive by their labour.[22]

It is, of course, true that the examples given by Petersen to illustrate both the conservative and the innovative types of movement arising from the interaction of man and his norms, involve emigration to foreign lands. Yet, it is possible to find appropriate parallels with population mobility within countries, particularly in the underdeveloped world. As Petersen stressed, the number of people involved in this type of movement is never very large. Whether conservative or innovative, the significance of free movement is not in its size but in the example it sets for later migration. In development terms, therefore, free migration has the effect of initiating the process whereby relatively neglected or virgin regions of a country come to be more closely integrated into its space economy. The massive movement of people which follows from the migration of the free moving, pioneering and enterprising individuals and small groups, simply 'becomes a style, an established pattern, an example of collective behaviour. Once it is well begun, the growth of such a movement is semi-automatic; so long as there are people to emigrate, the principal cause of emigration is prior emigration. Other circumstances operate as deterrents or incentives, but within this kind of attitudinal framework all factors except population growth are important principally in terms of the established behaviour.'[23] In its conservative form then, mass movement of population is the main factor in internal colonization activities within a country. It is, as indicated, directed mainly at the rural areas. Young males may predominate in this movement but this can only be temporary as the tendency would be for them to invite their families to join them to provide the extra labour needed for the various chores of settlement, or to take a wife and start a family.

In many underdeveloped countries, internal colonization movement takes place over wide areas, but in a situation of ethnic plurality. Host individuals or communities have no difficulty in responding to a demand for land from migrant groups, as long as this remains at the interpersonal and relatively informal level.[24] Difficulties arise when movements are formalized and made systematic by the government as a means of bringing about a more effective utilization of land. Such difficulties are, however, likely to be less serious in the circumstances of a population mobilized for the major transformation of rural structures, as in Tanzania, or where the government is prepared to extend the principle of eminent domain to unused or under-utilized land. This latter mechanism of land settlement is already being used by governments in many underdeveloped countries to achieve large-scale redistribution of population within the national territory. One of the best known of such efforts is provided in Indonesia where, even from the colonial period, a programme of transmigration had been in operation to redistribute the population from the densely settled central island (Java) to the relatively sparsely populated outer group of islands.[25] Other examples can be cited from Malaysia, Thailand, Brazil and Nigeria. Other than land settlement schemes, there are programmes and strategies which governments have used to achieve better integration between population and land resources. What is important to stress, however, is the critical importance of such integrative effort as part of the overall process of development.

Innovative mass movement also covers rural to urban migrations. The innovativeness in this context arises from responses which individuals have to make to the novel urban situation in which they find themselves. Especially within the modern industrial cities, the fact that upward social mobility depends on the labour or specialism which an individual can bring to the market place, means that individuals have to seek out opportunities which enable them to do this. How far they succeed, however, is not wholly within their control. It is related to the general socio-economic environment in the country and is usually greatly compromised in situations where artificial barriers of discrimination on the basis of race, sex or creed have been erected.[26]

In most underdeveloped countries, the lack of conjuncture between the rural and the urban economies has created special problems with regard to the integration of rural migrants into urban society. This is reflected in the phenomenon of serious urban unemployment and the concentration of such migrants in shanty towns or bidonvilles. For instance, lack of employment opportunities, particularly for females, has meant that many migrant communities are unbalanced in their sex ratio and therefore in their full adjustment to and integration into urban life. Although over time, some tendency towards sex equalization is noticeable, this is often quite easily disrupted with each new wave of employment, and the exaggerated response in the number of in-migrants. On the other hand, the development of both the informal sector economy and the shanty town must be seen as an attempt to create half-way or intermediate conditions of integration. The ability to survive, even within this sector, on more than a short term basis, or to strike a foothold in the shanty town community, requires innovative sensitivity to the urban conditions. Such ability does not open the door to full integration but keeps migrants in the anteroom safe from the full and harsh inclemency of modern urban life.

Mass population movement into different regions of a country represents a most significant way of achieving integration between, not only a people and its territory, but also different groups within the population. The latter type of integration, according to Pryor, involves three processes.[27] The first is the individual adjustment process which ensures that the individual is integrated into both the spatial economy of his destination and into the culture and societal network of the groups and institutions of the community into which he has moved. The second is the institutional adaptation process whereby the community and the economy adapt to the presence of strangers in a positive manner, involving a weakening of ethnicity and a reduction of discrimination and prejudice based on 'strangeness'. The third is the information diffusion process whereby the migrant becomes a medium of information transaction between his former and his present community, and they are thus placed in an interactive relation, at least potentially.

Even in its most unstructured form, the migration of people has a strong tendency to bring about some degree of spatial and national integration. This is reflected not only in the person of the migrant himself, but also in the series of activities which he sets in motion between the origin and destination of his movement, and sometimes other areas as well. Such activities involve exchange of goods and services, transfer of funds, transmission of new ideas, value systems, forms of organization, and artefacts.[28] All of these have the effect of reducing the conflict potential of differences and of increasing interaction between different regions of a country. How much advantage a country derives from such integrative tendencies depends on how much deliberateness and planning go into encouraging and guiding these spatial flows.

Population planning and national integration

The preceding sections have each considered various aspects of the dynamics of a population caught in the process of development. They have examined the effect of this process on population growth, population density with its implications for the human carrying capacity of land, changes in the quality of population, and migratory flows in response to differences in the resource potential of various localities. In each of these cases, it was shown that the effect of autonomous changes without any relation or reference to other societal processes has been to exacerbate problems of development – or to impair the full realization of benefits that may otherwise have been derived from a more deliberate and co-ordinated sequence of actions. It is this need to co-ordinate population processes with a view to achieving better integration of societal activities that makes population planning so critical in the development process.[29]

With regard to the problems arising from the autonomous changes due to rapid population growth, it is worthwhile to examine them within the conceptual framework provided by the theory of demographic transition.[30] This theory has been extrapolated from the experience of the developed countries of Europe and seeks to relate a sequence of changes in vital rates with changing socio-economic conditions. The theory recognizes four closely connected phases of demographic change. A first, or high stationary phase, in which both birth and death rates are high and relatively stable and the growth in size of population is very slight. High death rates occur because of the prevalence of famine, disease, and inter-communal strife and warfare. This stage is followed by an early expanding phase in which although birth rates remain high there is a significant drop in death rates. Population grows rapidly not only because of the increasing excess of births over deaths, but also because of the rise in life expectancy.

A third phase, the late expanding phase, sees a stabilization of the death rate at a fairly low level but, more importantly, a sharp drop in the birth rate. There grows a strong societal preference for the small family and a much higher level of material prosperity. A fourth phase, the low stationary phase, is achieved when both the birth and death rates now stabilized at rather low levels and the gains from births are just enough to cancel out the losses from deaths; consequently the population is stationary. There is an increasing proportion of aged in the population and the rise of concern to prevent an absolute decline in size of population.

It is generally agreed that most underdeveloped countries are in a sense in the early expanding phase. This is only in a sense because as already indicated, and unlike the situation in Europe, this change has taken place not as a result of major socio-economic transformation, but largely because of the extension into these countries of medical and sanitation technology from the developed countries. The result of this is the existence in most underdeveloped countries of a population with a significant proportion comprising the very young who are unable to fend for themselves. This large group has to be provided for and the greater their proportion in a population, the higher the burden of dependency – a situation of having many more mouths to feed than there are hands to provide the wherewithal. Table 31 shows that over the period 1950–75, the dependency ratio, defined as the ratio of the population under fifteen and over sixty-five to the population in the age range fifteen to sixty-four, increased in underdeveloped countries from 71.8 to 82.2, but made a sharp decline to 69.9 in the period 1975–85. The pattern in the major regions of the underdeveloped areas, however, shows that this decline was not uniform; indeed

Table 31 *Dependency ratios for different regions, 1950–85*

	Youth dependency: persons aged 0–14 as percentage of those aged 15–64			*Old age dependency: persons aged 65+ as percentage of those aged 15–64*			*Total dependency: persons aged 0–14 and 65+ as percentage of those aged 15–64*		
	1950	*1975*	*1985*	*1950*	*1975*	*1985*	*1950*	*1975*	*1985*
Underdeveloped countries	65.1	75.2	62.8	6.7	7.0	7.2	71.8	82.2	69.9
Developed countries	43.0	38.4	33.4	11.8	16.4	16.8	54.8	54.8	50.2
World	57.3	64.1	54.9	8.5	9.8	9.7	65.8	73.9	64.7
Africa	78.6	86.5	87.1	6.6	5.9	5.8	85.2	92.3	92.9
Latin America	72.2	75.2	65.7	5.9	7.4	7.7	78.1	82.6	73.5
East Asia	55.6	65.9	44.7	7.3	8.2	8.7	62.9	74.2	53.5
South Asia	68.3	75.8	67.7	6.5	6.6	6.8	74.8	82.3	74.4

Source: United Nations, *World Population Prospects: Estimates and Projections as Assessed in 1984* (New York, 1986), p. 34

the situation was of continued rise in dependency in Africa over the whole period.

It is this dependency burden that is at the heart of the concern about the degree of integration being achieved between a population and its resource base in the process of development. It also emphasizes the urgency in many underdeveloped countries for deliberate planning of their population. It is, for instance, quite obvious that in countries where the dependency burden is high, the per caput income will tend, other things being equal, to be low. In terms of its developmental implications, this fact has produced two opposing schools of thought. One school, represented by Ester Boserup, whose views have been discussed earlier, argues that such a high dependency burden itself provides the necessary stress on individuals, inducing them to the type of innovative behaviour compatible with development. A second school contends that the effects of the burden not only on individuals but also on societies – in terms of low savings rate, inadequate investment and employment opportunities, skewed composition of expenditure with too much emphasis on food, clothing and education – are such that development requires that a conscious effort be made to limit the number of children per family.[31] Particularly in the period after 1960, a massive campaign has been mounted to induce underdeveloped countries, especially those in Asia, to formulate, adopt and implement a deliberate policy of family planning as a means of ensuring the feasibility of their development effort. The situation by 1970 is revealed in Table 32. For underdeveloped regions as a whole, over 80 per cent of the population lives in countries with an official family planning programme compared with only about 55 per cent for the developed countries. Yet, in spite of this and after years of quite intense and relatively successful effort, it is now generally agreed that the growth of family planning acceptance in several of the programmes has tapered off and there is a serious search for new means of motivating the masses.[32] The cause of this failure was perhaps strongly underlined at the Bucharest Conference in 1974. At that conference, most underdeveloped countries spoke against engaging in an autonomous solution of a population problem which arose in the first place because of such selective emphasis with regard to development. Some countries came to argue that 'development was the best contraceptive'.

This position is, of course, unexceptionable provided that what is

Table 32 *Distribution of world population by government policy on population and family planning activities*

	Underdeveloped countries						
	Africa	*Asia*	*Latin America*	*Oceania*	*Total*	*Developed countries*	*World*
All positions	100.0	100.0	100.0	100.0	100.0	100.0	100.0
Use of contraceptives prohibited	—	—	—	—	—	—	—
Sales illegal	—	—	—	—	—	3.4	1.0
Sales allowed, advertisement illegal	—	—	—	—	—	2.2	0.7
Pronatalist incentives	—	—	—	—	—	8.9	2.6
Laissez-faire	37.4	3.6	39.3	7.3	11.9	30.1	17.3
Pronatalist	(8.1)	—	(36.9)	—	(4.9)	(22.8)	(10.2)
Neutral	(27.1)	(3.0)	(2.4)	(7.3)	(6.3)	(7.3)	(6.6)
Antinatalist	(2.2)	(0.6)	—	—	(0.7)	—	(0.5)
Official support of voluntary programme	24.8	1.4	10.3	—	5.6	10.6	7.1
Official programme	16.8	0.9	36.1	47.8	6.7	44.8	18.0
Official programme including motivation campaign	21.3	25.6	14.3	44.9	23.9	—	16.8
Official programme and economic incentive	—	28.0	—	—	21.2	—	14.9
Curtailment of rights and privileges with excess children	—	0.1	—	—	0.1	—	0.1
Restrictions on marriage	—	—	—	—	—	—	—
Restrictions on number of children	—	40.4	—	—	30.6	—	21.5
Involuntary fertility control	—	—	—	—	—	—	—

Source: World Bank Staff Report, p.73.

meant by development is clearly understood. If development implies the mindless pursuit of economic growth with the emphasis on increased commodity output and little concern with the population caught up in this process, then far from serving as a contraceptive, its effect is likely to be that of a stimulant. The mass poverty which such development generates in both rural and urban areas, is bearable only in so far as most people can find momentary escape in sexual relations. For development to begin the process of effectively depressing fertility, new structures must be created in both urban and rural areas which facilitate the transmission of a new image of family size that is deeply embedded in new forms of social organization and production relations.[33] Planning for the transformation of rural and urban structures in underdeveloped countries thus entails an associated planning for the size and density of population which the new structures can bear, particularly given the changed level of expectations in the society. Particularly for African and Latin American countries, with their relatively underpopulated tracts of land, such planning will involve considerable redistribution of population, both to effectively cultivate the new addition to arable land and to provide new services for the production from these areas, in the many small and medium-sized towns and rural markets.

Conclusion

Spatial integration involves taking a macro structural view of the whole process of development. It emphasizes that the population of a country, especially in terms of its dynamics, provides a veritable mechanism for achieving beneficial interaction with the environment and among the different pluralities involved. But as with other aspects of the development process, such benefits can only be derived to the extent of the deliberateness and planning that goes into managing this dynamics. It is contended that the social organizational framework developed in response to the spatial reorganization of the country provides the basis for managing the dynamics of population, as it does the productive activities of communities. This framework has significance because in a development context, it must serve as a channel for information flows linking grass-root activities to one another and to other levels in the hierarchy of relationships in the country. Population planning activity thus becomes one of the new and consequential adjustments about which information flows, through new and effective channels created in the process of the spatial reorganization of the country. Such flows are crucial in inducing new behaviour patterns not only with regard to fertility, morbidity and mobility issues, but also in the equally important areas of production and consumption activities.

11 Information flows

One of the more pervading characteristics of underdeveloped countries is the heterogeneity of conditions found in them. At whatever aspect of national life one cares to look, one is conscious more of differences than of a homogeneous national situation. Whether one looks at the people themselves, their environments, their cultures, their traditional economies, or social organizations, the compelling impression is one of diversity. This impression is stronger for countries with a large indigenous population, such as those in Africa and Asia, than it is for those in Latin America. The reason for the high degree of heterogeneity is partly historical – the fact that colonial powers in carving out spheres of trading influence for themselves, operated with scant regard to many of these factors. But the reason is also entailed in the fact of underdevelopment. Many of the communities that make up present national entities are still motivated by strong 'primordial attachments' of kinship, race, language, religion and custom.[1] In other words, many of these countries remain underdeveloped because they have not been able to integrate effectively a sufficient mass of population to achieve the type of change-generating interaction needed to move the social system to new and desirable heights.

This situation of plurality is conceptually often simplified in a dualistic model of underdevelopment. This model treats all precolonial plurality of conditions as a single entity characterized as traditional, and all newly introduced factors and processes as modern. Even with this simplification, the critical nature of the lack of deliberate effort to remove the dualism, as part of the requisite conditions for development, is only now being gradually appreciated as development efforts are increasingly frustrated because of internal political instability or a failure to activate appropriate images and responses to development among the general populace.

Information flow constitutes one of the principal instruments for reducing dualism and the high degree of heterogeneity found in many underdeveloped countries. Its significance is perhaps best appreciated through specifying what is involved in the concept of national integration. According to Weiner, national integration covers five major processes:[2]

(a) the bringing together, culturally and socially, of discrete groups into a single territorial unit and the establishment of a common national identity which overshadows or eliminates parochial loyalties;

(b) the establishment of a national central authority over subordinate political units or regions which may or may not coincide with distinct cultural or social groups;

(c) the linking of government with the governed such as to reduce opportunities for misunderstanding, frustration and disruptive conflict as a result of the gap between the elite and the mass, which arises as a result of marked differences in aspiration and values;

(d) the achievement of a minimum value consensus necessary to maintain a social order. These may be end values concerning justice and equity, the desirability of economic development as a goal, or agreement as to what constitutes desirable or undesirable social ends. Or the values may centre on means, on the instrumentalities and procedures for achieving goals or resolving conflicts;

(e) the capacity of people in a society to organize themselves for some common purpose or to create new organizations to carry out new purposes.

In these various contexts, it is generally agreed that information flow plays a crucial role in achieving integration.[3] Because of the primacy of the government as an agency for stimulating development, for generating the right attitudes and value systems and providing much of the needed information and knowledge, it is necessary to examine as an essential part of the development process, how best to structure information flow so as to ensure that it achieves the objectives of spatial integration and mass mobilization. In this context, it is useful first to define more clearly what is meant by information and to indicate the structures needed to make its transmission most effective.

The nature of information

There are various ways of defining information, some specific to a particular purpose, others more generalized. In specific terms, for instance, one can talk of 'development information'. This is defined by Quebral as 'the art and science of human communication applied to the speedy transformation of a country and the mass of its people from poverty to a dynamic state of economic growth that makes possible greater social equality and the larger fulfilment of the human potential. . . . It is the systematic use of communication to persuade specified groups to change their habits, life style or ways of thought.'[4] However, in order to fully appreciate the significance and diversity of the components of information, it is useful to generalize its definition.

According to Meier, information refers to the capacity to select from a set of alternatives.[5] It involves knowledge concerning the physical and human environment and is related to a potential for adaptation, regulation or control. Usually where information is concentrated, behaviour is more ordered and predictable. The necessity to make popular choices more ordered and predictable, for instance choices between wealth-generating and leisure-maximizing behaviour, requires that appropriate information be concentrated on specified groups towards this goal. In this way, the choices available can be more clearly discerned and their probable outcomes more easily identified and evaluated. Within a country with its network of social relations, the exchange and conservation of information provides the critical means not only to motivate but also to integrate various groups so that they can develop together as a cohesive, organized unit. Essential, therefore, to an understanding of the motivational and integrative processes at work in any territorial community, is a knowledge of the pattern and intensity of information flow in that space. For it is the dynamic exchange of information between component parts of a national system that creates the bonds of mutual awareness and interdependence which promote integrative behaviour. It is in this sense that Hagerstrand argues that all human action must be seen as taking 'place in a cross-fire of information'.[6]

Two types of information are identified. The first is direct real world information; the second 'indirect' system information.[7] Real world information refers to messages about elements and processes in the real world which can be directly observed. However, because of the large number of these elements, the number of attributes characterizing them and the complexity of the patterns of interrelations between them and their attributes, it is not possible to grasp, analyse or control real world situations except through simplified theories and models. System information on the other hand, refers to elementary messages or combinations of messages contained in theories or models, regardless of whether they are embodied in scientifically based theories and formal models, or in more loosely conceived knowledge, images and conceptions derived from experience and learning. The main role of system information is thus to enable individuals to establish relations between the messages from the real world system and the knowledge of it which they already possess. The ability to establish these relations is not automatic but is the product of training and learning. It is, in fact, the rationale behind all educational processes. Given the low rates of illiteracy and income of the masses in underdeveloped countries and the socio-economic attributes that go with them, the dissemination of the know-how of accepting and using a sizeable body of hitherto unfamiliar ideas and skills in very much less time than the process would normally take, becomes a primary necessity in putting the development process on line. It is for this reason that Quebral argues that 'the job of development communication is the process of development itself'.[8]

System information may also be viewed as part of a larger system of theoretical concepts, principles, empirical regularities, hypotheses and logical relations constituting coherent wholes and referring to larger segments of social reality. Such larger theoretical supersystems are referred to as sciences. Within this framework, system information can be seen as being generated from real world information through the application of particular methods and techniques of processing. Such conversion processes are called research. The methods and techniques used for converting real world information into system information represent a second-order set of system information which can be referred to as technical information (or know-how). Such information is increased through research, development and inventions. For a given country, it can also be augmented through deliberate borrowing from other countries. It is this type of information that is involved in the current concern with technology transfer.

By providing an interpretative and condensed overview of the exceedingly complex real world situation, the basic role of system information is thus to give man a greater measure of adaptive ability and control over his environment and society. A large part of system information, to the extent that it influences behaviour and guides decision-making, can thus be referred to as control information. Control information can be further subdivided into three:

(a) Preference information includes information related to societal goals, national objectives and preference functions, as well as value judgements. Much of the value information transmitted in the process of formal education, to the extent that it is directed at socializing the young members of society, belongs to this subcategory.

(b) Directive information comprises such classes of information as policy statements, plans, guidelines, decision criteria, rules, norms, directives and commands which may be geared to more or less specific situations. This type of information is of considerable importance in managing the development of large social systems. In underdeveloped countries, this embraces much of the information from government agencies directed at stimulating the process of development among different segments of the population.

(c) Prospective information relates to anticipation about future states and events, based on more or less scientific methods of forecasting or guessing, and constituting important input into present decision processes. This is the type of information collected by various statistical agencies and is particularly important in planning the direction and magnitude of change in large social systems.

A third category of system information is designated control signals. This comprises information that is used to trigger off control decisions and actions because of the feedback reactive attribute which they contain. In development terms, this type of information covers the reaction of the people to directive information, expressed either in forms of overt behaviour or through direct dialogue. The signal attribute of this category of system information refers to its potential to reflect various kinds of unwanted deviations such as, for instance, that between supply and demand (reflecting scarcities or surpluses) or between behaviour norms and actual behaviour. Its functional role, however, is to connect real world information with control information within the framework of control loops. Control signals can be derived from direct observations of the environment (in which case they may appear as real world information), but more often they are deliberately designed to connect particular subjects of real world information with particular subjects of control information.

Within a developmental framework, these various categories of information do relate to each other in a logical and functional manner. The nature of these relationships is shown in Figure 10. Essentially two types of relationships are depicted. One, represented by the unbroken lines, emphasizes how the different categories of information relate to and develop from one another. The other, indicated by the

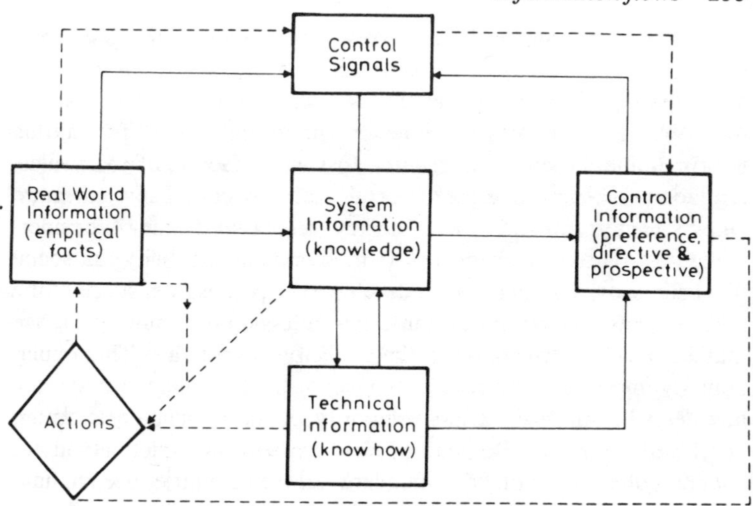

After T. Hermansen, p.11.

Figure 10 *Relationships between categories of information*

broken lines, shows how they are connected together within the context of controlling the development of the real world system. The critical feature of this other relationship is the link with the cell marked 'actions'. This emphasizes that increased information about the real world situation results in an improvement of our understanding. It also leads to changes in the behaviour of elements in the real world system which in turn generate changes in the structure of the real world system. It is this tremendous capacity of information flow to facilitate and bring about significant structural changes within a country that makes it so critical in the development process. However, the effectiveness of information flow in this regard depends, among other things, on the structure of its channels within a given national territory.

Network of information channels

In any country, the essential information-handling unit is the individual human being engaged in a continuous observation and evaluation of his environment and who is involved in a many-sided communication with his fellow men. Development, therefore, needs to equip the individual with a new set of system information which enables him to evaluate the real world information from his environment in a different way from the traditional, and to extend the scope and intensity of his communication with his fellow men. The channels through which this may be done are many and varied, formal and informal, and structured in relation to various factors of social differentiation such as age, sex, social class, occupational category, employment group and so on. The key element in defining the importance of a particular channel is its motivational capability. Indeed, it has been suggested that in the development process the success of a given channel of communication depends less on the amount of information disseminated, than on the capacity to motivate. This is particularly important with regard to unsophisticated illiterates and semi-literates who make up a large majority of the population in underdeveloped countries. Because of this concern with motivation, the channels of information flow in underdeveloped countries need to have most of the characteristics usually associated with the formal educational process.

The formal educational process is, in its own right, one of the most important channels for transmitting information to groups, selected generally on the basis of youthful age. It is the means whereby a given country undertakes the socialization of its young and their equipment with a wide variety of skills and competences. The purpose of information flows within such a system goes much further than this. In almost all countries, the role of formal education is to produce a people whose members, irrespective of the heterogeneity of their backgrounds, are 'united by more intensive social communication and are linked to centres and leading groups by an unbroken chain of connection in communications, and often also in economic life, with no sharp break in the possibilities of communication and substitution at any link, and hence with a somewhat better probability of social rise from rank to rank'.[9] We have already seen in the preceding chapter the tremendous progress made in the number of children brought within this structure of formalized information flow. For many underdeveloped countries, the evolution of national consciousness is as much the cause as the effect of development, hence the importance of integration for achieving both objectives.

Information flows through established educational channels are therefore important both in reducing the isolating effect of plural ethnicity and in replacing it by a new world-view and value system that accepts common nationality and development as desirable goals. The effectiveness of these channels in achieving these objectives can, according to Deutsch, be appraised in terms of how far they succeed in creating a community of complementary habits and facilities of communication. Complementary habits relate to the storage, recall, transmission, recombination and reapplication of relatively wide ranges of information which, taken together, constitute what anthropologists refer to as culture. In other words, an important outcome of the development process in a given country is the gradual evolution of a national culture which comes to be reflected in a convergence not only of production practices but also of consumption behaviour and style.

The significance of facilities of communication in producing a community of complementary habits largely explains why, in underdeveloped countries characterized by great ethnic diversity, emphasis in education had been to retain the language of the former colonial power. For many of these countries, the hope was that the access which this provides to a much wider store of information at the international level would help to achieve three things: create a community with a complementarity of communication habits; produce, along with that, a complementarity of acquired social and economic preferences involving the mobility of goods or persons; and encourage the rise of industrialism and the modern market economy. The last is expected to provide economic and psychological rewards for successful alignment with new national goals to men and women exposed to the risks of economic competition, uprooted by social and technological change and taught to hunger for success. The choice of a foreign language in these situations is also meant to reduce the conflict potential of having to decide among a number of competing local languages with all the anxieties of ethnic domination that raises.

How successful modern educational channels of social communication have been is perhaps best seen in their effect on rural youths and the massive out-migration in search of higher economic and psychological rewards from the process of development. Although this response itself reflects the failure of other channels of information to induce the appropriate type of motivation, it serves to underline the fact that the content of information flow is not neutral and must be closely evaluated in terms of the desired outcome.

Specifically with respect to the economically active population, who are the direct target of development efforts in underdeveloped countries, conceiving the channels of information flow in terms similar to that in the formal education process has a two-fold significance: context and scale. The issue of context relates to the need for not only appropriateness in the choice of channel, but also effectiveness in its capacity to motivate. Scale, on the other hand, refers to the size of the audience which can be reached and effectively motivated. In the case of formal education, in spite of the importance of books and audio-visual material, teachers are still regarded as the most critical channel for transmitting information. Their effectiveness is also closely correlated with face-to-face interaction with their students. For this reason, a limit is usually set to the size of a class with which a given teacher interacts.

These facts are important in considering channels of information flow to the economically active population in a country. This population usually divides up into broad occupational groups reflecting different sectors of the national economy. Among these groups, the farming community constitutes easily the most important. This is not only because farmers represent by far the largest proportion of the economically active population in most underdeveloped countries, especially in Africa and Asia, but particularly because their activities are critical for the success of any developmental effort. It is important to stress that within farming communities in different parts of the world, there exist various channels of communication. These channels have been important in the long history of these communities for the diffusion of successive items of innovation. Such channels have included kinship and friendship structure, marriage ties, market contacts and the settlement system. The first two channels are regarded as particularly significant in traditional societies, but their rather restricted span of access to new information has meant that their role in development is rather limited and sluggish. Viewed as a whole, communication within the farming communities in most underdeveloped countries is fragmented, slow and often unreliable. Few channels for interaction exist that link the components within the system and these do not work at high efficiency. A continuous flow of accurate two-way information as a basis for decisions becomes difficult to maintain under these conditions.

This is why in the development context, it has become important to deliberately create and institutionalize one or many agencies to provide the channel for this two-way information flow. In many countries, these agencies include the agricultural extension worker, the co-operative officer, the community development officer and so on. The extension worker is often regarded as the most important of these agencies. His role is to ensure a sufficient flow of information to the rural populace concerning the need for and the advantages of change, to teach the people the skills needed for this change through demonstrations, explanation and persuasion, and to encourage them to participate fully in all the relevant processes of decision-making.

One of the more depressing features of underdeveloped countries is the fact that in spite of the wide realization of the importance of extension workers, the story everywhere is of the gross inadequacy in the number of these agents. The consequence of this neglect or indifference is particularly highlighted in situations where private enterprise operates within the rural area in transferring information of a very limited range of innovations of interest to itself. The remarkable success that often accompanies such effort is a pointer to what could be done with a more comprehensive design for information flow among this occupational category. The experience of tobacco production in Nigeria, stimulated by the extension service effort of the foreign and privately owned Nigerian Tobacco Company, is a vivid example of what can be achieved.[10]

None the less, what is important to stress is the inadequacy and ineffectiveness of present channels of communication within the rural population. In Nigeria, for instance, even with the large proportion of relatively untrained staff being used, the existing ratio of extension workers to farm families is at least 1:2000 on average, and as high as 1:5000 in some places, compared to a ratio of 1:400-500 in developed countries.[11] On this issue, Yudelman's comment on the situation in Zimbabwe Rhodesia is perhaps typical for African and other underdeveloped countries:

> If there is to be any appreciable increase in African agricultural output, the process of converting ordinary native farmers into co-operators or master farmers has to be accelerated. Accomplishing this will require both an enlargement of the staff engaged in the extension service and a change in methods of extension. . . . There is no set formula for deriving an optimum ratio of extension workers to farmers, nor is there any magical formula for estimating the finances for these services. In Southern Rhodesia, the stated ideal was to have – in the higher rainfall areas – ten demonstrators, one African supervisor and one European land officer per 250,000 acres. Each demonstrator would be responsible for 25,000 acres, 3,000 of which would be arable. He would have to help five to six hundred families. This standard is comparable to the ratios in the United States and the United Kingdom . . . [however] in Southern Rhodesia, extension workers are dealing with a conservative and mostly illiterate group. They cannot rely on the media, such as pamphlets and radio, that are usually employed in advanced economies. Information has to be disseminated by word of mouth and this requires personal confrontations. The extension worker in Southern Rhodesia has to

provide more information than his counterpart in the advanced economies. He has to advise on prices and price changes and marketing procedures, to instruct producers on the timing of operations – to repeat simple demonstrations many times in order that they may be effective. In addition, if his work is to have any lasting effect he must follow up with repeated visits.[12]

The onerous nature of the responsibility that devolves on extension workers as channels of information flow in underdeveloped countries underscores the seriousness of the situation created by their limited number and scattered distribution. The same patchiness in distribution and spasmodic outburst of interest characterizes other forms of disseminating information to the rural masses. Adult education, functional literacy campaigns, community development projects, co-operative organizations – all show periods of intense interest and activity on the part of the government followed by one of apathy and virtual indifference. In such situations, middlemen operating in the rural areas have a field day and create their own intermediate channels through which filtered and highly distorted information gets to the population. For this reason, one often finds attempts by middlemen to ensure tight control of channels of communication with the rural population. Misra argues that this has been one of the most important stumbling blocks to the adoption of agricultural innovation in India.[13]

A new strategy for structuring information flow to the farming population in underdeveloped countries is thus called for. Apart from its spatial aspects, such a strategy must emphasize the importance of training the change agent to a level where he can serve as the appropriate channel of communication with the mass of farmers. On this score, there has been some concern as to whether the present strategy of using numerous low-level personnel to minister to large numbers of farmers is the most appropriate way of structuring information flow. Bearing in mind the similarity of the process of agricultural and rural development to that of education, it has been argued that a spatial approach which pays attention to scale and context may be more effective. On the issue of scale, for instance, Sam suggests that from experience in Burundi and Upper Volta, a catalytic element in rural development has been the Group Training Unit (GTU) method.[14] Under this method, instead of the traditional system whereby the agricultural extension agent tries to work with 400 to 1200 farm families but is rarely in direct contact with more than forty or fifty a month, the farmers themselves are organized into groups of twenty to twenty-five, taking close account of local situation. Each group is carefully encouraged to identify its natural leader who becomes a key link with the extension agent. The agent then works through the key member, meeting with the group once a week and dealing with different technical innovations for improved productivity. Sam emphasizes that this method combines the American concept of group dynamics with the Chinese technique of an agricultural production team. Along this same line, the Comilla Project in Bangladesh (then East Pakistan) insisted that extension work must be done through a 'model' farmer selected by each village co-operative, who came for a whole day each week to study the Centre's demonstration farm, to be coached by the Centre's experts and to receive oral and written instructions about current operations. In this way, according to Khan:

> The model farmers practised what they learnt in their plots and taught it to the members to whom they reported regularly in weekly meetings. Thus, every co-operative village got a trustworthy extension agent whose already considerable skill was constantly upgraded by the *thana* experts. Through this continuing teacher–student relationship, a handful of experts were able to extend their knowledge to two hundred and fifty villages, without asking for a regiment of half-baked, half-hearted, low-level government workers.[15]

Compared to the farming communities, the small indigenous industrial enterprises in most underdeveloped countries receive virtually no extension services from agencies of modernization. The flow of information within these occupational groups is in many countries left to be channelled at best through traditional institutions such as kinship groups, guilds and trade associations. In other countries, attempts have been made to encourage this class of activities to organize locally into co-operatives and, above these levels, into chambers of small-scale businesses. At present, where such organizations exist, the major interest has been how to get the government to extend more credit to them. Concern with increasing their access to technical and management information has been less evident. Part of the reason for the difficulty in structuring information to this category of activities is its very heterogeneous character. It embraces a wide range of traditional and modern small-scale enterprises including handloom, sericulture, coir, handicrafts, village industries and mechanized small-scale industries. Furthermore, although most underdeveloped countries have universities and colleges of technology, few of them have the type of institute concerned with the routine aspects of technology, able to encourage and co-ordinate the flow of inquiries to be directed to the appropriate technical specialists. The position has improved considerably since the 1960s when Industrial Research Institutes (in Latin America these are called Institutes of Industrial Technology) were set up in a number of underdeveloped countries. However, as Bass noted, many of these still show a strong bias towards academic recognition rather than the provision of technical information and services.[16]

Since many of these small-scale enterprises are in cities, towns and small market centres and tend to be locationally specific, especially in terms of their raw materials, there is a need to decentralize the system of providing information services to them on the basis of some hierarchical spatial principles. Indeed, as Bass also emphasized, the area served by such institutes should be restricted to that which can be effectively accommodated by direct contact.[17] In a large country with several dispersed industrial centres, policy-makers are faced with the choice between a large central organization with co-ordinated branches, or a series of independent operations. Argentina follows the first path, while Brazil and India tend towards the second. One advantage of several separate institutions is that a spirit of competition is created, which leads to diversified approaches to the provision of services.

At the area or district level, such institutions should not be sectorally specific but should serve all types of small-scale industry in the locality.[18] There is a limited scope for specialized institutions providing services for a single industry in a region, such as ceramics, textiles or woodworking. In general, the role of the institutions should include, apart from the dissemination of technical information, improvement of managerial practices, techno-economic evaluations, standardization, analysis and testing, product and process improvement, development of new products and processes, and 'trouble-shooting' of products and processes. This last implies physical inspection of the operations of a small-scale enterprise by the extension agency providing the information or service. In India, for instance, following on proposals from the International Labour Organization, the government set up a Development Commission for Small-Scale Industries whose functions included providing consulting engineers to these industries, keeping them in touch with the latest developments in technology and marketing, offering technical advice and demonstration of modern technical processes, preparing model schemes, designs, drawings and technical bulletins, conducting economic investigations, advising on management methods including marketing, distribution and surveys, conducting research and training classes and acting

Table 33 *Operations of the industrial extension service in India*

	1955–56	*1958–59*
Numbers seeking technical advice	3601	18,710
Visits outside by technical officers	7127	27,657
Numbers of parties advised to initiate firms	385	7978
Number of other assistances	2687	18,649
Number of mobile workshops in operation	–	47
Number of centres visited	–	1009
Number of artisans trained	–	6800
Designs and drawings	–	2609
Model schemes prepared	–	164
Technical bulletins prepared	–	70

Source: Development Commissioners for Small-Scale Industries, *Small-Scale Industries, Industrial Extension Service* (New Delhi, Ministry of Commerce and Industry 1959), p.17.
See also B. F. Hoselitz, *The Role of Small Industry in the Process of Economic Growth* (The Hague, 1968), p.163.

generally as an information channel.

The tremendous demand for such a service is shown in Table 33 by the very rapid rate of growth in the operations of this service in the three-year period 1955/6–1958/9. The importance of paying attention to the information needs of this category of occupations, is best appreciated against their crucial role both for agricultural development and the provision of employment, particularly in many small and medium-sized towns. In India, for instance, these small-scale enterprises provide employment for over 8 million persons in small and medium-sized urban centres, while in the case of China the number rises to approximately 18 million or about 36 per cent of the total industrial labour force.[19] In these countries also, there is a very rigorous system of licensing all such enterprises so that their number, size and growth can be constantly monitored and their need for information and other technical services frequently appraised.

The need of the larger enterprises for information is no less than that of smaller ones. Given their greater resources, it could easily be assumed that they would have no difficulty in securing this. However, two aspects of information flow at this level determine how easy it is for such enterprises to be served. One is the importance of face-to-face contacts; the other is the structural framework within which such contacts take place. Face-to-face contacts for the purpose of exchanging critical production information among large businesses often involve travelling over long distances. As Tornqvist reports in his study of large firms in Sweden, 'the most important contacts cannot be maintained with adequate efficiency by letters and telecommunications but demand direct personal contacts between personnel, and thus passenger transportation. That these contacts demand the personal attendance of often highly expert personnel is probably bound up with the fact that the contacts in many cases involve considerable elements of what we might call problem-solving, planning, keeping an eye on the course of events, pulse-feeling and reconnaissance.'[20]

Much of the information search of large firms is undertaken among themselves with little or no assistance from government agencies, although these firms also need to open direct channels to receive directives from the latter. Because of the intense and critical nature of information flow and processing for these organizations, Tornqvist suggests that this activity has a specialized existence within each firm. Indeed, most firms can be divided into two broad units:

(a) the administrative (or information or decision) unit which receives, processes and gives information;
(b) the operating (or production or manufacturing) unit which primarily receives, processes and issues materials and goods.

One of the most distinctive features of (a) is the need for very frequent personal contacts with counterparts in other firms as well as with other information units in government administration, banks and research outside industry. By contrast, the need for personal contacts between operating units in different firms and between them and information units outside the industry is extremely limited.

The ease with which face-to-face exchanges take place in underdeveloped countries is closely bound up with ownership structure and the fact that many of the larger firms are subsidiary or affiliates of foreign multinational corporations. The critical nature of information flow from such large enterprises is what is involved in the whole issue of technology transfer. It is what is needed to bridge what Streeten calls 'the communication gap' between rich and poor countries.[21] However, with regard to this type of information, Bhatt argues that there is a serious absorption problem which is qualitatively different from that faced by nineteenth century developing countries. According to him:

> Modern technology has long outgrown the stage when science learned from engineering and technological research and hence has become immensely

complex and knowledge-based, requiring high-level scientific and technological manpower for both its growth and operation. It is capital-, scale- and skill-intensive and, further because of economic development and increasing real income, it is geared to the production of sophisticated goods and services intended to meet *simultaneously* a wide variety of functional, aesthetic, comfort and status needs and wants.... The creative absorption of this technology to suit the specific needs of the poor countries requires a high degree of indigenous scientific and technological capability, which these countries lack.[22]

Bhatt, therefore, suggests that to facilitiate information flow at this level, underdeveloped countries should begin by establishing Technical Consultancy Service Centres in fields directly and vitally related to their development strategy. Such centres can then function as a vehicle for absorbing relevant modern technology, and serve as a communication link with foreign sources of technology. They can also:

(a) help in making appropriate technological choices and in diffusing relevant modern technology;
(b) support and improve machine-building capacities by providing machine industries with designs as well as links with the production structure;
(c) identify technological research problems and thus link research with industry;
(d) generate a wide variety of project ideas in diverse fields;
(e) guide the planning process by providing norms for input, skill and capital coefficients.

Some centres have already been established in the more industrially-developed underdeveloped countries, particularly those in Latin America and a few in Asia, notably India. In a recent study of four of such institutes in Latin America, it was suggested that their potential may be greatly reduced by barriers impeding access to and communication of technology and technical information.[23] The problem of obtaining outside information is acute, not only because of a reliance on existing technology and information and the scarcity of resources and professional personnel, but also because of the degree of isolation from the international technical community. However, the greater the diversity of experiences of personnel of such institutes, the greater the chance that needed information will be available via personal experience and contacts.

Information flow and popular mobilization

For a government, however, while it is important that different actors – individuals of different ages and sexes, farming households, small, medium and large-scale firms, industrial groups, and so on – are provided with the information vital for their efficient performance in the process of development, it is equally essential that channels be developed to mobilize all of these actors on the basis of directive (control) information and control signals. Or, as Soja puts it, 'a primary function of any politically organized area is to integrate effectively its territorial components, to create a community of interests which accommodates innovation, sustains development and promotes the general welfare of its adherents'.[24]

Of vital importance for effectiveness are the channels of communication which governments establish to transmit and receive information about major changes in the needs and habits of their population, or in the other relevant conditions of their life. The need for such channels arises because of the very nature of power which a government tends to exercise, especially in the process of development. According to Deutsch:

> Power is exercised by force or persuasion. Force may be applied directly, as destruction, violence, or their threat; or indirectly, as the withholding of economic necessities, or the withholding of knowledge. . . . Coercive power is [however] effective in regions near its sources, or against minor centres of social organization. It is ineffective or unpredictable over extreme distances or against major aggregations of mankind. It may, at most, destroy, but it cannot compel, and it cannot govern. . . . What is true of the limits of power to coerce seems equally true of the power to persuade; as well as of the wider ability of politics to relate, to include, to organize. All these powers or abilities depend on existing social structures and facilities of communication. . . . What matters, therefore, is the distribution of individuals and groups that can be persuaded – and kept persuaded – within any given time. This distribution is in part a function of the general distribution of social communication, past as well as present, and it is, therefore, necessarily uneven. . . . This dependence on the limited social bases of power, and on the limits of its sway, was already true for the 'night-watchman' states of nineteenth century *laissez-faire* liberalism. It is even more true of the emerging welfare states and production planning states of our own time. The slow but widespread shift in emphasis from law to administration, and from local government to regional development authorities, points in the same direction.[25]

For many underdeveloped countries this 'slow but widespread shift' is as yet hardly noticeable in the structure of governance and its insubstantial manifestation is no doubt a critical factor in the persistence of underdevelopment. Continued emphasis on law and order in the administration of communities in these countries means not only a limitation to the type of information flow between government and the people but also the persistence of channels of communication inappropriate to the task of development. The typical channel of communication between government and the masses of the population is provided in most countries by the structure of local government.

In many underdeveloped countries, this structure still reflects much of their colonial history. Using the situation in Africa for illustration, one finds that in French-speaking African countries there are three levels of administration below the national government. The lowest is the village where the formal structure is provided by a village chief *(chef du village)*. Although chosen within the community and usually on a traditional hereditary basis, he was dependent on the colonial administration for the recognition of his tenure. The next level is the canton administered by a *chef de canton*. He was responsible for implementing the policies of the colonial government and maintaining its authority. The *chef de canton* was usually salaried and as such was expected to have some degree of general education. The third level is the district or province *(cercle, subdivision, sous-préfecture)*. Here, the formal structure of administration is provided by a *chef de subdivision* (or a *commandant de cercle*, usually shortened to *commandant*) at the provincial level. The *commandant* exercised very wide powers as general administrator, supervising other civil servants and chiefs, controlling development programmes, taxation and the police, and presiding over the customary tribunal and the *conseil de notables*, whose members were nominated by him and which had a consultative role.

With independence, the central government in these countries continued to rely upon the same agents to carry the main burden of local administration and development. Even though in some of them the government conceded and extended the idea of elected local councils at regional, urban and village levels, the actual powers of these councils have remained limited. In the independent, English-speaking African countries, the position is not substantially different. Although the institution of elected local councils is generally more widespread, their actual powers are greatly circumscribed and their efficiency highly compromised by their interpenetration with the local organization of the governing party and their vulnerability to corruption.

What is therefore most critical is the perception of these various agencies of governments of their role in the development process. In many underdeveloped countries, development is often conceived simply as investment being handed down from higher to lower levels of government. As such, the task of the administrator was to see that the money was well spent or that the project was executed as expected. This was apart from his normal duty of maintaining law and order, ensuring that taxes were collected and seeing to it that other agencies of government such as education, health, agriculture and water supply, discharged their obligations faithfully. There was little direct involvement of the populace in all this, except occasionally when sectors of government activities such as community development or co-operatives needed them to justify their own existence. In short, the flow of information within this structure was largely assymetrical, a predominantly one-way flow from above downwards.

Not unexpectedly, the perception of the nature of development is quite different between the agents of government and the population. In a study of Muranga District in Kenya in 1973, some 6000 individuals were asked what they would first of all like to see improved in their village.[26] The answers gave the following preferences:

	percentage of sample
Communications, power and water	34.5
Health	23.4
Education	15.6
Administration and law enforcement	11.4
Industry and commerce	8.0
Agriculture and veterinary services	7.2

What is remarkable about this range of preference is the low order of preference which the villagers accorded their own productive activities as an area calling for government assistance. In short, the local image of government is as a body providing services and not as one mobilizing them to change or improve on their own productive organization and technology.

This limited perception of the role of government at the local level is itself part of the problem of development. It has led political scientists and administration specialists to suggest that underdeveloped countries may have to replace the present system of public administration inherited from the colonial past with that of a development administration. Development administration 'embraces the array of new functions assumed by developing countries embarking on the path of modernization and industrialization. Ordinarily it involves the establishment of machinery for planning economic growth and mobilizing and allocating resources to expand national income . . . [it also seeks] to foster industrial development, manage new state economic enterprises, raise agricultural output, develop natural resources, improve the educational system, and achieve other developmental goals.'[27] There has been serious concern in the literature as to the usefulness of this new concept of administration and what in fact differentiates it from other kinds of public administration. Swerdlow suggests that the only way to make the concept most useful is to apply it not only to goal-oriented, change-minded administration but also to those countries which have as a top priority the change from low productivity to much higher productivity levels, sustained over a reasonably long period of time.[28]

One of the critical issues in development administration is the role of the bureaucracy as a machinery for transmitting information between the government and the governed, and mobilizing the latter to carry out programmes or projects serving developmental objectives. Bureaucracy, according to Schaffer, is a particular form of machinery for decision-making and allocation.[29] It is expert, universalistic, professional, computative, depersonalized, disenchanted and routinized. Within it as an organization, particular patterns tend to be reiterated. These patterns encourage the emergence of a hierarchy, because the acquisition of expertise and the known commitment to the reiterated and fairly exclusive use of past cases as precedents to determine present ones, make possible appellate supervision. This is what gives rise to the familiar pyramids in the public administration of many underdeveloped countries. Furthermore, rules about data storage or a filing system, feedback and search, limit the information available to guide choices which have to be made in the process of coping with new situations. The choices will, accordingly, tend to be repetitive or, at the most, adaptive. The bureaucratic model is clearly not meant to operate on a performance criterion of efficiency or 'output'. Its emphasis is on repetition and reiteration rather than on innovation. Its style of decision-making carries to the greatest height the equity of the rule. Consequently, bureaucracy encourages the emergence of a wide gap in perception and communication between the decision-maker and the community. What the official can see, hear, or listen to is only a part of the 'real' situation. He sees a 'case', a case that can (or cannot) be brought within the scope of a rule, and not the whole human situation which gives rise to it.

Yet, it is clear that what is required in underdeveloped countries is an administration which is concerned with innovation in circumstances 'in which there are unusually extensive needs and peculiarly few resources and exceptionally severe obstacles to meeting the needs'.[30] Such an administration will be organized to produce a different sort of pyramid. The executive will be wide and flat, with few levels and more functional specialization in supervision, as against the present kind of administration which tends to be narrow and high, with more levels and less specialization. The task could then more clearly be seen as building a new sort of community, representing the locality in the centre, reinforcing or altering local leadership, institutionalizing programmes or advancing a particular project and recruiting resources of taxation or labour. Such an administration would by definition generate new sorts of information from the field to headquarters and would constitute a different type of communication channel between those concerned with controls and others concerned with projects, between leaders and the people they represent. As a channel for communication, especially in terms of access to and community perception of development, such an administration will conceive of its task as entailing much 'directive education'. Such a conception does not need to make administration politics and would be wrong to seek to recruit support by ideology. Rather, it will make administration part of the response to the demands of development and there is no reason why such a response should not be made comprehensible by education.

The crux of the issue, then, is how to establish an administrative structure and machinery which would facilitate the mobilization of the masses of the population in the process of development and ensure their real and active participation in decision-making. Mobilization is seen here as the critical factor in the administration of development because it focuses not only on the direct 'delivery' of services by the agencies of government to clients or beneficiaries, but also on the expectation of responsive behaviour from the latter, which may be induced but cannot be coerced by public authority, such as improved cultivation practices, free labour for public works or the adoption of birth control methods. Mobilization is also important because of the low income and low status condition of the masses in most underdeveloped countries, the very limited financial and administrative resources available to their governments which preclude the extension of services to intended beneficiaries on a one-to-one basis, and the serious information and confidence gaps between the government and the intended beneficiaries. The effects of this gap are greatly complicated by the problem of ineffective interaction especially through bureaucratic channels.

What is required for such mobilization is a spatial organization based on small territorial units of communities of the type already indicated in earlier chapters, which can ensure the formal equality of individual members, accountability to a local group and opportunities for the participation and the exercise of influence by individual members. Such an intermediary structure is what Esman refers to as 'constituency organization'.[31] Esman suggests that such a structure can be formed from various bases ranging from local government units, to co-operatives, farmers' associations, women's groups or communes. He argues that because constituency organizations are essential vehicles in underdeveloped countries for providing the mass public with services which require individualized attention or responsive behaviour, they must be conceived, both in the design and implementation of action programmes, as integral components of the administrative infrastructure. Furthermore, governments interested in reaching the mass public must invest in building and sustaining these organizations and be prepared to cope with the political consequences of mobilized mass constituencies.

Already, various underdeveloped countries have been engaged in establishing such a framework for mobilization, albeit with varying degrees of comprehensiveness and political commitment. In an in-depth study of sixteen countries (all but three of which were in Asia) by Cornell University's Centre for International Studies, some very important conclusions emerged as to the relation between this type of organizational structure and levels of development achievements.[32] According to Esman, the four important findings of the study are:

(a) Successful rural development, as measured by established indicators of agricultural productivity and social welfare, is closely associated with vigorous locally accountable institutions. Countries in which such organizations are weak do poorly by most criteria of rural development; thus constituency organizations appear to be a necessary condition of rural development.

(b) Effective local institutions cannot function in isolation. They must be linked through two-way flow of information, influence and resources with centres of power at the district, regional and national levels where policies are determined and resources are allocated. These linkages and exchanges are likely to be more reliable and productive if they occur through multiple channels –

political (party cadres), associational (federated farmers' organizations) and private (commercial enterprises) as well as the usual bureaucratic networks, because excessive reliance on any single channel can lead to distortion and blocking of communication.

(c) Constituency organizations perform similar functions and seem to be equally feasible in political systems as different in their ideologies as China, South Korea, Sri Lanka, Israel and Egypt. Regimes that undertake seriously to organize their population, particularly the peasantry, and link them systematically to agencies of the state are not distinguished by ideology from those which do not, but rather by their willingness to risk the political consequences of a mobilized population.

(d) In all these cases, successful constituency organizations are not traditional structures, but formal institutions deliberately sponsored and supported by public authority, ranging from Chinese communes and Taiwanese farmers' associations to Israeli *kibbutzim* and Egyptian co-operatives. Some of these modern organizations, however, do incorporate traditional, informal arrangements for mutual assistance, based usually on kinship ties.

Spatial restructuring of the type represented by constituency organizations thus constitutes a major means of mediating information and resource flow between the administrative agencies of government and the masses of the population. Apart from this, it ensures that administration becomes more responsive to local needs, preferences and priorities by providing the relevant information on local conditions and reasonably accurate feedback on the impact of programmes provided by government. Such organized units also make it possible for local groups to acquire, operate, manage, and maintain facilities for their mutual benefit thus reducing the high degree of social alienation that at present characterizes popular attitudes to social amenities. They can also make more effective claims on behalf of their members on services, funds and other concessions from the government at all levels.

Developing a control information system

Such a fundamental restructuring for purposes of ensuring full mobilization of the population in the process of development has tremendous implications for decision-making at various levels of government. As such, it also requires the installation of a system for collecting, recording, storing, processing, retrieving and sometimes displaying, a wide range of information generated at the level of this basic territorial unit and the other levels of the hierarchical organization of government. Complex as this may appear, what is required is a very simple frame for reporting performances and problems, and a simple system for transmitting this to the next level of administration where a degree of collation is done. The important thing is to keep the main objective of the process in mind and reduce the degree of its bureaucratization and routinization.

This is crucial because in most underdeveloped countries, there already exist organizations such as the Central Office of Statistics, the Bureau of the Census, the Survey Departments, the Geological and Meteorological Services, whose functions are to establish and operate some type of information systems. The activities of these various organizations are characterized by their emphasis on their own specialized needs and ends and the scant attention they pay to mutual co-operation and co-ordination. The latter is made particularly difficult because of the differences in the spatial units adopted for the purpose of collecting, recording or analysing information. The result has been great uncertainty in appreciating changes going on within the country or in evaluating how changes in one set of elements relate to those in others. More than this, many of these agencies suffer from the fact that, partly because of the rather restricted range of information they collect (most Statistical Offices provide comprehensive information mainly on external trade and population census), and the tardiness with which they analyse and present these to the public and the policy-makers, they are not highly regarded as important elements in the decision-making processes in underdeveloped countries.

This problem of how to assure effectiveness in the two-way flow of information in a country has led to a somewhat paradoxical position. This is, that while it is important to ensure simplicity in the flow of information, the task of co-ordinating and processing it for decision-making may require that a government takes advantage of available sophisticated systems of information analysis. Such a system would need to be explicit in terms of the locational characteristics of each territorial unit providing the information and make it possible to distinguish the relevant characteristics of each item of information. Essentially, the information generated in a country can broadly be classified into two: stock and flow information.[33] Stock information depicts the variable attributes of elements and processes taking place within a territorial unit, while flow information refers to displacements or movements of elements of different kinds out of or into the unit. The terms 'stock' and 'flow' information thus refer to a certain horizontal system of relations between territorial units. Flows taking place within such units, for example, between various economic activities within a unit, are from the point of view of a spatial information system viewed as variable stock attributes.

The need to develop a system of information that takes account of the total population spread out across the entire national space in small territorially defined units, has led to consideration of the possibility of using a geo-co-ordinate or grid-square system as the basis of information collection in underdeveloped countries. This system is conceptually very simple. Each element having a definite location in geographical space or for that matter each territorial unit, is assigned its locational reference within a grid-square system taken from a uniform map covering the entire national area. This reference, together with the time notation of the observation, is filed with a number of typological and variable characteristics. From such a spatially oriented file, data can be aggregated and analysed in a completely free and flexible way and presented in accordance with the particular problems of interest. None the less, it must be obvious that although this system is not particularly expensive to establish, its functioning depends on the availability of large computers and on extensive systems of reporting to keep the registers and information files up to date. Various means and procedures can, however, be formalized which would be sensitive to costs and simplify the reporting network.

Whether a geo-co-ordinate system of information management is adopted or not, there can be no doubt that the development process requires the co-ordination and control of highly diffused sets of information in order to be able to influence the behaviour of widely dispersed decision-making elements in a predetermined direction. An appropriate system of information for this purpose may differ from country to country, but given the fact that in most underdeveloped countries the overriding concern is the mobilization of the people to engage in their own development, it will be necessary to review their existing information systems and try to avoid some of the mistakes of present-day developed countries. More important, this review must be based on a strong appreciation of the critical importance of a two-way flow of information in the development process. It must also recognize the need to create a system which can ensure that this flow can take place efficiently and effectively and which can guarantee that decisions are made on the basis of as much information as possible.

Conclusion

Clearly, an information system is being canvassed for underdeveloped countries which, to be functional, requires the use of one of the most sophisticated technologies at present known to man. The dependence on large computers for storage, retrieval and analysis of information would seem to be at variance with the concern with the growth of indigenous technology discussed in earlier chapters. Yet, the paradox is a false one and the variance is only apparent. For the crux of the matter is the need to conceptualize the development process in a country as a matter of deliberate choice which can be achieved with policies and instruments of different degrees of sophistication. What is important is the strategic value of particular instruments and policies. The flow of information has been shown to be a most critical factor in the successful mobilization of a country for purposes of development. To be effective with regard to this flow and in the context of the time constraint imposed by the process itself, it may be necessary for a central authority to seek a means for a quicker digest and dissemination of the necessary control information.

Within the development context, information flow is not only a means to facilitate appropriate decision-making but also a requirement for achieving greater national integration. The dissemination, in particular, of preference information, it has been shown, plays a considerable role in making diverse people within the same political borders accept broadly the same set of goals and be prepared to work towards the same general objectives. The dissemination of similar directive information, particularly in a situation in which administration entails some notion of directive education, ensures that popular response is more enlightened and goal-directed. Undoubtedly, all of this will depend on structures or channels of communication which are appropriate to the task at hand. How far a given system is appropriate can be measured in terms of its output or product, as the in-depth study of sixteen countries in Asia and elsewhere has tried to show. In other words, the effectiveness of information flow has a direct relation to the physical output from territorial units or communities, and is reflected in the criss-cross pattern of movements of goods and services within the country.

12 Movements of goods and services

The flow of information within a national territory creates the conditions which favour increasingly intensive interaction between different areas, manifested in the movement of goods and services between them. The intensity of this interaction and the volume and diversity of the goods and services which interchange, thus become a measure not only of national integration but also of the level of development within the country. The bases of this interactive exchange of goods and services derive from regional variations in ecological conditions and natural resource endowments, uneven distribution of productive factors and their indivisibility, as well as regional variations in human resources, traditions, taste and consumption patterns.[1]

These conditions are necessary for the movement of goods and services to take place. They are, however, not sufficient. For transactional movements to come into existence requires three other conditions: accessibility, market exchange relations and a price mechanism. Accessibility is to a large extent determined by transport facilities. The volume of movement between two places is in fact often in proportion to the quality, speed and cost of transportation facilities available. This relationship easily becomes circular, with improved facilities leading to increased volume of movement and the latter necessitating a further improvement of facilities.

But before such a situation can develop, the communities involved must accept the need for market exchange relations or trading between them. These relations must not be confused with other types of exchange relations which one encounters among different societies and which provide the basis for some form of economic integration among their populations. Polanyi, indeed, identifies three distinct modes of economic integration among human societies, namely reciprocity, redistribution and market exchange.[2] In the first two of these, exchange relations are of a very special type. In reciprocity, exchange is usually directly between individuals or families, with the amounts exchanged being approximately equal (in the long run) among participants. This type of exchange relation is characteristic of primitive groups which are egalitarian in their social structure with no established order of dominance and paramountcy. It is common, among groups, as in pre-colonial eastern Africa, where long distance trade or urban agglomeration have not developed. Redistribution, on the other hand, involves a flow of goods but essentially to support the activities of an elite. It is the mode of integration of a rank society and although it may exhibit features of modern exchange relations such as market or price, neither of these determines the supply or demand for production factors. This is the characteristic mode for a large part of Europe in the medieval period, the pre-industrial societies of Asia and some parts of Africa up to the present century.

Market exchange relations exist when the price mechanism not only determines whether or not goods shall be produced or exchanged, but also serves to co-ordinate the productive activities of large numbers of individuals acting independently. To function effectively, the price mechanism requires that such individuals should be able to respond appropriately to price signals, their responses focusing on the relation between prices and the potential profits from their activities.

This responsiveness to the price mechanism is perhaps the most critical of the sufficient conditions for large-scale movements of goods and services to take place. It manifests itself in a subtle but real calculation of the comparative advantage that accrues to the individual in producing particular goods and services in particular regions *vis-à-vis* other regions of the country. Comparative advantage relates not so much to what a given region can or cannot produce, but rather to what activities can earn the greater profit from using regional resources to produce goods for export and importing others forgone in the process. This decision process in response to price signals ensures the maximization of the gross product of the region and represents an economically rational allocation of resources.

This issue of the increase in gross product arising from the exchange and movement of goods and services within a country is not neutral in terms of its beneficiaries. Indeed, it is possible that various elements in the exchange relation, such as the price at which goods are sold, the transport rate, the exchange regulations and institutions, can all be so structured as to create a situation of 'unequal exchange' among the various participants in the process which may eventually act to destabilize the basis of the relations. Indeed, it has been argued that the integrative effect of the international flow of goods and services during the colonial regimes, and even the present neo-colonial period, is of this unequal exchange type and that the current economic crises of production in most underdeveloped countries arise from this fact.[3]

In considering the economic effects of the movement of goods and services within a country, it is thus necessary to be clear as to what type of integrative relations they subsume and consequently how far they contribute to the overall goal of development. It is worth stressing that this means assessing how far the pattern and structure of the movements contribute to enhancing the productive capacity of individuals in all parts of the country and ensuring them an equitable share in the benefits deriving from this increased mobility. To underscore the distinction between the types of integration, it will be necessary to examine first the exploitative type achieved under the colonial regime which is still dominant in many underdeveloped countries today, and second, to consider what changes are needed to transform this to serve a more developmental role.

Colonial spatial integrative strategy

It is now commonplace to describe the objective of colonial expansion as one of exploiting the resources of their erstwhile dependent territories. From the very beginning, therefore, transportation development was accorded the highest priority not only because of its obvious effect in facilitating the evacuation of the commodities produced, but more because of its capacity to integrate the local with the international economy through the imposition of a price fixing market mechanism. As Lord Lugard so aptly pointed out at the height of the British colonial adventure in Africa, 'the material development of Africa may be summed up in the one word – transport'.[4]

The major mode of transport in colonial territories up to the Second World War was the railway. Railway construction in many territories began in the last quarter of the nineteenth century and served to link interior areas with their ports. Because of their tremendous capacity for moving heavy, bulky goods such as agricultural commodities and mineral ores, they became the major spines of the transport network in most countries. However, the practice was seldom to establish a railway network as such but only to use the railway to 'open up' the territory; hence the tendency for the railway, especially in Africa, to show on maps as single lines of penetration. For a long time, this was the only major investment of the colonial administration in most territories and it was therefore important that it paid its way. In consequence, railways tended to enjoy a high degree of monopolistic power and to preclude competition, particularly from the growing road network. Thus, it was only towards the end of the colonial regime when political independence was virtually assured, that axial road arteries running parallel to the rail lines were allowed to be improved or developed.[5] This overwhelming dependence on railways which served no more than a penetration function was particularly hard on land-locked countries. Years after independence, such countries continue to suffer severely from rather restricted outlets as the case of Zambia so clearly demonstrates.[6]

Invariably, the termini of these railways were seaports and the colonial period was one of major seaport development in many dependent territories. Indeed, even in 1967, it was claimed that in Africa only 6 per cent of external trade was overland trade with neighbouring territories.[7] The rest was trade going out through the seaports. An important feature of the colonial economy was its concentration of port functions on one or just a few ports in a country, leaving the others gradually to decline. Part of the reason for this was, of course, the high cost of providing modern port infrastructure and operating the facilities and equipment. Another reason was that the physical characteristics of many pre-colonial ports were unsuitable for the modern ocean-going vessels and to improve them would involve expensive earth moving and the creation of almost completely new artificial harbours. This was certainly the case in Takoradi, Abidjan and Lagos on the West African coast.

Roads in the colonial economy were given relatively less importance although it was not always easy to restrict them to this role. They served either as extension of the rail into regions whose profitable resource potentials were discovered after the rail had been planned and built or, more frequently, as feeder roads to the railway. Feeder roads had special significance, however, only in those regions where export commodities could be produced and it was here that they received all the attention of the colonial administration, particularly with respect to building bridges over non-fordable crossings. These areas stand out on most maps by their dense network of roads and contrast sharply with other areas whose productive activities were oriented largely to home consumption.

This pattern of colonial transportation development thus had tremendous implications not only for the movements of goods and services but also for the levels of development of different rural regions and the urban centres that service them. Specifically, by ensuring greater cash income to farmers in the export crop producing areas, these areas were transformed into the major markets both for food commodities from areas not suited for export crop production and for imported commodities from abroad. Within each country, there developed two types of rural economy – an internal exchange economy and an export economy. Each set in motion its streams of commodity movements and established different organizations and institutions to facilitate these movements.

The example of Nigeria would perhaps help to illustrate the major features of these two movements. With the decisions at the Berlin Conference of 1884, Britain, like other European powers, began the process of 'effective occupation' of this large chunk of the African continental area. The instrument for such 'effective occupation' was the railway. Starting from Lagos in 1895, the railway reached Ibadan in 1900, Ilorin in 1908 and Kano in 1912. An eastern branch from Port Harcourt in 1913 reached the coal fields of Enugu in 1916, and joined up with the western line at Kaduna in 1926. An extension from Zaria through the cotton-growing areas to Kaura Namoda was completed in 1929.

The effect of this skeletal network of rail lines was dramatic in terms of export production. According to Harrison-Church, in the twenty-five years before the Second World War, the railway enabled the export of groundnuts to increase 200 times, of cocoa thirty times, of tin twelve times and the internal trade in fish to develop a hundred-fold.[8] One reason for this, as Hodder shows in his study of tin mining on the Jos Plateau, was the remarkable shortening of travel times. Before the railway reached the Jos Plateau, the journey from the tin mines to the coast took thirty-five days and cost £29.10s a ton. With a connection to the coast at Port Harcourt in 1927, not only was the journey reduced to less than thirty-five hours, but the cost came down to £8 per ton and exports rose from 10,926 tons in 1927 to 13,069 tons in 1928.[9] Over the next fifty years, exports of agricultural commodities from these areas were to rise remarkably, particularly with the expansion of the road transport system. However, in spite of various regulations in favour of the railway, it gradually lost its importance to the roads in conveying commodities such as cocoa, palm oil and palm kernel, rubber and timber, from areas within 150 miles of the coast where the competition from road transport was not easy to check. The same story of a tremendous increase in export production characterized the development of the railway and the colonial road system in other underdeveloped regions. In Latin America for example, although the railway was initially established by domestic or national capital, it was soon taken over by the British and used to expand production rapidly from the nitrate and copper mines in Chile, the coffee plantations in Brazil and the ranches of Argentina as well as to promote other forms of raw material export in Paraguay, Guatemala and Mexico.[10]

The other consequence of colonial transport development was the broad pattern of regional specialization that it induced within countries. Using the Nigerian example again, the transport development soon revealed northern and southern zones of export agricultural activity and an extensive middle belt not particularly favoured in this respect except for small quantities of benniseed and soya beans (Figure 11). Large parts of the middle belt were, however, excellent food producing areas. Indeed, it would appear that the same factors which militated against their participation in export crop production seemed to favour them for local food production, both of the southern root crops of yam and cassava and the northern cereals, particularly sorghum. Within the flood plains of the Niger and Benue rivers and their numerous tributaries, the middle belt also produced large quantities of rice and consequently became something of a bread basket for the country. Yet, given the export emphasis of the colonial economy, very little encouragement was given to transport development in the middle belt and this left it as the most backward region in the country. Again, examples of regional differentiation and disparities resulting from selective transport development can be catalogued from other underdeveloped countries. In India, for instance, regional transport surveys have drawn attention to wide disparities in levels of development due to differences in transport facilities, for example in Andhra Pradesh between the coastal, Rayalseema and Telengana districts; in Gujarat, between Saurashtra and Kutch and the north-eastern hilly tracts, and the central and southern districts; or in Rajasthan, between the eastern and the western districts.[11]

None the less, significant internal movements of commodities were generated between different parts of the country in the wake of the rudimentary colonial network of transportation, generally as a subsidiary activity to the major export movements. In Nigeria, much of this internal flow comprised food items for many migrants retained their traditional dietary habits and thus constituted a ready market for food commodities from their home area. Moreover, the gradually rising level of urbanization created new and expanding markets in different parts of the country, such that local specialization in produc-

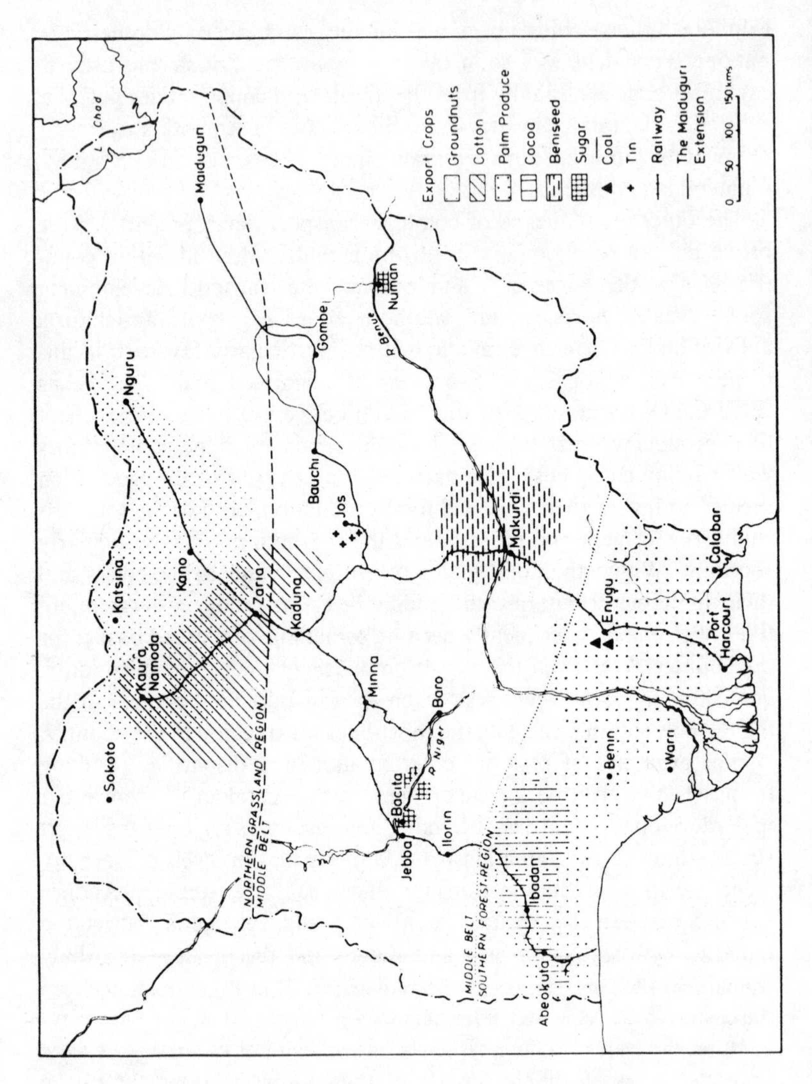

Figure 11 *Transportation and regional specialization in Nigeria*

tion of a specific food crop took place in certain areas. Indeed, according to Hay and Smith, by 1964, some 501,000 tons of local foodstuffs were involved in inter-regional trade in the country, valued at some £13 million.[12] To this should be added another £24.5 million-worth of other locally produced agricultural commodities such as cattle, kola, and timber and another £27.9 million-worth of locally manufactured goods.

The profit from this extensive internal trade and movement of goods across the country went only marginally to the farmers but more to the myriad traders and middlemen who interposed between them and the final consumers. The chain of traders begins with the innumerable itinerant buyers who go round the farms buying small quantities of commodities from the farmers in their inaccessible hamlets and villages. These they sell in the small rural markets on tracks along which only few trucks care to ply. Other middlemen buy from the itinerant buyers for sale in rural markets with greater accessibility. Eventually, the rural traders meet the urban traders at periodic markets on major roads, who then more vigorously continue the process of bulking of the commodities for the urban and regional markets. At the regional markets, a reverse process begins of breaking the bulk of imported or locally manufactured commodities and selling smaller lots through another chain of retailers until the rural consumer eventually buys. Not unexpectedly, by this time the price for the imported or manufactured commodity bears little or no relation to that paid to the farmer for his crops in the first place.

The multiplicity of traders in the internal exchange system has received a variety of comments in the literature of underdevelopment. According to Bauer, for instance,

> The number and variety of intermediaries have been much criticized by official and unofficial observers. They are condemned as wasteful and are said to be responsible for wide distributive margins both in the sale of merchandise and the purchase of produce. These criticisms rest on a misunderstanding. The system is a logical adaptation to certain fundamental factors in the West African economies which will persist for many years to come. So far from being wasteful, it is highly economic in saving and salvaging those resources which are particularly scarce in West Africa (above all, real capital) by using the resources which are largely redundant and for which there is very little demand; and it is productive by any rational economic criteria.[13]

Hunter also noted that India and Pakistan, as well as West Africa, are fortunate in having a massive corps of skilful indigenous traders while East and (to a lesser degree) Central Africa have benefited greatly from immigrant traders.[14] The whole of South-East Asia has benefited from the overseas Chinese traders, though some countries have spent much effort trying to expel them. On the question of the exploitative role of the traders, Jones observed that reports from Kenya, Sierra

Leone and eastern Nigeria seem to indicate that this occurs only in areas poorly served by roads or where the total production for sale is small. 'Correction of this kind of situation', he concludes, 'must lie in improved transport, if that is the problem or in increased local specialization in production to make the amounts offered for sale large enough to attract competitive buying'.[15]

The consensus would seem to be that the multiplicity of small traders is not itself inimical to the interest of producers except where transport is not well developed. To the extent that this latter condition characterized the rural areas of a large number of underdeveloped countries in the colonial period, and does so even today, it would seem to emphasize the importance of this problem.

But it was not only the farmers who suffered but also the consumer. Particularly within the urban centres, various institutions have been developed by traders to control the market. In the kola nut and cattle trade between northern and western Nigeria, for instance, the role of the landlord system in controlling trade and traders and regulating prices and credit has been studied in detail by Cohen among the Hausa in Ibadan.[16] A similar situation is reported from Ghana by Polly Hill.[17] Such institutions in general make it easier to form sellers' rings to prevent price competition and maintain prices. According to Bauer, 'It is said that in some markets if a newcomer enters the meat trade and does not conform to the prescribed prices, one of the Hausa traders quietly takes up a position close to the newcomer and sells at a specially low price to drive him out of business.'[18] Other forms of restrictive practice are found among retailers in urban markets, especially where they are well organized into unions, all of which are directed at operating effective price rings.

In Latin America there is growing concern that food retailing in cities has become disadvantageous to the mass of poor consumers, due to various restrictions which the small-scale retailers practise because of the near monopoly position guaranteed them by the present policies of the government. Indeed, in a study of Cali, Colombia, it was found that the cost of food was about 9 per cent higher for the mass of the poor than would have been the case if they could have shopped in the large supermarkets where the rich 15 per cent of the population do their buying.[19]

Compared to the situation in the internal exchange economy, the export trade in most underdeveloped countries shows a high degree of organization and rationalization, dating from the colonial period. In Nigeria, for instance, government intervention in this sector began with the setting up of the Produce Control Board during the Second World War which was transformed into Commodity Marketing Boards in 1947. The purpose of this intervention in the case of cocoa was described in the Colonial Office White Paper as follows:

> to secure the most favourable arrangements for the purchase, grading, export and marketing of Nigerian cocoa, and to assist in the development by all possible means of cocoa industry of Nigeria for the benefit and prosperity of producers.[20]

This intervention made the government the sole buyer of cocoa and similar export commodities. A network of centres for buying, grading and storing was developed throughout the export producing areas in close proximity to the railways or their ancilliary roads, and all the produce found its way to a central bulking depot at the port of export.

Marketing boards do not necessarily have to be monopolistic in their operation and among underdeveloped countries a variety of types can be found. Abbott and Creupelandt, for instance, identified six types of agricultural marketing boards in underdeveloped countries.[21] These are:

(a) advisory and promotional boards such as those for tea in Ceylon and tobacco in Zimbabwe Rhodesia;

(b) regulatory boards such as the Sisal and Tea Boards in Kenya and the export control offices in certain French-speaking African countries;

(c) price stabilization boards such as the Pyrethrum Board in Tanzania and the *caisses de stabilisation* in some French-speaking African countries.

(d) price stabilization and trading boards such as the special boards, supply institutes or departments of development banks in many Latin American and Near Eastern countries, set up in the 1950s to maintain buffer stocks of basic foods such as grains and beans and to stabilize internal prices for producers and consumers through trading alongside other enterprises;

(e) monopoly export marketing boards of the type described for Nigeria and other British West African countries as well as Senegal (groundnuts), Cameroon (cocoa) and Morocco (cocoa), and India (coffee).

(f) monopoly and processing boards usually for commodities primarily produced for domestic processing and consumption, for example maize in Kenya, Zimbabwe Rhodesia and Zambia.

Although there is no doubt that these various boards have been able to achieve specific improvements in the marketing operations in underdeveloped countries, they have, since the early 1950s, come in for some major criticisms. This is particularly true of the monopoly export marketing boards whose reserves were often appropriated by governments for other development purposes, thereby reducing the real income of the primary producers. It was also claimed that many of these boards created unnecessarily high costs and wastage in attaining the objectives for which they were set up. But in spite of the many criticisms, there is no doubt that, particularly during the colonial period, the establishment of marketing boards provided a high degree of organization and marketing efficiency for export commodities quite lacking for the non-export commodities that were mainly for internal exchange.

The same degree of organization was noticeable for the trade in imported manufactured goods which was dominated and organized largely by big, foreign firms. Bauer estimated that in Nigeria, for instance, these firms controlled 85 per cent of the import trade in 1949, while Levantine and Indian firms accounted for another 10 per cent leaving only 5 per cent for the African firms.[22] The large European import houses were vertically integrated in the sense that, besides importing, they also operated wholesale and retail establishments, particularly in the major urban centres in the country. But not all their imports reached the consumers through subsequent stages of their organizations. They also sold to independent middlemen, both wholesalers and retailers. None the less, the organizations of the European firm tended to be geographically extensive and elaborately integrated in contrast to those of the Levantine, Indian and African firms which invariably were limited to the ports and to a few other commercial centres.

The distribution network for imported manufactured goods below the level of the larger firms was again characterized by a multiplicity of retailers. Here, there may be some justification for the fact that, given the generally low purchasing power, most consumers could only afford to buy very small quantities of a particular commodity, hence the need for retailers at different levels of transaction. Whatever the justification, what this situation meant in real terms for the rural population was that, particularly in areas with poor transport facilities, they were made to sell their produce at very low prices while constrained to buy their requirements of imported articles at inordinately high prices. This inequity in distribution is essentially a function of the pattern and quality of accessibility available to rural communities in underdeveloped countries and is to a considerable extent a major factor of the pervading air of poverty associated with them. It is clear, therefore, that not only the transport system but also the institutional arrangements governing trade within colonial territories were greatly influenced by the overriding objectives of a colonial economy. That such systems and institutions are inadequate for self-centred development can hardly be gainsaid. Yet, before considering an appropriate alternative it will be necessary to identify the major characteristics of the colonial transportation system and their impact, not simply on export production but also on the overall developmental situation in different parts of an underdeveloped country.

A model of transportation development

In viewing the pattern of transport development in a number of underdeveloped countries, Taaffe, Morrill and Gould identified what they regarded as an ideal typical sequence.[23] This sequence sees the expansion of a transport network as a spatial diffusion process revealing certain broad regularities. Using the pattern observed in the development of the network in Ghana and Nigeria principally, but also secondarily in Kenya, Tanzania, Malaya and Brazil, they proposed a four phase sequence of development.

The first phase, which is pre-colonial, is described for a hypothetical underdeveloped country (see Figure 12). Each small port has a small trading hinterland with which it is linked by short and inconsequential routeways. There is little connection between these routeways except by the occasional fishing boats and irregular arrivals of trading vessels.

The second phase coincides with the period of colonial rule and begins with the emergence of the first major penetration lines from the sea coast to the interior. It is marked by the differential growth of coastal ports and the enhancement of the economic importance of the inland transportation terminals. The local hinterland begins to expand and to force the construction of diagonal routeways, all focusing on the ports. Railways provide the major penetration lines and their construction was motivated by the following needs:

(a) to connect the colonial headquarters on the sea coast with an interior area for political and military control;
(b) to tap areas of exploitable minerals;
(c) to reach areas of potential agricultural export production.

The third phase is marked by the greater emphasis on lateral

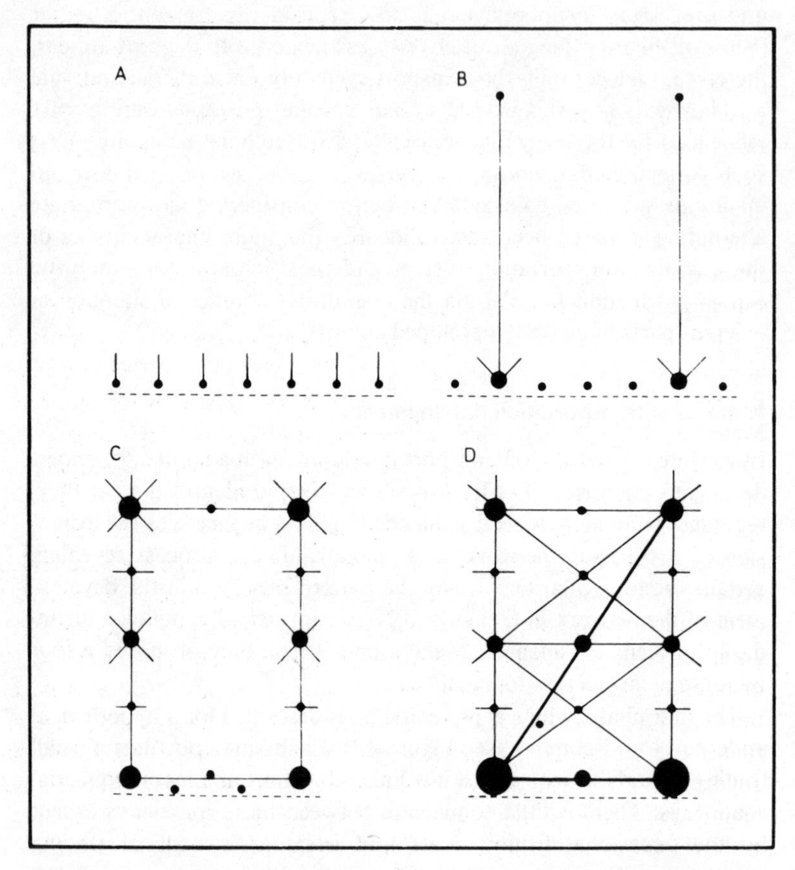

Figure 12 *Ideal–typical sequence of transport development in under-developed countries*

interconnections based on feeder routes. The development of these feeder routes results not only in the expansion of areas tributary to particular ports but also in intense competition between these sea coast terminals giving rise to the phenomenon of trade capture and expansion. The expansion process also encourages the growth of intermediate centres particularly in locations between the sea coast and interior terminals and reveals the broad outline of an integrated system. Theoretically, this phase of lateral interconnections should continue until all the ports, interior centres and major intermediate nodes are linked.

The fourth phase is marked by an emphasis on high priority

linkages. By this point, most lateral connections have been established and a fairly complete and coherent network of routes has emerged. A new process of concentration, similar to that noticed in the second phase, begins to take place. This is revealed, first by the steady rise in the importance of road traffic which first complements the railroad, then competes with it and finally overwhelms it. Second, it manifests itself in the disproportionate growth of road traffic between certain interior centres and the port. The result is a compelling need for improved transport facilities which sets up a spiralling effect of more traffic calling forth still further improvement. Consequently major national trunk routes or 'main streets' become strongly discernible. Taaffe *et al.* see this phase as a logical, though weak, extrapolation of concentration processes noted in the earlier stages of transport development but they suspect that in underdeveloped countries such high priority linkages are likely to emerge not along export trunk routes but along routes connecting two centres concerned with internal exchange activities.

It is difficult to follow the logic of this last statement, particularly given the fact that this fourth phase developed from earlier phases in which export trunk routes had received the greatest attention in the colonial and post-colonial space economy. The contradiction can only be explained by the fact that, as they admit, this fourth phase is in part based on 'highly generalized evidence from areas with well-developed transportation systems', that is, the developed countries. Its main role is thus to underline the difficulty of extrapolating a colonial route network into a future characterized by a more self-centred development. This then is the fundamental weakness of this model which has received considerable attention in the literature as providing some perspective of future patterns of transportation development in countries engaged in the process of socio-economic transformation.

It is against this background that one must view the growing reservations about the role of colonial and neo-colonial transport systems for inducing processes of self-reliant development in a country. McCall, for instance, argues that:

> Colonial transport investment was primarily a tool to promote economic, social and political underdevelopment, by giving access to more profitable sources of raw materials and extending the market for the metropole's manufactured goods. . . . In no case did the foreign builders intend the railways to serve the interests of the population through whose territory they passed . . . [thus, the] initial decisions made by technologically advanced colonial powers have had very long term effects on the subsequent develop-

ment and structure of the space economy, through the operation of geographical inertia, cumulative causation or deviation-amplification. Regional imbalances and lop-sided spatial structures are concomitant with core–periphery effects, the scale economies of capitalism, accumulation of capital, 'chains of exploitation' and so on.[24]

Besides this broad critique there are other issues relating to the development impact of colonial and post-colonial transportation which merit notice, particularly as they affect the vast rural majority in underdeveloped countries. These are: the choice of technology in constructing the routeways, the problem of costs and rates, and the control of transport resources. The choice of road construction technology in the colonial and post-colonial economies has not been a neutral technical decision made on the basis of social, environmental and engineering criteria. In many underdeveloped countries, there have been built-in biases in favour of capital-intensive road construction technologies which have caused large areas of the country to remain poorly served. These biases are usually justified on the basis, first that there are profound obstacles to the wider use of more appropriate technology, and second, that there are practical problems for engineers resulting from the fact that labour-intensive techniques are comparatively slow, expensive, difficult to manage and incompatible with high quality. Muller has recently demonstrated that these criticisms are greatly exaggerated and misconceived.[25] In his view perhaps the greatest obstacle to a rational evaluation of an appropriate technology for road construction in underdeveloped countries comprises the technocratic attitude and narrow specialization of consultants, the worship of modernism and the general technological dependence of their decision makers.

The factor of technological dependence is perhaps the most serious since it means that decisions about physical and technical parameters tend to be left to consulting engineers or multinational contracting firms. The latter invariably take their technology from an international market that is increasingly capital-intensive and usually in standard packages developed for road building in advanced industrial countries. Even where local engineers are involved, their usually foreign training makes them hardly better placed than the foreign consultants and this often results in the drawing up of contract procedures which hinder the development of alternative technologies. Governments themselves also make such development difficult since they often place great premium on speed of construction, the minimization of cash outlays and supervisory personnel, and the avoidance of the risks of using new techniques. The effect of all this is to limit the areas that can be served by good roads usually to inter-urban stretches, and particularly to deny the vast rural population access to tolerably good roads.

The determination of transport costs and rates also provides opportunities for various forms of discrimination against the rural population. This was particularly true with the colonial railway and the monopoly power which it could and did exercise. Wolff noted, for instance, that in Uganda in 1905, the Colonists' Association persuaded the railway to subsidize agricultural import and export rates to their members but not to the more numerous native producers.[26] Similarly, the Kenya colonists in 1911 got the railway to raise the freight rates on imported blankets to increase the native need for cash and so impel them to enter the wage labour market. However, to balance the picture, there was the decision of the Nigerian Railway in 1935 to stop carrying imported kola nuts from Ghana as part of the strategy to encourage local production of the commodity.[27]

None the less, it is generally true that transport rates and licensing regulations in many underdeveloped countries are still designed to facilitate the cheap evacuation of agricultural products or mineral exports. Because of the rail pricing policies based on 'values of services', imported manufactured goods for internal consumption are usually made to bear higher rates. This means that the rural consumers in particular pay considerably more for high value manufactured goods than urban purchasers. The situation hardly improves for the rural consumer even with improvement in the road network, because the main monetary gain due to road improvement in underdeveloped countries shows up as decreased transport costs for traders and middlemen rather than lower prices to consumers. McCall thus concludes:

> Road and rail in the neo-colonial situation are complementary in their operations and in their price structures, the first exporting bulky, low value agricultural commodities at the lowest possible rate, and the second distributing imported manufactures at a relatively low cost to wholesale (regional) centres. It is notable that the next stage of distribution to the village community has not been assisted by investments in the transport system.[28]

In evaluating transportation investment in underdeveloped countries, it is thus necessary to take account of differential impact on social classes within specific regions. This impact is closely related to the issue of what agency or agencies control specific transport modes. In most underdeveloped countries, the government usually exercises control over the railway in terms of its physical elements – notably the infrastructure (running surfaces, terminals, marshalling yards, and freight loading equipment), the rights of way and the vehicles – and also its operation, involving the determination of the characteristics of flows, and the orientation of all costs and returns associated with these. The manner in which this power of control has been used to discriminate in favour of export crop producers and colonists has already been indicated.

Road transportation, on the other hand, tends to be controlled by both government and private entrepreneurs. Usually, the construction of roads and the provision of equipment for them are the responsibility of government. Private enterprise, however, owns most of the vehicles and in general the operation of them. In many underdeveloped countries, driving commercial vehicles has been a significant means of entry into modern sector employment. Owning a vehicle has also been a key step in creating local entrepreneurs, most of whom, however, tend to be urban based. Indeed, the World Bank recognized that 'the transport industry is . . . one of the best starting points for an independent business career in developing countries'.[29] However, given the generally poor conditions of most rural roads, transporters find themselves almost in a monopoly situation which they exploit to the disadvantage of the rural population.

A quite typical situation is that reported by Van der Tak and de Weille in their follow-up study of a World Bank financed road project in Iran.[30] Their results show that, although following this project, transport costs for agricultural produce dropped by 20 per cent, the prices paid to the farmers remained practically unchanged. They found that the monopoly position of transporters and middlemen allowed a difference of 25–50 per cent between the farm and the Teheran prices, whereas the transport costs accounted for only 5–10 per cent of the margin. Indeed, they concluded that the road has had no development effects. That is, no increase in agricultural production has been due to the road, and an overhaul of the marketing structure might have had far greater development effects than any road improvement. Similar report of widespread disillusionment with road projects is relayed from Yemen by Dunant and others in their appraisal of the feeder road project financed by the World Bank. One of the frequent remarks made to them by the people in the area was that only merchants and government officials would benefit from the project. Some 75 per cent of those questioned, although already living within one and a half kilometres of existing roads, did not expect any benefit from the project. Dunant *et al.* commented; '[They are] already experienced . . . and know that the situation did not improve since road construction'.[31]

Transportation for self-centred development

It is thus clear that transportation as an instrument for moving goods and services within a country cannot be regarded as neutral in respect of its development or national integrative effect. The ideal sequence model of transportation ignores most of the crucial aspects of the role of transportation in development, noting only the growth in network density and volume of traffic on the major routes. It misses the fact that one major weakness of colonial and neo-colonial transportation systems is their inadequate interlinkages and intermodal connections and their limited integration into the total spatial economy of these territories. To extrapolate such a system further into the future is, of course, to perpetuate those forces which have helped to distort and sharpen spatial and social inequalities in underdeveloped countries.

None the less, in seeking to restructure the transportation system towards the goal of self-centred development, it is important to bear in mind that transport holds a contradictory position in the process of development. On the one hand, as is emphasized by economists, it is not productive in itself but it is responsive to forces generated in the production and consumption sectors. The need for transport is thus always a derived need, and to study it must be to study sectoral activities. On the other hand, transport is an infrastructural element with profound implications for overall development – most importantly for the spatial distribution of power and control.[32] The transportation needs of rural and urban areas of a country and the relation of transport to the spatial distribution of power and control thus become the critical elements of interest.

The rural areas of most underdeveloped countries have remained relatively backward because of their isolation. Yet, it is generally agreed that the transportation needs of rural residents tend on the whole to be simple and modest and to involve for a vast majority no more than access from the village to market centres. Johnson suggests that for rural development three types of roads are needed to meet different needs: the 'commuter route' needed to carry daily traffic to and from work and to allow for the clustering of the population into villages; the 'farm-to-market roads' to permit access to district markets and avert monopolistic or monopsonistic conditions; and 'truck roads' to allow each functional area to obtain goods and services from other areas.[33]

The serious inadequacy of the present situation is perhaps best revealed in the case of India. Here, it was shown that in 1967 only 11 per cent of the 646,000 villages were connected with the rest of the country by all-weather roads and one out of three villages were more than five miles from a dependable road connection.[34] Such isolation of many villages clearly impedes the spread of new attitudes and techniques as well as the physical movements of goods and services. The position is not too different in many other underdeveloped countries and the need to improve accessibility for the rural population is one of the most recognized and pressing needs of development.

None the less, road construction and transport development at this level present many difficulties. In the first place, because of the highly dispersed nature of rural settlements, the mileage of roads that needs to be constructed tends to be considerable. For the same reason, the traffic to be carried on these roads is small relative to those on major roads leading to the rural markets or to other towns. Usually, the traffic comprises no more than the farmers' households bringing inputs to the farm, conveying surplus output to the market, attending to infrequent social occasions or getting the children to school. Given their low income, the more frequent mode of travel is by foot or bicycle. In those areas where draught animals are available, they are also used to convey loads or to draw carts and thus constitute an important component of rural traffic. In many underdeveloped countries, however, a recent, very important and growing component of rural traffic is the lorry, usually one in the five to seven ton range.

The various characteristics of rural feeder roads – the extensive mileage, the relatively low volume of traffic and the limited vehicular size – have implications for their construction. Essentially, the implication is that the technology of construction need not nor cannot afford to be sophisticated or capital-intensive. Such roads, in fact, should and can be built by labour-intensive technology, preferably with the labour of the farmers themselves. What perhaps requires the greatest attention on such roads are the bridges crossing small streams that become swollen and impassable in spate. For these, there are various technological options which still permit labour-intensive technology. However, one important concomitant of a labour-intensive road building technology is the constant problem of maintenance. This is largely a matter of supervisory vigilance, but it constitutes an important aspect of any rural feeder road development.

The strategy for the construction and maintenance of rural roads calls attention to the vital role of mobilized communities in the development process. The issue goes back to the determination of appropriate territorial units into which the rural population should be grouped and the type of functions which should be performed at this level. Indeed, as Buchanan noted about China, most of the local roads serving purely local needs were built with the labour of commune members and although no estimate of the aggregate length of these feeder roads exists, he observed that they formed 'a close-meshed network, acting as vital links between the team, the brigade and the commune and the rest of China'.[35] In the slack period of the autumn months of 1959 alone, Emerson estimated that more than 7 million Chinese were mobilized for road building and short distance transportation.[36]

This type of public work using unemployed or under-utilized labour is an important aspect of non-monetary capital formation upon which underdeveloped countries must increasingly depend if they are ever to move their societies out of the present trap of poverty. It would, of course, be a misrepresentation of the facts not to emphasize that this form of capital formation is not unknown in underdeveloped countries. Indeed, Zimmerman quotes various writers to show that about 25 per cent of capital formation in the rural areas of India and Pakistan is currently of this type.[37] What is needed then is the organization and motivation to use this strategy to cope with other aspects of the development process.

In a self-centred development, rural feeder roads should form the backdrop against which the improvement of the rest of the national transportation network must be seen. Their frequent neglect in the scheme of things has been in part responsible for the backwardness of most rural areas of underdeveloped countries. None the less, they are no more than headwater channels guiding and directing traffic streams into larger and larger routeways. These routeways are in increasing order of traffic importance, a fact which implies progressive technological sophistication in their construction and a higher degree of regulation of the mix of transportation modes and traffic. At the highest level, the road transport system is essentially inter-urban and the varying importance of stretches within it reflect different magnitudes of interaction between urban centres of differing sizes. A few underdeveloped countries such as India, Malaysia, Nigeria, Thailand and the Philippines have proposals for a co-ordinated national network of roads, but none of them is marked by any special attention given to rural roads.[38]

This is not the place to go into a lengthy discussion of road transport development in a country. Suffice it to say that for self-centred development, attention must also centre on the extent to which the road network serves different sections of the population, particularly the underprivileged such as the women, the children and the poor. Transport planning should show greater appreciation of the needs of these categories of people both in the provision of facilities and in the time scheduling of their operations.

Provision of facilities should begin with the design of the roads themselves. In many underdeveloped countries, particularly in Africa, the great majority of the population are unable to purchase any form of vehicle. Yet, there is usually no provision made for pedestrian walks nor for pedestrian crossings on the roads. The problems faced by pedestrians, especially in the urban areas, are many, not least of them being the high probability of being knocked down and killed by inattentive motorists and cyclists. Cyclists in turn have so little provision made for their needs that they have to weave and dodge their way, often dangerously, between motor vehicles in undue haste. An appreciation of the movement needs of these various groups could make conditions in many cities more liveable and less harassing than at present.

The mode of transportation is equally of importance. As already indicated, the poor and the low income groups, particularly those in the rural areas, are left to the harsh mercies of small-scale transport entrepreneurs. Transport development thus requires significantly greater attention to the provision and disposition of mass transit facilities. This means that where it is impractical to establish mass transit transport facilities or to get communities to take a hand in the provision of their own transport requirements, it will be necessary to create an organizational framework within which the operations of the numerous private operators can be regulated. Regulation is particularly necessary with regard to the scheduling of services on various routes. Again, for the rural areas in particular, the provision of regular, even if infrequent, service will save many a farmer a long and sometimes futile wait at the road-side. The same concern for regularity needs to be extended not only to inter-urban but also to intra-urban services. Some countries have already established highway authorities but many have tended to concentrate on the actual construction of routeways. Moreover, such authorities tend to be highly centralized as agencies of the central government and have not been able to pay due regard to transport needs in different parts of the country.

The developmental role of other modes of transportation, notably railways, river boats, aircraft and ocean vessels, is so obvious as to require no elaborate treatment in a book such as this. What needs to be emphasized is the importance of each of these modes, particularly the railways and the river boats, to be more effectively integrated into the national sectoral activities and to be better connected with one another. The objective of integration is best served not only through expansion and reorientation of the routeways as appropriate, but especially through operating a more flexible and responsive pricing system.

However, as Singh observed for India, two main sets of facts need to be better provided to ensure effectiveness in coping with the transport needs of communities in underdeveloped countries.[39] These are data pertaining to traffic flows and requirements, and those bearing on transport costs. At present, the factual and estimate basis for planning in the various branches of transport, with perhaps the exception of the railway and the major ports, is extremely precarious in most underdeveloped countries. There are few studies of commodity flow or transport survey covering the whole country, and work on transport costs is at best at an exploratory stage. Moreover, the various agencies concerned have continued to operate on their own without much attention to combining services in meaningful ways or placing their common tasks above their separate short-term gains. The resolution of problems such as balancing alternative development from the point of view of the economy as a whole, minimizing investment and operational costs, and developing appropriate pricing policies, is of central importance for efficiency in the provision of transport facilities. Similarly, the regular collection and flow of information on traffic streams and costs will go a long way to assist co-ordinated development and operation of transport services.

Organizing the flow of goods and services

The provision of adequate transportation does not, however, solve the problem of development, especially for the rural producer, nor does it necessarily ensure effective integration between rural and urban activities. This is because of the basic differences in the rhythm of activities in both of these geographic domains. In the rural areas, productive activities, notably agriculture, tend to be greatly controlled by environmental factors and result in output being highly concentrated within a short period of the year. In some countries, such concentration occurs more than once a year. Particularly in the irrigated rice producing countries of Asia, two or three harvests a year are often realized. Most farming communities thus face the problem at least annually of how to cope with this sudden surfeit of output.

The options open to them for dealing with the situation have always been three-fold: store, process or sell. The storage of agricultural surpluses has a long history in most rural communities of underdeveloped countries and many have evolved ingenious ways and methods of preserving agricultural commodities for substantial periods of time. In West Africa, for instance, Morgan records three main types of facility depending on whether storage is to last temporarily for a few weeks, for a short term period of some months, or a long term one of several years.[40] The facilities comprised woven grass bins, clay bins or granary pits respectively. In most communities, social rules of varying degrees of elaboration govern how or where each type of crop is stored, who owns which store or can open the storage bin and in what circumstances this can be done. In the Casamance, Senegal, de Sapir reports that among the polygamous Diola, each wife, apart from the farmer himself, has her own bin (*buntungab*) constructed as a separate, round, room sized, two level structure of puddled adobe with a conical thatched roof. The storage area is on the top level and her kitchen underneath so that the kitchen smoke can keep out rats and other pests, and dry excessive moisture from the rice crop.[41] In the case of divorce, the wife takes all her stored crop or as much as she can. In general, secrecy pervades the attitude towards the stored crop. The door can be opened only early in the morning or late in the afternoon when others are not around. Strangers may never enter the *buntungab* since it is believed that this will result in the stored crop 'disappearing' or being 'bewitched' away.

In many parts of the underdeveloped world, neither the storage facilities nor the technology for preservation against pests and diseases are developed enough to cope with harvest surpluses of a volume much bigger than for a subsistence economy. Thus, a major problem facing rural areas in most underdeveloped countries is the development of appropriate storage facilities and preservation technology, particularly for the different varieties of food crop being produced.[42] This problem has been discussed in Chapter 5. Its significance here is that it affects the flows of goods and services in the country. With an inadequately developed storage system, farmers have to get rid of the rest of their harvests above the amount that can be stored or put away for present consumption. The simultaneity in the existence of such surpluses among thousands of farmers creates a glut which, in the absence of adequate storage facilities and government policies, leads to a drop in prices and a heightened demand for transportation with the consequent sharp rise in transport rates. The resulting dilemma in which the rural population in some countries find themselves during this period is an important factor in their general poverty.

Processing of part of the surplus harvest, particularly by the women members of rural households, has traditionally provided some way out of this dilemma. Processed commodities in general last longer and, although requiring their own storage facilities, are on the whole easier to accommodate. Traditional processing facilities are themselves rather small in scale and can cope with only a limited portion of the surplus harvest. Besides, since in some countries this is again women's work, various rules and customs govern the transaction between a man and his wife or wives in this situation. Modern processing facilities, on the other hand, entail capital investment of a type which cannot be allowed to remain idle during slack periods without seriously compromising their profitability and economic rationale. To be successful, their installation requires the rationalization of agricultural production to ensure a fairly steady supply of raw materials spread over many months. Even this requires some measure of improved storage facilities and is more likely practicable in spatially reorganized rural areas.

Processed agricultural materials constitute a significant portion of the goods moving across a country and helping towards effective economic integration of its different regions. The processing itself has a positive and additional advantage in generally making the product more transferable over greater distances. It also makes it better placed to bear the cost of transportation. In short, it can be asserted as a general rule that the greater the degree of processing of agricultural commodities in a country, the wider the span of inter-regional exchanges based on them and, consequently, the higher the degree of national integration.

Whether in processed or non-processed form, most commodities move across a country as material elements of trade. Farmers sell their surpluses and so do the processing and manufacturing enterprises of different sizes and complexity. But the organization of the trade in the produce of the millions of farmers, who are the majority of the population in most underdeveloped countries, is critical for self-centred development. It has already been indicated that, up to now, most of the farmers have had to relate to the rest of the economy through a long chain of middlemen operating along a hierarchy of retailing and bulking markets. It was also indicated that, at least with regard to export crops, the colonial and post-colonial administrations in many countries designed an alternative marketing system which did not do away with middlemen but kept their number within reason. The elements of this new system can be identified as follows:

(a) the establishment of quality standardization and grades for the various commodities;
(b) the provision of premium prices for the best grade and discounted prices for poorer grades;
(c) the wide publicity given to these various prices for each season;
(d) the setting up of various types of marketing organization to ensure quality and stabilize prices;
(e) the licensing of middlemen or buying agents;
(f) the determination of transport rates, handling charges and permissible trading margins;
(g) the establishment of storage facilities with adequate protection against pests and diseases.

It is obvious that there is very little about this system to prevent it from being extended to cover the whole of the agricultural economy. This has, in fact, been attempted to a certain extent in a number of underdeveloped countries, particularly in Latin America and Asia. For many African countries, however, part of the problem of extending this system to cover commodities involved in internal exchange is a certain inertia about taking part in the rather innovative enterprise of determining quality criteria and methods of grading. It is instructive to realize that this was precisely what the colonial administration had to do with regard to a whole series of exportable, tropical but unfamiliar commodities such as cocoa, palm kernels, palm oil, rubber, cotton and groundnuts. The fact that in spite of the wide variety of these commodities, some solid, some liquid, quality criteria and grades were established, emphasizes that there is no insurmountable difficulty inherent in the operation.

Standardization and grading, particularly if accompanied by premium pricing, not only provides incentives for farmers to improve their production but, more importantly, establishes a frame of reference operative throughout the country. This means that demand for and supply of a given commodity can be expressed in universally recognizable units. Not only does this facilitate the ease of trading in such commodities across a national space, but it will also greatly enhance the efficient operation of processing industries in the country. Prices can also be publicized of universally recognizable units and so reduce the possibility of exploiting rural producers.

While it should be possible for producers to form themselves into marketing co-operatives and sell direct to government, this should not conflict with providing opportunities, through licensing, for a limited number of middlemen. Nor should there be any conflict between co-operatives or even middlemen, wanting to construct and maintain their own separate storage facilities and those which the government statutorily should build and maintain as a buffer to help stabilize prices for the farmers.[43]

None the less, unlike export crops, the bulking and movement of goods within a country has a dual role in the sense of the breaking of bulk in an attempt to reach the ultimate consumer especially in the urban centres. It is particularly in this field that the role of middlemen should remain paramount. If the government were to become the major buyer of the produce from the rural areas, it would then be the main seller to wholesalers and processing firms. From the former, a chain of smaller wholesalers and retailers would develop to reach out to all urban centres and consumers. Even in this case, however, the licensing of traders and commercial firms would be required, if only to facilitate the regulation and operation of activities in this field. At any rate, governments in underdeveloped countries need to pay attention to the provision of various market services to mediate in the critical interaction in the development process between producers and distributors. A few countries are already showing an appreciation of the importance of such government supported services, including Mexico, Bolivia, India, Malaysia, Thailand, Malawi, Ethiopia, Nigeria, Kenya and Tanzania.[44]

Whether with regard to the bulking of produce from the innumerable farms, or the breaking of bulk and distribution to urban centres and neighbourhoods, attention will need to be paid to issues such as catchment area and marketing field. The bulking of commodities at central points throughout the country requires the use of various optimization techniques to minimize unnecessary costs of transportation from the farms to the bulking centres. Some of the countries mentioned above, notably Bolivia, Malaysia, Mexico and Pakistan, have in fact proceeded to construct warehouses and market outlets to improve the farmer's bargaining power. The same consideration will influence, in somewhat different ways, the retailing operations of various traders and wholesalers, although here the resolution may have to be the outcome of a spatially competitive process. What the relation of the catchment or market area should be to the basic territorial unit of organization of the administrative divisions of a country will be something for different governments to decide. Generally, it can be assumed that the stronger the organizational emphasis on co-operativization, the more likely it will be for the boundaries of the commericial and administrative activities to coincide.

Conclusion

The movement of goods and services across a country represents a major aspect of the integration of the country. It denotes a progressive move to interdependence based on increasing regional specialization and a better appreciation of the benefits derivable from the full exploitation of the principle of comparative advantage. Such increasing interdependency also ensures that the forces making for interregional transmission of growth within a country become so much more powerful as eventually to outstrip those making for international transmission and resulting from initial colonial relations.[45] Moreover, the general theory of political integration indicates that transactions involving the exchange of goods and services between different areas of a country constitute one of ten major factors of the national integrative process.[46] None the less, Deutsch calls attention to instances where although people from different cultures lived in one society and exchanged goods and services over a long period of time, the information flow and by implication, national integration was minimal.[47]

At any rate, as has been emphasized, for most underdeveloped countries, conditions are still less than adequate or appropriate for full integrative effects to result from the movement of goods and services within the national territory. For one thing, the network pattern of major transport routes still reflects the outward orientation of their colonial past. A restructuring of this pattern to bring it more in accord with national objectives and priorities will be an important first step towards achieving a more self-centred pattern of national integration and development. In this connection, new emphasis will have to be placed on the transportation need of the vast majority of the population living in the rural areas and requiring no more than all-weather feeder roads. Some importance must also attach to the convenience of the populace and this can be achieved to a high degree by thinking of the different categories of users of routeways during the construction process, and by paying greater attention to the disposition and time scheduling of various transportation modes. All of this is critical because transportation is an important tool of power and control and it is necessary to ensure that it is not used to impair the capacity of individuals to cope with the changing circumstances of their lives. Indeed, Adelman and Dalton emphasize the strong association between the availability of feeder roads and village income.[48]

Equally important in enhancing the capacity of the masses for development is the organization of the disposal and distribution of goods and commodities. Here again the situation of the rural population has been shown to be more critical than that of urban residents and the appropriate organization must thus be one that protects them from exploitation in the hands of middlemen. The effect of this organization must be one that not only facilitates the expansion of the internal market for their goods but also stimulates greater productivity.

Both improved transportation and the establishment of better trading organization should bring about a greater integration of economic activities within a country. In this matter, however, it is impossible for a country to operate as a closed system. Self-centred development does not imply autarky; rather it entails the reorientation of the objectives of productive effort to serve primarily national interests. Export and import of goods must continue and be part of the overall movement of commodities within a country. But their priority in the whole scheme of things must be re-evaluated. Such a re-evaluation highlights the need for reappraising the external relations of a country, particularly those with trans-national organizations whose presence in underdeveloped countries is precisely geared to foster the import–export trade.

13 External relations

In the preceding chapters, underdeveloped countries have been treated as if each one of them were a closed system with few or no relations with other countries of the world (except in their 'colonial' status). Indeed, the impression could easily have been formed that any of these countries could be developed without giving serious thought to its external relations. Yet, just as improved transportation and communications as well as the enhanced movements of people, goods and information have made for closer integration within a country, these same factors on a global scale have brought about greater interdependence among the countries of the world.[1] Hence, no view of the development process is complete without consideration being given to how this affects and is affected by the external relations of a country.

The external relations of any given country tend, on the whole, to be characteristically complex. They are, according to Northedge, the product of environmental factors both internal and external to the country and in consequence may, at least theoretically, be conceived as limitless and embracing virtually the whole universe.[2] In practice, however, this environment is greatly circumscribed by the range of interests and the limitations of power of the particular country. For underdeveloped countries and in the special context of development, the most critical range of interests is that relating to productive capabilities, encompassing specifically economic and technological issues. These interests, in recent years, have been pursued within the limitations of power imposed by the dependency of these countries on the advanced industrial countries of the world. The fact of dependency thus provides a first dimension for examining the problems of external relations of underdeveloped countries. A second dimension is provided by attempts at regional and inter-regional co-operation as a means of increasing bargaining power and reducing some of the more negative effects of their interaction with the developed countries.

Problems of external dependent relations

The underdeveloped countries of Africa provide the best example for illustrating the problems of external dependent relations for self-centred development. Even after the formal termination of their colonial status, many of these countries continue to show evidence of a greatly impaired and demoralized capacity to act fully in their own interests. One of the reasons for the continued impairment of this capacity is the colonial fragmentation of the continent into a disparate medley of trading areas. A continent whose population in 1975 was less than 400 million was put into 56 'national' entities with an average national population of about 7 million and a median population of less than 4 million. The position in Latin America is only slightly better, although for Asia many of the countries show a larger population concentrated on limited productive land.

The small size of most of the underdeveloped countries of Africa has constituted a major obstacle not only to their independence of action in their external relations but also to their ability effectively to undertake their own development. The evolution of the external relations of most of these countries can be generalized as involving five stages. These stages marked the transition of each of the countries from being a primitive 'reserve' outside the world market to becoming a truly underdeveloped economy, dominated by and integrated into the world market.

The first stage is that of an exploratory trading economy. The external relations of the country were based on the identification by foreign European powers of the range of raw materials available and their decision on which of them to exploit. This was the stage when many colonial trading companies were established and made easy monopoly profits without much risk or great investment.

The second stage witnessed relatively substantial infrastructural investment mainly in transport (ports, railways and roads) and in limited social overheads especially related to urban development. It also involved the installation of water supply, electricity and sanitary equipment as well as the establishment of schools, hospitals and housing.

The third stage was one of accelerated colonial economic growth. This occurred particularly at the end of the Second World War and was marked by greater investment in both social overhead capital and directly productive activities. The aim, however, was to rebuild Europe's losses in the war through planning the 'development' of the colonies and ostensibly investing in the welfare of the people.

The fourth stage was characterized by rapid growth in the bureaucratic machinery of exploitation due particularly to the granting of political independence. Political independence and the wide acceptance of a strategy of 'planned' development encouraged a pattern of public expenditure with a high propensity to importation. The economy soon reached a stage where imports began to outstrip exports and the country in consequence came increasingly to depend on 'aid', technical assistance and capital inflows from the metropolitan countries.

The fifth stage is one of stagnation when the 'dualistic' structure of the economy becomes most marked and is characterized by a heightened dependence relation on the metropolitan country. This stage also shows in sharp relief the most glaring difference in the distribution of growth between the various sectors of the economy and in per caput income between social classes. Most importantly, the stage is marked by the stagnation or even decline of the agricultural sector which still provides employment for the vast majority of the population.

Samir Amin provides a detailed analysis of these five stages in the evolution of the external relations of nine French-speaking West African countries.[3] Referring to all five stages as reflecting 'outward-directed' growth or growth based on external demand and external financing, he shows how, on the eve of the Second World War, most of these countries had hardly gone beyond the exploratory trading stage. Real colonial exploitation by then had touched only Senegal where most of the infrastructural investments were concentrated. The other eight territories remained as minimally exploited colonial 'reserves'. Thus in the period between 1920 and 1940, the external trade of Senagal alone accounted for nearly 60 per cent of the total for French West Africa. This total itself had doubled in real volume during the period, giving a growth rate of 3.5 per cent per annum, a rate two to three times lower than the post-war rate. Public expenditure during the same period grew at a rate of around 2 per cent per annum while the role of external finance was negligible, averaging less than 2 per cent of total public spending.

The third stage of accelerated economic growth opened with the post-war plans of the *Fonds d'Investment pour le Développement Économique et Social des Colonies* (FIDES). (In the English-speaking countries, the comparable event was the Colonial Development and Welfare Act which created a special fund for colonial 'development'.)

The flood of investment made possible under this programme went not only to improve the infrastructure (for instance, most West African ports had additional berths constructed during the period) but also to bring about rapid expansion in export production. The organization of marketing was reformed through the creation of marketing boards and a new price incentive system for export crops was instituted. Indeed during the period 1948 to 1960, economic growth in French West Africa paralleled the growth in external financial contribution. Thus, while the real annual growth rate of export earnings was 4.7 per cent, that for all income from abroad was 6.8 per cent.

The reason for this large volume of external financing was that well before 1960, the accelerated rate of colonial exploitation had produced a crisis in the public finances of the various territories. Economic growth had lagged behind current public expenditure and there was a need to meet a growing proportion of that expenditure out of the budget of the metropolitan country. So, even before their political independence, the economy of these countries was already stagnating and unable to provide the necessary resources for their administration.

The position did not improve appreciably as the countries entered the fourth stage beginning with political independence. If anything, the gap between the rate of growth in current government expenditure, which was increasing, and the rate of growth of the economy, which was slowing down, continued to widen. The position was further exacerbated by the illusion of 'planning' which simply encouraged greater bureaucratization of the economy. Efforts to increase taxation, widely practised in the decade immediately following independence, merely postponed the day of reckoning. The situation reflected the paralysis of national savings which, despite the growth in gross domestic product remained unable to bring about an automatic and spontaneous transition from growth stimulated from abroad to internally generated and self-financing growth. During 1960-70 local savings accounted for no more than 1 per cent of the region's gross domestic product. This fact illustrates better than anything else the completely dependent nature of the economy of the country and its total lack of internal dynamism.

The final stage in the evolution of the present pattern of external relations of these nine countries came with active 'industrialization' which emphasized most clearly the dualistic structure of the neo-colonial economy. In Senegal, industrialization has a much longer history than elsewhere in West Africa. Starting from the 1930s, industries were developed here particularly as a result of 'refugee' capital being removed from France during the Second World War.[4] Many of these industries were installed to supply a market comprising the whole of French West Africa. Profitability was the strict criterion on which this industrialization was based and was the only basis on which it continued to attract French capital. This had the effect of confining it within narrow limits, on the one hand, to industries which were given geographical protection in the form of transport costs and so had to be in Africa (vegetable oil processing, packaging, canning, cement), and on the other to industries in which cheap labour was a decisive advantage (canning, sugar processing, brewing, tobacco, matches, textiles). The overwhelming importance of the criterion of profitability within a structure of relative prices imposed from outside, also meant that the industries had very little connection with one another but were geared either to final consumption or to agriculture, of which in some cases (vegetable oil processing) they were no more than an extension. Furthermore, these industries import not only all of their equipment but also 35 per cent of their raw material. Under these conditions, it is easy to appreciate that their development would quickly reach its ceiling and that they would be able to make only a small contribution to the country's balance of payments when they are not aggravating it.

In the meantime, agricultural production was gradually moving towards stagnation. Even in the area of the export production of groundnuts, where much governmental effort was concentrated, the pattern of development showed a remarkable rate of growth of 8.8 per cent per annum from 1885 to 1914 followed by 2.7 per cent for the period 1918 to 1940.[5] There was then stagnation until 1950, when growth began again at the very high rate of 7.7 per cent per annum during the decade 1950–60, but then fell to a rate of 4 per cent for the following decade, 1960–9. Not only does growth seem to have remained at this level for some years, but there are already signs of a regression, shown by a reduction in the area sown for the first time in the history of the country, except for the period of the Second World War. The record harvest which produced a sale of 1.1 million tons in 1965–66 was followed by three mediocre harvests: 0.79 million tons in 1966–67; 0.83 million tons in 1967–68; and 0.6 million tons in 1968–69.[6] Even in the 1980s, the position has hardly changed substantially, production recording 0.57 million, 0.68 million and 0.59 million in 1983, 1984 and 1985.[7]

What is perhaps more significant from the point of view of the individual farmer is the continuous decline in his purchasing power in spite of increases in his productive capacity. This is not a question of decline in the price of groundnuts which, expressed in constant values, has not in fact dropped with the clear regularity so often alleged. Rather, it is a function of how the labour productivity and earnings of the Senegalese worker compares with that of his trading partner in France. Amin expresses this relationship as the double factorial terms of trade, defined as the terms of trade multiplied by the rate between the labour productivity index for Senegal and France.[7] This expresses the change in the quantity of labour contained in one 'basket of imports' exchanged for the variable quantity of groundnuts obtained from a constant quantity of Senegalese labour. Taking 1938 as 100 and based on real wage rates in the bush, the double factorial terms of trade for Senegal have deteriorated at the very high annual average rate of 4.2 per cent, falling from an index of 320 in 1911 to 186 in 1920 and 53 in 1957. This deterioration reflects a growing inequality in trade at the expense of the Senegalese peasant, who received hardly a seventh of what he received less than a century ago, in terms of the value contained in the products exchanged. No wonder that the World Bank itself noted that the economy of Senegal had stagnated. In spite of a growth rate at current prices of 7.3 per cent for industry during the 1960s, agriculture, livestock rearing and fishing recorded only a 3.2 per cent growth rate and so did the gross domestic product.[8] Indeed, the GDP has tended on the whole to decline from 4 per cent in the period 1959–64 to only 1.9 per cent in the period 1965–69. In per caput product terms, the position is even more serious since there has been a growth, at current prices over the decade, of only 1.1 per cent per annum: 1.9 per cent in the period 1959–65, and only 0.15 per cent in 1965–9. A stationary per caput product means serious decline. Not only does the gap between the underdeveloped country and the developed world widen under these conditions, but the stationary real product also produces serious internal distortions between social groups, between regions and particularly between urban and rural areas.

The Senegalese experience indicates the crisis of stagnation inherent in peripheral capitalist development in underdeveloped countries. Various writers have tried to show evidence of this type of dependent development, especially in Latin America and to a less extent in Asia.[9] However, as Palma indicated, the stagnationist element in this characterization need not always occur, particularly as capitalist development is always marked by the swing of the business cycle.[10]

Hence, with changes in the terms of trade in favour of agricultural or mineral products, some underdeveloped countries are able to take advantage of the favourable situation and rapidly accelerate the rhythm of their economic development. This has happened in the history of many Latin American countries and is certainly most evident today in the case of OPEC countries. This internal dynamism is, however, usually short lived. The inexorable logic of the existing strategy of development with its concomitant pattern of external relations leaves the underdeveloped country still unable to chart its own development path, and condemns it to sharper internal contradictions due to the incomplete nature of the process of capitalist penetration of its economy.

One of the major mechanisms for maintaining this pattern of external relations is the monetary system operative in underdeveloped countries. For some of these countries, particularly French-speaking African countries, political independence had not led to the establishment of national central banks. In the case of those countries in West Africa, their international monetary transactions continue to be handled by the Central Bank of West African States (BCEAO). This bank, a carry-over from colonial times, provides a common currency for seven of the nine states (that is, excluding Mali and Guinea). In 1962, the BCEAO was reconstituted and the seven countries concluded a co-operation agreement by which France guaranteed the convertibility into French francs of the CFA francs issued by BCEAO. The member countries undertook to keep their external reserves in an operations account opened by BCEAO at the French Treasury, with which a special relationship was also established. This relationship included the maintenance of a fixed exchange rate between the CFA and the French franc, unlimited access to French francs, and freedom of transfer between France and these countries. This relationship limits the control these countries have over their own currencies.[11] In a period of great instability in the exchange rate of the franc, most of these countries have suffered with France and their capacity to import outside of the franc zone has been considerably impaired.

The position is not much better for many other underdeveloped countries which have their own central banks but continue to hold their reserves in the major international currencies. The only advantage such countries have had is the freedom to switch their reserves from one international currency to another, or to decide whether or not to devalue their currencies against that of the metropolitan country. Nigeria, for instance, decided in 1976 to diversify the foreign currencies in which its external reserves were held and in consequence adversely affected the parity of the British pound sterling in which most reserves had been held until then. This, however, did not protect the reserves from the monetary instability that has characterized most currencies in recent years.

In the growing development literature, therefore, considerable attention is being paid to the consequences of this aspect of independence.[12] The pressure of underdeveloped countries to have a more stable international currency in the form of special drawing rights (SDR) on the International Monetary Fund is only the first step in the move to reduce the level of dependency. Certainly, in the long run, in so far as the SDR replaces a reserve currency, these countries are freed from the obligation of transferring real resources to erstwhile metropolitan countries that is implicit in the present system. None the less, there are still various difficulties in the whole issue of international monetary relations as it affects development.The increasing sophistication of the continuing dialogue between the Group of 24, representing the underdeveloped countries and the Committee of 20, developed countries, is hopefully likely to result in some form of resolution.[13]

The role of multinational corporations

The attainment of political independence created in many underdeveloped countries the illusion of having secured a greater measure of control over their own destiny. To some extent, the potential for this is inherent in political independence but it is realizable only in so far as a country has a loud and strong voice in the decision-making process governing the volume and direction of economic activities within its territory. This is a more difficult and complex challenge, partly because of the intricate external linkages of the erstwhile colonial economy that these countries have inherited, but particularly because of the overwhelming role of powerful and wealthy multinational corporations in mediating the pattern, scope and content of these linkages. These corporations control most of the direct foreign investment in underdeveloped countries. In 1971, for instance, the total book value of foreign investment was estimated at over $150 billion, of which about a third went to underdeveloped countries. Of this amount, the United States accounted for over 55 per cent, followed by Britain with about 20 per cent and France with about 7 per cent.[14]

In terms of external relations, the multinational corporations not only strengthened the process of exploitation of underdeveloped countries but also deepened the pattern of their technological dependence on the advanced industrialized nations. Their continued exploitation is particularly notable in the concentration of most of their investment in petroleum and other extractive mining activities. In 1969/70, for instance, these two activities represented 86 per cent of total multinational investment in western Asia, 74 per cent in Africa, 55 per cent in southern and eastern Asia and 22 per cent in Latin America.[15] The exploitation of such minerals as copper, tin, bauxite, nickel and iron-ore has been undertaken not only to meet current needs of the developed countries but also to stockpile against future needs and the possibility of changing policies in the underdeveloped countries concerned. The pattern of investment, however, shows variations over time and between different regions of the underdeveloped world. Using the United States direct foreign investment as an example, Table 34 shows that for Latin America, it was initially concentrated in the extractive industries but progressively its emphasis shifted to manufacturing. On the other hand, for the rest of the underdeveloped world investment is still largely in the extractive industries.

But perhaps the most far-reaching impact of multinational corporations on the external relations of underdeveloped countries is, as has been shown in earlier chapters, on the relative ease or otherwise of real

Table 34 *Sectoral breakdown of US direct foreign investments, 1950–1973 (per cent)*

	1950–60	*1960–67*	*1967–73*
Latin America			
Mining and Smelting	14.5	2.3	6.2
Petroleum	42.1	6.4	14.4
Manufacturing	21.3	64.8	44.9
Other	22.2	26.5	34.5
Other underdeveloped countries			
Mining and Smelting			3.5
Petroleum			56.5
Manufacturing			10.1
Other			29.9

Source: A. K. Helleiner, 'Transnational enterprises in the manufacturing sector of the less developed countries', *World Development,* vol. 3, no. 9 (September 1975), p. 643.

technological transfer. The primary vehicle of such a transfer has been largely through the strategy and programme of industrialization adopted in most of these countries. This strategy was based on the principle of import substitution while the success of the programme came to depend a lot on the investment decisions of multinational corporations. Given the predominantly profit maximizing objective of these decisions, it is easy to appreciate the emphasis given to highly sophisticated and capital-intensive production methods where the real need of underdeveloped countries is for labour-intensive technology. The inappropriate nature of the technology imported makes it difficult to bridge the wide technological gap between developed and underdeveloped countries, psychologically deepens the feeling of technological impotence and materially underscores the reality of economic dependence. In his study of the situation in the Andean countries of Latin America, Mytelka found that licensing technological packages by multinational corporations to subsidiaries or national enterprises in underdeveloped countries has been the major factor limiting technological assimilation and reducing the potential for adaption and extension of imported technology. To quote him:

> Ownership structure, product sector and licensing interact with choice of machinery imports and research and development activities in such a way as to produce a 'technological dependence syndrome' in which opportunities for 'learning by doing' are consistently missed.[16]

In terms of their actual operations, the roles of multinational corporations have not always been compatible with the best interests of underdeveloped countries. Various studies undertaken by the United Nations Conference on Trade and Development (UNCTAD) reveal that, on at least three grounds, the activities of these corporations have been prejudicial to the development of underdeveloped countries. First, these corporations often impose export restrictions on their subsidiaries and affiliates in underdeveloped countries. Such restrictions can take the form of market sharing whereby the subsidiaries and affiliates are only permitted to export to certain countries and are precluded from others; or it may involve the retention by the parent company of the primary responsibility for export activities and an insistence on its prior approval before any export can be made by the subsidiary company. Sometimes, the restriction requires that the subsidiary exports to or only through special firms. All of this can work against the trade and development of underdeveloped countries, especially where the subsidiaries involved occupy a prominent position in the economy and its growth and expansion are constrained by being barred from exporting, especially to nearby markets which may have been allocated to the parent or another affiliate company.

Second, multinational corporations often tie the import of raw materials and intermediate goods required for use in domestic production to sources of supply within their own corporate structure. Such tied purchasing arrangements not infrequently frustrate the logic of import substitution industrialization, since they prevent the growth of backward linkage relations involving the development of local sources of raw materials. The refusal of the consortium of multinational corporations to develop the ample bauxite resources of Ghana as a complement to their investment in the Volta hydro-electricity scheme is a good case in point.[17] Moreover, this tying of imports to specified sources enables multinational corporations to exploit the technique of transfer pricing to the disadvantage of underdeveloped countries. Prices which the corporation charge for imports supplied by their subsidiaries or affiliates are to a large extent determined arbitrarily in the absence of world market prices. In general, therefore, transfer pricing is used as a means of minimizing the tax liability of the corporations as a whole through limiting or increasing profits made by their subsidiaries for tax purposes or to reduce the impact of custom duties or to accumulate surplus funds in 'safe' currency areas.[18]

Finally, multinational corporations engage in tremendous surplus creaming through easy movement of funds out of underdeveloped countries. Apart from the use of arbitrary transfer pricing, they utilize a variety of other methods for repatriating high profits out of these countries. One of these methods is through payment arrangements between subsidiaries and affiliates. For instance, payment by affiliates to a subsidiary in an underdeveloped country for goods purchased from a subsidiary may be deferred. On the other hand, the subsidiary may be instructed to make immediate payment for all purchases made from affiliates, instead of being allowed the normal thirty or ninety day payment period.

It may, of course, be argued that any underdeveloped country can well protect itself against the machinations of multinational corporations. Such an argument misses the real essence of the problem posed by these corporations. Apart from the fact that most of these practices are carried out on an informal basis, without written or justiciable records, so much secrecy surrounds them that even developed countries, when they are on the receiving end of these operations, have not always succeeded in getting the information necessary for taking appropriate action.[19] Moreover, in terms of resources most underdeveloped countries are weak relative to the multinational corporations and find themselves in a pathetically helpless position to stand up to them.[20] Besides, the capacity of multinational corporations to suborn or co-opt influential members of the government of underdeveloped countries to their service is currently a matter of international concern. Some hope is being raised by the enactment of a code of conduct to which the multinational corporations must subscribe. The critical issue here is that of surveillance and the extent to which the interests of such an intelligence agency necessarily coincides with those of underdeveloped countries. Nevertheless, such a development represents a move in the right direction and underdeveloped countries face a challenge of using it for their own ends.

The metropolitan interests

The role of multinational corporations is, however, only a part, though not an insignificant one, of the reciprocal interests of metropolitan industrial countries in their external relations with underdeveloped countries. These interests go beyond the exploitation of the resources of the latter to embrace the protection of their own domestic economies against the negative consequences of economic growth in underdeveloped countries. It also includes the transfer to those countries of consumption patterns, institutional and organizational arrangements, educational, health and social systems, and the more general values, ideals and lifestyles current in the metropolitan countries.

The protection of their own domestic economies often compels developed countries to follow policies which run counter to their public profession of commitment to the development of countries of the Third World. Where the export agricultural commodities from an underdeveloped country can substitute for domestic crops, quantitative import restrictions, variable import levies, tariffs expressed in *ad valorem* or specific terms, or a mixture of both, are imposed to severely limit the amount that can be imported. Yet, for some underdeveloped countries, such commodities provide a very high percentage of their total foreign exchange earnings. In 1983, for instance, food and animal feeds accounted by value for some 19 per cent of total export from underdeveloped countries. This ranged from over 35 per cent for Latin America, 21 per cent for Africa, 19 per cent for Asia and under 4 per cent for the Middle East. This restriction of access to the markets of developed countries quite naturally exerts a downward pressure on both the volume and the price of agricultural exports from underdeveloped countries and is currently the subject of discussion and negotiation in many international forums.

Besides such direct actions, there are other indirect ways in which events in the metropolitan countries adversely affect the development of Third World countries. The continued economic growth of developed countries gives rise to structural changes in their economy which impact negatively on underdeveloped countries. For example, tremendous scientific and technological progress in the petrochemical industries has given rise to a rapidly expanding output of new and varied types of synthetic materials which are substitutable for natural products usually imported from underdeveloped countries. The consequent reduction in demand for the specific natural product weakens the bargaining power of the Third World countries concerned for a more equitable price. Furthermore, structural changes in the economies of the developed countries greatly affect the composition of their industrial output. Increasingly, greater emphasis is attached to chemical and engineering industries, while those industries which depend largely on agricultural raw materials, notably textiles, clothing, timber, and vegetable oils, have become less significant. Moreover, highly sophisticated technology now ensures that production processes use less and less raw materials per unit of manufactured output. These improvements in technology clearly affect the international trade of underdeveloped countries in certain mineral ores such as tin.

It is thus clear that directly or indirectly, the interests of metropolitan countries often do not permit any altruistic commitment to the development of underdeveloped countries. What is, however, more serious is the manner in which current consumption patterns, institutional and organizational arrangements and other social systems in underdeveloped countries, make it difficult for their leadership to conceive of a realistic way out of the present *impasse* in their development. The most effective blinkers limiting the search of underdeveloped countries for alternative developmental paths are provided by the educational system and a world view fashioned largely on those of the western industrialized countries. An aspect of this world view of direct relevance to a discussion of external relations of underdeveloped countries, is provided by the present conventional wisdom regarding the basis of international trade relations between countries.

According to international trade theory, two countries will engage in trade because of their different relative factor endowments and each will export the commodity that uses relatively intensively its relatively abundant factor. Such trade will result in complete equalization of factor prices so that the immobility of factors between countries does not prevent the maximization of world output and hence the realization of international economic efficiency resulting in mutual benefits to all concerned. Yet, it is a well-known fact that international trade has failed not only to bring about growth in the economies of most underdeveloped countries but also to stabilize their share of world trade. Thus, while in 1970 their exports accounted for nearly a third of the world total, by 1972, their share had dropped to less than 20 per cent (although it had risen between 1980 and 1985 to around 25 per cent) and this might have been worse if the petroleum exporting countries had not maintained their relative position over the period. By contrast, the developed market economy countries, with which virtually all underdeveloped countries traded, increased their share of world exports from 60 to 72 per cent during the same period. This phenomenon whereby of two international trading groups one continues to prosper from the gains from trade while the other suffers only adverse consequences due, not to a fall in the actual volume of commodities exchanged, but largely to price movements, poses serious problems for the current basis of external relations of underdeveloped countries.

An equally important aspect of these relations is predicated by a significant assumption in international trade theory. This is that the two countries trading have identical consumption tastes in the strict sense of identical homothetic utility functions. Indeed, one can argue that this is the whole thrust of the colonial venture of the industrialized countries, namely to stimulate a taste for foreign products in underdeveloped countries so that the incentive for economic growth would always be in terms of more foreign goods that can be acquired. On this issue, a somewhat positive role for the monopolistic competitive advantages enjoyed by multinational corporations in the markets of underdeveloped countries has even been projected. This is based on the argument that consumers in underdeveloped countries cannot maximize their satisfaction unless they involve themselves in a perpetual learning process concerning the new goods and services being placed at their disposal. This process entails the acquisition of knowledge about the new technologies of consumption based on new, capital-intensive technological discoveries. Since these consumers have limited time to devote to this learning process, it is economical for the producer to undertake the adaptation of commodity consumption to rising incomes so as to minimize the associated educational process for the consumer. Moreover the producer, in this case the multinational corporation, has both the resources and the economic incentive to do so. From this point of view, the argument concludes, monopolistic competition in underdeveloped countries can be seen more 'as a rational and dynamic mechanism of social adjustment of consumption patterns to economic development rather than a socially undesirable imperfection of market structure, which ought to be reduced or eliminated by social policy in the interests of protecting the consumer from "exploitation" '.[21]

If the effect of the operation of multinational corporations, and of current patterns of international trade and external relations, were to be considered wholly in these benevolent terms, the question then is: How far can it succeed in bringing about the development of underdeveloped countries, especially as it involves and implies the acquisition of tastes, styles and standards of living similar to or comparable with those currently enjoyed by the developed countries? This is today perhaps the most fundamental question that every underdeveloped country has to answer as it confronts the apparent intractability of the development problems which it has to resolve and the limited set of options open to it. In this regard, it is important to bear in mind what can best be described as the new 'development ethics' that is pervading the activities and aspirations of governments in most of these countries, and is supported and reinforced by the international community. This ethic obliges every government to strive to improve the socio-economic conditions of all strata of its population and to mobilize them to participate actively in their own development. In other words, the development of these countries is increasingly seen as involving the acquisition of certain tastes, styles and standards of living, not simply by a wealthy minority but by the large majority of the populace. This, indeed, has been the fundamental premise on which the whole volume has been based. The crucial question then is: how far will this be possible if those tastes, styles and standards of living are to be of the type currently enjoyed in the metropolitan countries of Europe and North America, countries with whom they largely engage in international trade and maintain the most intensive external relations?

The constraints of future global resources

The present standard of living in the developed countries of Western

Europe, North America and Japan is based on an industrialization programme that depends considerably on the large scale consumption of non-renewable resources. The implication of this fact for underdeveloped countries can best be illustrated by considering the situation with regard to the consumption of two critical resources, notably steel and energy. Collectively, the developed countries consume nearly 500 million tons of steel per annum corresponding to nearly 600 kilogrammes per person.[22] By contrast, the underdeveloped countries, excluding China, consume as a whole only about 25 million tons of steel each year, corresponding to about 20 kilogrammes per person per annum. What is perhaps more remarkable about the situation is that, although consumption in the developed countries is some twenty-five-fold greater than in the underdeveloped, the overall growth rates for both groups of countries are almost the same, that is, nearly 6 per cent per annum. Similar rates of growth are observed for other widely used metals such as copper, zinc, chromium, silver and lead. Indeed, for other metals notably aluminium, niobium, vanadium, titanium and molybdenum, the rates of growth of consumption are even higher. This means then that following the current pattern of international trade and development, the gap between developed and underdeveloped countries will become an unbridgeable chasm by the year 2000.

The position with per caput consumption is even more serious. Given their significantly lower rates of population growth, per caput steel consumption in the developed countries is growing even more rapidly than in the underdeveloped. According to Oyawoye, assuming that the developed countries can be persuaded to reduce their growth rate of steel consumption from 6 per cent to 2 per cent annually, and the underdeveloped countries can step up their growth rate from 6 per cent to 10 per cent, this will only marginally reduce the gap by raising the consumption of the underdeveloped countries by the year 2000 to half that of the developed countries in 1975.[23] However, such a seemingly favourable but most unlikely event would then have brought total world consumption of steel to 1300 million tons in the year 2000, as against the 2150 million tons which the developed countries alone would have consumed in that year if the present rate of 6 per cent growth were maintained.

The same situation is revealed when one considers energy consumption. According to Marois, 5000 million tons of coal equivalent is consumed each year by developed countries compared to about 200 million tons in underdeveloped countries. This corresponds to a per caput consumption of 6 tons and 0.3 tons respectively. Again, the rate of increase in the two groups of countries in spite of the twenty-five-fold difference is remarkably close, amounting to 5.2 per cent and 5.6 per cent respectively.

In terms of ability to replicate in underdeveloped countries the tastes and living standards entailed in these consumption figures, even in the long run, it is necessary to consider the sources of minerals that have made them possible. In spite of views popular in the underdeveloped countries, many of the minerals needed for the present level of consumption in developed countries are produced in those countries. Table 35, for instance, depicts the position with respect to ten minerals including petroleum. With regard to iron-ore, for instance, it shows that developed countries provided about three-quarters of current world production. For bauxite, copper, tin and manganese, the proportion is under 70 per cent but for zinc, lead and phosphate, it is over 70 per cent. Even with respect to energy, developed countries produce nearly two-thirds of the world's energy resources (measured in coal equivalents) although they consume over 83 per cent of the world's total, that is they consume, in addition, nearly half of what is produced in the underdeveloped countries. This is, of course, not to minimize the vital contribution of underdeveloped countries to the mineral markets but to emphasize that some of the myths about their rich mineral wealth do not, at least at present, stand up to close examination. The only reservation that needs to be made in this connection is that in many of the underdeveloped countries the level of mineral exploration and survey is still far from adequate in giving a fair idea of their mineral potential.

None the less, if the present trend of global consumption continues such new discoveries would not necessarily serve the needs of underdeveloped countries. Most exploration and exploitation remain largely in the hands of big foreign enterprises who are busy moving what is mined to stockpile it elsewhere. This means that the availability of these minerals for establishing in underdeveloped countries a pattern of development similar to that of present developed countries would pose serious problems for peaceful international relations.

The logical conclusion to the finite nature of global resources is that underdeveloped countries have only two development options open to them. One is to continue along the existing development path based on the notion of achieving a pattern and style of consumption similar to that of the developed countries; the other is to accept the non-viability of this course of action and to settle down to fashioning new develop-

Table 35 *Share of underdeveloped countries in the production of major minerals, 1972*

	Total output (thousand metric tons)	*Output of underdeveloped countries (thousand metric tons)*	*Underdeveloped* countries' share of total output (per cent)*
Bauxite	65,314	25,343	38.8
Coal	3,042	532	17.5
Copper	6,631	2,414	36.4
Crude petroleum	18,598	11,187	60.2
Iron-ore	768,359	184,907	24.1
Lead	3,492	460	13.2
Manganese	21,452	8,332	38.8
Phosphate	93,612	23,021	24.6
Tin	299,602	175,807	73.4
Zinc	5,551	982	17.7

Source: United States Department of the Interior, Bureau of Mines, *Minerals Yearbook*, vol. 3 (1972). See also Helen Hughes, 'Economic rents, the distribution of gains from mineral exploitation, and mineral development policy', *World Development*, vol. 3, nos. 11 and 12 (November–December 1975), p. 823.

* The percentage share refers to major producers only. In each case about 10 per cent of production is produced by other countries and could not be identified.

mental goals more consistent with their local resource endowment and socio-cultural disposition. Both of these options have far-reaching implications for their external relations. The first choice involves an acceptance of their current dependency status and a need to adjust to the dominating role of multinational corporations while, maybe, searching for ways and means of minimizing some of the deleterious effects of their present methods of operation. Such a choice also accepts an outward directed orientation for the economy and would seek to achieve 'development' through trying to attract more and more foreign enterprises and foreign aid. Socially, such a choice implies a commitment to satisfying mainly the luxury needs of the wealthy or advantaged minority of the population.

The second choice focuses on the economy itself and on the vital importance of transforming those fundamental structural elements which at present impair its capacity to respond effectively to new societal goals and provide for the basic needs of the vast majority of the population. Tastes and consumption patterns are strongly conditioned and influenced both by the strength of the productive sectors of the economy and the value preferences of a society. The ability of underdeveloped countries to bring about such fundamental structural transformations to their economy and society, however, depends on how far they can exercise their sovereign rights in determining who and what to allow into their countries or keep out of them. In such a situation, external relations, particularly international trade, will continue to play a vital role in development, but its priorities in each underdeveloped country will be determined by the needs of that country rather than directed and manipulated by powerful external agencies and organizations with interests not always congruent with those of the country.

The concept of selective closure

The idea that the development process requires a country to go through a phase of constrained isolation from the exigencies of intense international interaction during which it can incubate new and more viable structural forms, is one which is only gradually being appreciated among the development specialists. Writing on this issue, Dudley Seers, for instance, noted as follows:

> . . . we do not as yet understand much about what self-reliance implies for development strategies, but some of the economic aspects are obvious enough. They include reducing dependence on imported necessities, especially basic foods, petroleum and its products, capital equipment and expertise. This would involve changing consumption patterns as well as increasing the relevant productive capacity. Redistribution of income would help, but policies would also be needed to change living styles at given income levels – using taxes, price policies, advertising and perhaps rationing. In many countries, self-reliance would also involve increasing national ownership and control, especially of sub-soil assets, and improving national capacity for negotiating with transnational corporations.
>
> There are other implications as well, especially in cultural policy. These are more country-specific, but as a general rule, let us say that 'development' now implies, *inter alia*, reducing cultural dependence on one or more of the great powers – i.e. increasing the use of national languages in schools, allotting more television time to programmes produced locally (or in neighbouring countries), raising the proportion of higher degrees obtained at home, etc. . . .
>
> Of course, an emphasis on reducing dependence does not necessarily mean aiming at autarchy. How far it is desirable, or even possible to go in that direction, depends on the country's size, location and natural resources; on its cultural homogeneity and the depth of its traditions; on the extent to which

its economy needs imported inputs to satisfy consumption patterns which have to be taken – at least in the short term – as political minima. *The key to a development strategy of the type suggested is not to break all links which would almost anywhere be socially damaging and politically unworkable, but to adopt a selective approach to external influence of all types.*[24]

Selective closure thus implies a greater deliberateness as to what type of external relations a country should maintain during the process of development and what sorts of external influence it should allow to permeate its society. The concept quite clearly challenges that of unrestricted exposure implied in the notion of development as modernization or, more correctly, westernization which was very much in vogue in the social sciences in the 1960s. What it emphasizes more than anything else is that in all fields of endeavour a country must, first and foremost, tap and draw upon its own internal and traditional resources and it is only to the extent that these can be further improved, developed or complemented from external sources should external relations be encouraged and sustained. It implies what the Chinese figuratively refer to as 'learning to walk on two legs' – the traditional and the modern.

The logic behind the concept is, of course, not difficult to appreciate. If development implies 'societal transformation', then like all 'transformation' acts, it cannot be undertaken in the full glare of international exposure and intense external interactions and impact. A country needs some degree of solitude to be able to look into itself, to discover the inner resources and capabilities of its peoples and to use these to fashion a new society which can operate in the modern world more effectively. Such a phase of solitude is transitional but vital to a development process. It is like the isolation and solitude needed to hatch an egg into a chick. During such periods, the link-up into an international communications and trade system will become highly selective and limited, while the internal communications and trade system is being greatly enhanced.

On the face of it such restricted interaction may appear inimical to an international world order of growing interdependence and trade but in fact it is not. It is rather the alternative to watching many of the underdeveloped countries grind gradually to economic stagnation and bankruptcy, or accepting that their external relations would be those of client states, forever pleading for moratoria on foreign debt servicing or conceding that the only course out of the present impasse in development strategy is through violent social revolution arising from internal frustration.

In operational terms, a policy of selective closure would thus have three main strands to it. First, it would stress the predominant use of local resources in all sectors of national life. Starting particularly with food, it would require a drastic reduction of the dependence of the minority urban and elite population on the importation of luxury food and drink items from the outside world. The only exception would be where such items are required by a cross-section of the population for purely nutritional reasons and where it can be shown that no local substitutes are available. Similar constraints must be exercised with regard to the manufacturing of raw materials, even where the resulting products are not of the same standard of refinement as imported ones. Constructional materials provide an important area for selective closure. Policies in this field, however, must go hand-in-hand with a commitment to technological choices which are, at least in the initial stage, more consistent with abundant local resources. For example, for the numerous small bridges that would need to be built in rural areas, stone bridges using not only local rocks but much local labour, would be more realistic than steel bridges requiring a large foreign exchange outlay.

The second strand in this policy of selective closure relates to changes in the composition of trade commodities and consequently in the relation with multinational corporations. This is likely to be particularly significant with regard to the pattern of current imports. A change in production priorities in a country based *ab initio* on available local resources, is likely to represent a reversal of the current strategy of import substitution industrialization. This would not mean necessarily the termination of relationships with multinational corporations. It would, however, mean that some were more welcome than others. For example, multinational corporations involved in car assembly industries may be replaced by those engaged in the manufacture of various types of agricultural machinery and equipment. Given the tendencies towards a conglomerate structure among many of these corporations, such a shift of emphasis need not have disastrous consequences for their operations. But it would certainly be more advantageous and more directly relevant to the current needs of many underdeveloped countries.

The third strand of the policy of selective closure relates to the whole issue of foreign aid from governments of the developed countries. This is a most complex issue whose detailed consideration goes beyond the concerns of the present volume. However, it is useful to note that opinions in the developed countries range between those who think aid is a vital and necessary contribution to the development efforts of underdeveloped countries and so should be continued, and those who see it as 'largely a myth; at best a wholly inadequate payment for goods received, at worst another name for the continued exploitation of the poor countries by the rich'.[25] This second view has been very critical of the rather patronizing element in the whole relation between aid-givers and aid-receivers and has called attention to the fact that nearly all aid programmes, whether or not this is specified, have been conditional on underdeveloped countries pursuing certain policies. As far as the Western world was concerned, such policies usually mean conservative economic policies and respect for private enterprise, usually foreign private enterprise. In general then, it was argued that two types of 'aid' should be recognized: disaster relief and so called development assistance.[26] The former is temporary emergency aid aimed at directly alleviating acute human misery resulting from natural catastrophe, war, and the like. It is given in a selfless and humble spirit and is not designed to 'develop' the recipient country. On the other hand, although nations have learned from each other throughout recorded history, especially through trade and other forms of personal contacts, development assistance is a phenomenon involving the virtual imposition overnight of entire complexes of institutions, technologies, attitudes – indeed, ways of life – by one society on another.

It is, in fact, noteworthy that opposition to foreign aid has come not only from the radical left but also from the conservative right in the developed countries. Arguing against the continuation of foreign aid, Bauer asserts:

> Foreign aid is a system of gifts; to call recipients partners in development patronises them still further by treating them as minors ignorant of simple realities. To patronise them thus is not only distasteful but also paradoxical, since the advocates of aid usually consider the people of less developed countries (practically all of them non-white) at least as the equals of those of the west, if not their moral superiors.[27]

His main argument, however, particularly with regard to the West, is that for these countries to accept the virtue of the free market economy and then to urge that the concept of aid is right and should be extended, represents a fundamental contradiction inimical to the interests of both groups of countries.

Arguments as to the wisdom or otherwise of giving 'aid' will continue to be heard for quite some time to come. What will be critical for the development effect of underdeveloped countries, however, is the question of initiative and purpose. Where the initiative to give aid and assistance comes from the developed countries, it would be unrealistic to assume that, even with the best of intentions, their interests in so doing will coincide with or converge towards those of the recipient country. On the other hand, where the initiative comes from the underdeveloped countries, the interests served are likely to be closer to the development needs of those countries. This is not to say that underdeveloped countries have not in the past exercised their initiative in seeking foreign assistance in a manner that served the interest of a small self-seeking class. But foreign assistance sought within the framework of a policy of selective closure is more likely to be related to specific areas of development where the use of local resources and competence has been exhausted.

Perhaps the most crucial area of continued foreign assistance would be with technology transfer. A policy of selective closure places great emphasis on the identification, cataloguing, rationalization and development of traditional technology. It also requires the acquisition and adaptation of modern technologies of different complexities and sophistication. The deliberate search for and importation of such technologies may not be straightforward, or of a type which multinational corporations currently operating in a country can or care to provide. Successful negotiations between a developed and an underdeveloped country to make such technologies transferable would be the best form of aid and assistance.

The concept of selective closure thus implies a conscious redefinition of the external relations of a country with a view to terminating those relations which deepen dependency and weaken self-reliance. It also involves establishing new and more discriminating relations, especially those calculated to enhance the capability of a country to develop its resources in a self-centred manner. This concept would, in particular, affect the long standing privileged position of erstwhile metropolitan powers in their former colonial territory. It would certainly influence the manner in which multinational corporations continued to operate in underdeveloped countries. As such it poses a serious question of the ability of most underdeveloped countries, even where they accept its validity, to adopt it as practical policy.

The imperative of regional organizations

It is easy to appreciate that the larger a country, the greater the probability of it possessing a wide variety of natural resources. Moreover, the more abundant and varied the resources of a country, the greater its chances of successfully implementing a policy of selective closure during the process of its development. This is not to say that smaller countries such as Cuba have not successfully transformed their socio-economic structures behind selectively closed doors. But much of this was possible because of the peculiar circumstances of confrontation and the checkmate in big power conflict situations which enabled it to receive certain types of assistance and support in the process of transformation. Such a situation can hardly be regarded as normal and most underdeveloped countries would need to draw strength from their internal rather than their external relations.

It is, however, a fact of the situation that the majority of underdeveloped countries are small in area, in natural resources and in population. While these things do not necessarily prevent them from being able to engage in the deliberate structural transformation of their economy and society, it certainly limits the extent to which they can operate a policy of selective closure, in terms both of satisfying the varied needs of their population and bargaining effectively with external agencies and organizations about the modalities of their involvement in the development process. These facts, besides others, give a greater imperative to the need for closer co-operation and regional integration between underdeveloped countries, particularly those of Africa.

Of course, co-operation of various kinds already exists among underdeveloped countries in the three major regions of Africa, Asia, and Latin America. The United Nations has been successful in fostering varying degrees of co-operation through continent-wide economic commissions. The United Nations Economic Commission for Africa, for instance, has responsibility for stimulating development in all African countries through joint African efforts. Recognizing, however, that it would be unrealistic to work towards the integration of all African countries into a single economic unit, it has divided the continent into regions and the countries within each region should strive towards working as an integrated economic unit. The subregions have been identified on the basis of geographical contiguity and ethnic and social considerations. Thus, four regions were identified in Africa, namely East, Central, West and North Africa. South Africa is at present excluded from this effort because of its apartheid policy and the denial of elementary human rights of self-determination to the majority of its population. Similar efforts at subregional economic integration are being undertaken for Asian countries, the Middle East, and Central and South America.

The history of these efforts shows that the move towards integration is, for many reasons, a protracted and difficult one. In Africa, for instance, attempts at regional co-operation date back even to the colonial period. The colonial administrations grouped their territories for certain specific purposes into regional blocs. Thus, there was French West Africa and French Equatorial Africa, each operating as a kind of federation and with a separate monetary system. The British also organized their West and East Africa possessions in much the same manner. However, in terms of deliberate efforts at integration, the most notable instance was provided by the British East African Common Market which dates from 1917 when Uganda and Kenya formed a Customs Union.[28] The Customs Union came to include Tanganyika in 1927 and between all three territories there was complete freedom of exchange, accompanied by a common currency, a common external tariff and a common income tax. In 1947, this Union was upgraded to the East African High Commission which added the administration of common services to the operation of the *de facto* common market. This change, in fact, was an attempt to consolidate the three economies into a closely knit economic unit. With the independence of Tanganyika in 1961 and the increasing problems of equalizing industrial growth and the general rate of development in the three territories, the High Commission was replaced by the East Africa Common Services Organization (EACSO). Its main purpose was to rationalize the functions of the various organs responsible for the operation of such joint services as railways, post and telecommunications and shipping, and to administer matters relating to general transportation, finance, social research and economic relations. EACSO was eventually replaced by the East African Community in 1967 which established for the first time a legal framework for co-operating in all fields of development, and provided co-ordination of industrial development and general development policies. After ten years of turbulent existence, the Community was dissolved in 1977.

The East African Community was the most outstanding effort to preserve and improve upon earlier colonial attempts at regional co-operation in Africa. Its eventual collapse underlines how difficult regional co-operation can be and how carefully and realistically it has to be approached.[29] Many of the other regional efforts did not even last a year after the political independence of one of the constituent members. These early failures must, however, be evaluated in terms of the circumstances of their establishment. Most of them started as unilateral efforts by a colonial administration in which the varying interests of the constituent parts were not considered or negotiated. Their objectives were more towards improving the efficiency of colonial management rather than enhancing the overall welfare of the peoples of the co-operating countries. They did not have to confront problems of different ideological orientations or different openness to external influences and interventions. All of these have to be taken into account and resolved when dealing with independent countries. The potency of these issues probably explains why in the case of East Africa it was not possible to resolve them within the colonially derived structures of the Community.

This, of course, partly explains the variety of forms in which regional co-operation currently exists in different parts of the underdeveloped world. There are co-operations on the basis of individual sectors or groups of sectors. These often involve agreement on common policies and joint action, but leave the individual country a wide area of independence with regard to the overall direction of its economy. Real co-operation and integrative effort begins when the countries involved have to start operating as a single unit for significant areas of their economic life. Even in this respect, there are gradations in the degrees of integration. Thus, countries may agree to free trade arrangements, to forming customs unions, to becoming common markets or to end up as economic unions.[30] In a free trade area, the member states agree to suppress or gradually eliminate all trade restrictions between them. Tariff and other trade restrictions between member countries and third parties remain unaffected by this arrangement and members are free to impose any levels of tariff against non-member countries. The principal goal of a free trade area is to gradually achieve a state of perfect competition among the industries of the member states. A customs union, on the other hand, involves an agreement not only to remove tariffs levied on imports from member countries but also to establish common external tariffs on imports from non-member countries. In a common market, in addition to free trade arrangements and a common external tariff envisaged in customs unions, there is implied free movement between the member states of the factors of production, such as capital, labour, managerial skills and enterprise. Although the countries may not have a common currency, there is often free convertibility between the currencies. Finally, an economic union involves an agreement among member countries to integrate all their economic activities and to undertake joint decisions in all aspects of economic development policies. To achieve such a union requires harmonious political relationships and identity of political ideologies. Clearly, for economic unions to be successfully established, it is vital to create conditions approximating a full political unit.

It is obvious that these various forms of co-operation represent an ascending order of regional integration. The flexibility and opportunities which they offer underdeveloped countries for a gradual process of mutual understanding and ever increasing commitment, are critical for the arduous task of self-reliant development. Creating larger economic collectives not only moves these countries to higher levels of development but ensures that they learn the rules of international co-operation with compeers devoid of the psychosis of real or assumed exploitative relations. Such achievements should reinforce confidence among the underdeveloped countries in their ability to hold their own in an increasingly sophisticated world.

Conclusion

The external relations of underdeveloped countries thus constitute an important element in their strategy of development. These relations, however, have to be re-examined and redefined within the framework of a commitment to self-centred and self-reliant development. It has been emphasized that particularly with respect to the developed countries, most especially the erstwhile metropolitan ones, a period of more restrained relationships marked by selective closure to some influences from those quarters, may be vital for incubating the fundamental structural transformations necessary to set a country on the path of cumulative and self-sustaining development. Such a policy is not likely to be easy, not least because of difficulties that may be encountered from multinational corporations whose short term interests may appear to be threatened. Yet it must be emphasized that multinational corporations are too sophisticated as organizations not to appreciate when their long term advantage will best be served by going along with change. That appreciation will be greatly enhanced if underdeveloped countries make it easier for them to be compliant through increased and determined efforts at regional co-operation and integration.

But as has been emphasized the road to such an outcome is rough and difficult. Regional entities of different degrees of integration are already emerging in various parts of the underdeveloped world and they each provide useful lessons as to what to do and what not to do. In Latin America, for example, the Andean group of countries, comprising Bolivia, Chile, Colombia, Ecuador and Peru, which was set up to deal with the restrictive practices of multinational corporations, deserves much wider study and appreciation in other emerging regional groupings.[31] The Andean countries were particularly concerned with enhancing domestic technological activities. They therefore harmonized their policies and concerted their efforts to curb some of the more notorious restrictive practices of multinational corporations, including those involving the use of imported technology. The member states undertook not to authorize, save in exceptional cases, agreements which prohibited or limited the export of manufactured products requiring the purchase of raw materials, intermediate goods and equipment from a determined source, or which contained restrictions relating to sale and resale prices, obligations to pay royalties on unused patents and trademarks, prohibitions on the use of competing technologies and restrictions on the volume and structure of production.

Similar efforts at concerted self-reliance can no doubt be identified among other groups of underdeveloped countries. What is lacking is a more deliberate concern among underdeveloped countries for learning from each other's successes or failures. Part of the reason for this is linguistic. Many of the publications on the Latin American experience are in Spanish or Portuguese while most underdeveloped countries of Africa and Asia are either English or French or Arabic speaking. Another reason is, of course, the continued dominance of the metropolitan countries in the publications field, particularly in Africa. A more important reason is the very limited trade relations and tenuous cultural contacts among underdeveloped countries.

The United Nations and a number of specialized international agencies have so far been the main forum for their interaction. Interaction at this level, however, is strongly overlaid by the rules of diplomatic behaviour and precedence. Often it is also greatly influenced by the specific combative context of protecting their mutual interests against the developed countries. Thus, while the advantageous effects of increasing interaction between underdeveloped countries cannot be underestimated, it is still true that the benefits to be derived from wider and more intensive contacts among groups of them in different stages of development, are still waiting to be fully explored. The forging of new links and the intensification of external relations across the three continents of Africa, Asia and Latin

America – in other words the establishment of stronger south–south as against the present north–south relations – is bound to be of overwhelming significance for the direction and effectiveness of the development process in many of these countries in the decades ahead.

14 Conclusion

The spatial perspective emphasizes that what is involved in the development process essentially concerns the land and people of a country in their entirety. The stress here is on 'entirety', the importance of appreciating that development is not just about a project area but about a total national territory, not just about a project community but about the whole national society. The various arguments for this reorientation in the conceptualization of the development process have been offered in different parts of this book. Perhaps the most important of these arguments is that the social structural relations which constitute a major object of transformation during the development process cannot be tampered with effectively in a piecemeal fashion. Laws and societal rules which imply new disciplines, new attitudes and new behaviour patterns cannot effectively be introduced into a society if they apply only to a minority who can always evade them by merging back into the antecedent social structures.

Equally important for the idea of development as involving the entire land and people in a country is the obvious fact that land and people are two important resources which many underdeveloped countries have in abundance. Particularly in the case of Africa, the relatively low density of population indicates that there is an abundance of land *vis-à-vis* population. The same is almost true of Latin America although in the case of Asia there are countries such as India and Indonesia where population appears to be in more abundance than the land. Yet, the achievements of China underline the fact that a large and dense population is not an automatic barrier to development properly conceived. Indeed, it is this fact, as well as many other contradictions in the strategy of development which most underdeveloped countries have been pursuing over the past three decades, that has helped to put into sharper relief the critical question as to what the development process involves.

The investment–mobilization distinction

In the early two decades of adopting conscious and deliberate development strategies in underdeveloped countries, the major thrust, as discussed in earlier chapters, made no distinction between economic growth and economic development. The latter was simply treated as an extension of the former or sometimes even as synonymous with it. Techniques and strategies that were appropriate to the one were applied indiscriminately to the other. There is today, no doubt, a more widespread appreciation that the pervasive failure of development in most underdeveloped countries to influence the lives of the majority of the population in a beneficial way, is largely the result of this earlier insensitivity.

Yet, it is true that because of the strength of intellectual habits, much rethinking on the development issue continues to be couched in growth rather than development terms. The distinction between the two is critical. Economic growth has tended to be concerned with linear increments in a set of variables in a society which are characterized by being measurable especially in monetary terms. On the other hand, development, while it embraces such a growth, goes beyond it to involve changes in the relations between various classes in society and between them and the environmental resources on which they depend. More than this, development involves changes not only in the overall size of the system but also in its complexity.[1] To continue to think of development as economic growth implies giving primacy to investment planning as the most important process of bringing it about.

On the other hand, the thrust of the present volume is the need for re-evaluating this somewhat simplistic approach to development. Again and again, events in the real world have underlined the fact that what is missing or in short supply in underdeveloped countries is not so much investment capital as the ability to muster and mobilize the society to make full use of the capital resources available to it. Or, as Lockwood puts it,

> . . . the great bottleneck in economic development is usually not a dearth of capital resources, or even of skills. This may be serious at the outset. If it persists as a major obstacle, however, it is apt to be because of resistances encountered in constructing a social framework which will provide incentives and opportunity for human enterprise in new forms, thereby releasing the productive capabilities latent in most peoples of the world.[2]

The mobilization of the total population of a given country is thus the most critical factor in the construction of a new and more developmental social framework. Since most underdeveloped countries, at least in Africa and Asia, start from a position where the majority of the population live on and off the land, the importance of mobilizing the land as part of the construction or restructuring effort is critical. The role of geography, of regional and settlement planning, becomes of considerable importance in this endeavour. Equally vital is the emphasis on organizational and institutional design and planning.

The effort at mobilizing the people as a whole becomes itself an element in the development of a country, since to accomplish it requires tremendous innovative capacity on the part of the leadership. Such innovation almost axiomatically must involve using traditional systems of organization in a modernizing context and therefore transforming them in a manner still comprehensible to the vast majority of the population. The authenticity which remains an essential part of such transformed traditional systems represents an essential part of self-reliance and self-centred development. The important point, however, is that what is transformed is of the society, it is part of its tradition which by being transformed remains traditional not alien. Even where this involves ideas, values, techniques and technologies from other peoples and other lands, these can become part of the traditional resources of a people if they are imaginatively adapted and assimilated. On this issue of tradition and authentic innovation, it may perhaps be useful to quote from Eliot. According to him,

> Every nation, every race has not only its own creative, but its own critical turn of mind . . . Yet, if the only form of tradition, of handing down, consisted in following the ways of the immediate generation before us in a blind or timid adherence to its successes, 'tradition' should positively be discouraged . . . Tradition [however], is a matter of much wider significance. It cannot be inherited and if you want it you must obtain it by great labour. It involves the historical sense; and the historical sense involves a perception, not only of the pastness of the past, but of its presence . . . The difference between the present and the past is that the conscious present is an awareness of the past in a way and to an extent which the past's awareness of itself cannot show.[3]

The mobilization of a people to construct a new social framework which could induce a tremendous release of latent productive capabilities involves great labour, not only in gaining a strong awareness of the past but its important role in the conscious present. It is this that ensures the modernization of traditional structures and the traditionalization of modern forms. It is this more than anything else that makes development of a people self-centred and authentic to themselves and that does not imply frustrating attempts at 'westernization' misconstrued as their own modernization.

Societal mobilization also encourages other forms of innovative realism, particularly in the area of artificially and externally imposed standards. For example, to educate all the children of a country or to provide them with health facilities would quickly underscore the impracticality of conceiving a school, at least initially, as an architectural structure of distinctive beauty or character in a community. Nor could all teachers be defined strictly as college graduates or someone who has necessarily completed his training. For different countries or even different regions of a country, innovative adjustments have to be made not only to provide the present cohorts with education but also to engage in a systematic and progressive upgrading of local skills. The same thing can be said of health services. To wait until there are adequate numbers of Western trained doctors before a significant segment can have access to health care, is to be guilty of inhibiting the release of vitally needed and latent productive capabilities. Yet, as is well known, the common ailments from which the vast majority of the population suffer most of the time, can be dealt with by individuals whose training does not involve the eighteen years currently required by modern medical authorities. Indeed, it can be argued that without such a commitment to the whole populace, China could not have invented that most romantic of its officials, the 'barefoot' doctor.

Mobilization as development strategy also induces an active appreciation of traditional technologies and knowledge and of the need to consciously build up their inventory. If one were again to use the case of health services, it is a fact of common experience that there is no community which has not developed its own traditional system of health care. Even if this comprised no more than the use of so called 'witch-doctors', it is well known that in the area of curative practice, they too depend a lot on herbs and various organic materials. In a study of materials used in traditional medicine, a mobilized society would be forced to incorporate the resources represented by these traditional forms of health service, to use them in a more systematic way to reach out deliberately to a larger proportion of the population, to rationalize and upgrade the quality of service of their practitioners, and to assess the nature of the knowledge, technologies and competence entailed in their form of practice. Such an assessment becomes the critical factor indicating how such traditional forms relate to current knowledge and in what ways they can help the advancement of human knowledge. But the immediate challenge can be better appreciated when it is realized that while in 1969, a country like Britain spent about $100 per head of its population on health services and the United States some $300, Ethiopa spent only around $0.50 for the health care of its 22 million people.[4] The expenditure for health in Britain represented only 5 per cent of the country's gross national product (GNP); the Ethiopian expenditure represented perhaps 0.6 per cent of its GNP. None the less, even if the Ethiopian expenditure for health care were to be multiplied to a figure equivalent to Britain's 5 per cent of GNP, total expenditure for health care would still only be around $3.00 per head. Nothing can better underscore the importance of imaginative adaptation and use of traditional forms of health care to complement the more expensive modern system.

The same thing goes for the mobilization of traditional technologies in the production system. The need to ensure that everyone is gainfully employed requires an effort to keep alive some of the traditional labour-intensive technologies, not as a permanent feature of life but within a programmed framework of phased obsolescence. Again, the deliberate nature of such a policy should entail a conscious assessment of items in traditional technologies which can provide a useful addition or breakthrough to the fuller utilization of the environmental resources of a country.

Mobilization strategy with its emphasis on inventory, evaluation, and systematic utilization of traditional forms and processes does not, however, imply a rejection of advanced technologies or organizational methods. What it does is to insist on a reasonable and structured balance so that a society can engage in overall transformation without losing complete touch with its own authentic resources or with external possibilities. Indeed, within the context of deliberate policy, the need for such a balance may necessitate a positive attitude to the 'brain-drain' phenomenon. In such a situation, an underdeveloped country could encourage some of its most talented young men, whose specialization and skills are in fields not relevant or vital to its immediate needs, to emigrate to the developed countries where such skills can be utilized and further developed in the hope that when conditions are right such individuals will return.[5]

The ideological smokescreen

It would, of course, be naïve to assume that development through investment or mobilization planning represent two discrete modes of societal reorganization. Investment planning can be assumed to imply a capitalist approach to development with its presumed emphasis on free enterprise and freedom of the individual, while the mobilization approach may be assumed to imply socialism with its strong concern with collectivities. This impression may be further reinforced by the fact that much of what passes for economic planning in most underdeveloped countries today, influenced as they are mainly by ideas of the Western capitalist world, is really no more than investment planning.[6] Yet, it is easy to appreciate that neither capitalism nor socialism can become the dominant mode of societal organization without involving both investment and mobilization planning. Indeed, it has often been claimed that much of the problem of underdeveloped countries is not that their economies have been penetrated by capitalism but that the penetration has not been total and complete. The result has been that they remain peripheral and only partially integrated to the world capitalist system. Indeed, as Bradby puts it:

> The traditional modes are not destroyed so thoroughly or dominated so powerfully as at the parallel stage in the metropolis. This is because in many societies a break so radical as the expulsion of a peasant from his land just is not possible, so that men can pass from one mode to the other, and back again; and secondly, because capitalism has for a long time occupied only the superstructure, and has not bothered itself with developing agriculture and integrating it into the market. This stage corresponds to the present 'neo-colonial' period.[7]

In other words, even if a capitalist path to development were to be preferred, the task of mobilizing the people and the resources still remains.

The important point being made here is that the need for mobilization is not predicated by the choice or preference for a specific ideology, but is a condition of any serious development effort. Mobilization for development forces a country to conceive or visualize the type of society into which it wishes to evolve. Given the initial cultural differences within and between underdeveloped countries, it is unlikely that the preferred future society will be the same everywhere. But whatever the preferences of a given country, it will have to work out a consistent framework of social relations which would offer adequate incentives and opportunities for the majority of its population and which, in particular, would help in releasing their latent productive capabilities. Articulating the basis and content of such a framework constitutes the ideology of development of the country. To be meaningful and effective such an ideology must derive its strength and substance from the culture and peculiar social circumstances of the given country. All that has been attempted in the present volume is to highlight the fundamental features and the essential nature of any development process.

The need to emphasize the country–specific nature of development ideologies. arises because of the popular reaction to the word 'ideology'. Not infrequently, when the word is used it is immediately assumed to imply either socialism or communism. This is understandable inasmuch as these two ideologies have been the best articulated. None the less, as one surveys the advanced industrialized countries one after the other, it is not easy to find any two which are alike. Each has its own distinctive features. Although they can be broadly categorized into either a capitalist or a socialist camp, the differences among the countries of one group are sometimes as significant as the differences between groups.

Yet, what is most striking about both groups of countries is the high degree of control they have over their own fortunes and the comprehensive nature of their mobilization as a country and as a people. This, it has been stressed, is the real essence of development. It is the critical factor in the efficient organization of production and in the systematic effort at raising productivity. More significantly, it is the precondition for the emergence of an authentic national culture and for the release of the creative energies of a people in all fields of human endeavour.

The question of class analysis

In all of this discussion so far, the issue of class structure and its relation to the development process has not been introduced directly into the analysis. Yet, it is recognized that there is a large school of thought which regards class structure as the most critical element in the understanding of the current situation in most underdeveloped countries and hence central to any analysis of the problems of their development. Indeed, Buchanan asserts that the character of the bourgeoisie is of decisive importance in any analysis of the contemporary social geography of the countries of the Third World,[8] and Slater argues that:

> As of the present, geographers, and also regional economists, largely remain ensnared in the belief that one can scientifically examine the development and

organization of a space-economy in the setting of an implicitly harmonious social order, where there would seem to be no internal structural contradictions . . . Not only has the existence of social classes and the contradictions between them been generally overlooked in geographical analysis, but also questions concerning class consciousness, class struggle, and political movements have been predominantly ignored in spatial studies.[9]

It would be easy to plead ignorance of the methodology of class analysis or to contend that such analysis belongs more properly to another discipline. One can also easily admit to a naïveté in confronting the development problem within a framework in which class analysis and class struggle are not central. But these are easy options which should be rejected. While not underestimating the importance of the class structure in any society, there are weighty reasons why in the present work it was deliberately decided not to give it much attention.

Perhaps the most important of these reasons derives from a basic distinction which needs to be made between the philosophy and the methodology of development. The philosophy of development concerns itself with the principles and general conceptions of this area of human activity. It attempts to set forth hypotheses and interpretations of the reality of societal development and, not unexpectedly, provides a variety of viewpoints and paradigms of analysis. One of the most notable interpretations has been that of Marx which sees development as a dialectical process resulting from the resolution of the contradictions inherent in modes of social production as mediated by different social classes. By contrast, the methodology of development is concerned more with the principles of procedure, the strategies and tactics of bringing about a desired mode of social relations and production. It seeks to identify theoretical or practical ways of accomplishing particular tasks or goals.

Strictly, of course, both the philosophy and the methodology of development do interpenetrate and it is not possible to consider one without some appreciation of the other. What is possible, especially in a heuristic context, is to put emphasis on one rather than the other. In this particular volume, the preferred emphasis has been on the methodology of development. This has meant concentrating on what has to be done and how to do it, given a government that is committed to development. It has not concerned itself with how that government comes to be and what stage of class struggle resolution it represents. It is my belief that the development literature has had its full share of philosophical analysis, and contributions continue to be made. Moreover, in terms of the methodology of development, it is obvious that a major contribution has been by economics and that this particular set of methods has tended to concentrate on investment planning, on the allocation of scarce resources between competing ends, rather than on the full mobilization of the people and natural resources of underdeveloped countries. This particular aspect of development strategy, it is contended, has a very geographical and spatial dimension and constitutes the main rationale for this volume.

A second reason why class analysis has not featured in this volume has to do with the time span inherent in the type of resolution to the problems of underdevelopment which such analysis contains. When so much of what constitutes the elements of underdevelopment in a country is elaborated in terms of class interests and privileges, the solution usually proffered is class struggle and revolution. In some underdeveloped countries, there is no gainsaying the fact that revolution is likely to be the vital key to open the door to other possibilities. But the cry to revolution as a basic ingredient of the solution to the problems of underdevelopment has often been made possible because of the focal salience of an individual, usually representing an historically entrenched class, whose overthrow could mark the end of an era or a system of social relations. This is true, for instance, in recent years of King Farouk of Egypt or the Emperor of Ethiopia. Their overthrow was seen as necessary before any major socio-economic transformation could begin. In many underdeveloped countries particularly those in Africa, no such focally salient individual embodying a system of social relations exists. Predicating the prospect of transformation on class struggle and revolution in such a situation has the characteristics of shadow boxing. In most African countries, for instance, land reform does not imply taking land from one class and redistributing it among members of another class. What it involves is a reorganization of the present pattern among their present owners.

This is not to imply that classes are not gradually evolving in these countries or that in some other regions of the underdeveloped world they do not represent part of the social reality. Rather it is to say that the class structure varies considerably among underdeveloped countries, that its potency for engendering the type of struggle that would lead to change needs to be evaluated for each individual country. Moreover, preoccupation with class analysis may divert attention from other types of social structures, notably bureaucracy which could be a more crucial factor in fuelling class formation, and inhibiting the full participation of the population in the development process.[10]

The point to be emphasized is therefore two-fold. First, conditions for social revolution as a prologue to socio-economic transformation are not present in every underdeveloped country, at least not yet. To regard the existence of such conditions and the resolution of the conflict arising from them as a prerequisite for development, could imply waiting until doomsday. Second, even where a revolution has to take place, the thrust of the present volume is about what to do afterwards. This book is about the development process and assumes that a country is prepared to engage in a self-centred and self-reliant transformation of its social and economic structures.

This brings us to the third reason for not emphasizing class analysis in this volume. In many studies of underdeveloped countries, there is often confusion in treating the elites as if they constitute a class everywhere. Indeed no society, however primitive, can do without elites. Such groups of individuals are needed to co-ordinate and harmonize the diversified activities of members of the society; to symbolize its moral unity by emphasizing its common purposes and interests; to combat factionalism and resolve group conflicts and to protect the society from external danger. Elites are thus a minority set apart from the rest of society by their pre-eminence in one or more of the activities connected with these goals.

In modern times, and in underdeveloped countries, there is no single comprehensive elite but rather a complex system of specialized elites linked to the social order and to each other in a variety of ways. Indeed, so numerous and varied are the elites that they seldom possess enough common features and affinities to avoid marked strains and tensions among themselves. None the less, one important factor which differentiates them from the masses, apart from their different skills and talents, is the social weight and social significance that are attached to their activities. On this basis, Keller distinguishes between the strategic and the segmental elites.[11] The strategic elites are those which claim or are assigned responsibilities for and influence over the society as a whole. They are thus the prime movers and models for the entire society and are variously referred to as the ruling elite, the power elite or the top influentials. The segmental elites, by contrast, have major responsibilities for subdomains of societal life.

Keller further distinguishes the strategic elites from other groups which had similar responsibilities in the past – namely ruling caste, aristocracy and ruling class. The major distinguishing characteristics between these groups lie in their recruitment procedures and their methods of maintaining themselves as a group. Ruling castes, for instance, recruit their members through biological reproduction and set themselves apart from the rest of the society by means of religion, kinship ties, language, residence, economic standing, occupational activities and prestige. An aristocracy, on the other hand, monopolizes the exercise of key social functions through limiting these to families bound by blood, wealth and a special life style, and supported by income from landed property. For the ruling class, although members are recruited on the basis of wealth and property rather than blood and religion, and although variously differentiated and specialized sectors may be distinguished, they are all bound together by a common culture and by close interaction across segmental boundaries.

As against these traditional groups, the strategic elites have no single stratum which exercises all key social functions. Instead, they are differentiated by the specializations associated with the performance of these functions. Recruitment into their ranks is thus not by blood, wealth or property but rather by merit and skills, particularly based on education appropriate to their specialized tasks. Hence, strategic elites are marked by their diversity as well as by impermanence.

In underdeveloped countries, even where, as in some Asian countries, there has historically been a ruling caste or aristocracy, or as in Latin America, a ruling class established in the early decades of their colonial history, the expansion of education and the upsurge of democratic sentiments among the people have had the effect of weakening the hold on power of these groups and strengthening the emergence of a strategic elite. For most African countries, the colonial period weakened the traditional elites and through education and exposure to modern influences, replaced them by a strategic elite. The development path which all of these countries will follow is thus going to be determined not so much by the traditional caste or ruling class but by the vision of their emerging strategic elites and their understanding of how best to translate this vision into reality. To attempt to analyse the situation in many underdeveloped countries through casting these elites in the mould of a social class which needs to be displaced in a revolutionary struggle, is to ignore the current dynamic situation within the societies concerned.

Conclusion

The second half of the twentieth century will go down in human history as the period when mankind sought severally and together to improve the conditions of living and the quality of life of its four billion-odd members organized within different national borders. The period began with political independence for many of those countries which were still under colonial rule and was followed by increasing concern about how the successor national governments were succeeding in improving the lot of their populace. In the first quarter century ideas as to what to do and how to set about it were dominated by the analysis and considered views of scholars from the developed countries. Even before that quarter was half-way through, it was already becoming obvious that development based on these views and suggestions was moving underdeveloped countries farther and farther away from their societal goals. The Conference of Ministers of the Economic Commission for Africa meeting in Tunis in February 1971, for instance, reflected the general feeling among underdeveloped countries of the unimpressive nature of their achievement during the First Development Decade.[12] According to them, their failure could be put down to three factors. First, the inadequate integration of their national economies, especially in its physical, organizational, economic and socio-psychological aspects. Second, the limited relevance of the development of science and technology in the developed world to the problems facing their countries, in particular problems bearing upon the transformation of their socio-economic structures. Third, the over-elaborate or otherwise unsuitable character of the models which had been adopted for formulating and implementing their economic development plans. Virtually all of these models were of foreign vintage, the product of an outward-orientation of their economies and, more importantly, of their intellectual and educational efforts.

In the hope, therefore, of achieving better results in the Second Development Decade, the Conference commended to each of its members the importance of ensuring that they first, effectively marshall their national and external development resources; second, mobilize all sectors of their population for participation in activities which should lead to the integration of the traditional sector, at present the less productive sector, with the modern dynamic sector; and third, promote structural changes to reduce the almost exclusive dependence on external factors for the initiation of the processes of transformation and development.

As the Second Development Decade draws to its close, it is difficult to show that these recommendations have been taken seriously to heart in many underdeveloped countries in Africa or elsewhere, or that they have started to give rise to improved conditions. On the other hand, it can be argued that the implied changes in strategy represented by these recommendations cannot be undertaken without the intellectual foundations on which such reorientation needs to be based. One of the most exciting aspects of the present decade is the growing literature by scholars from underdeveloped countries analysing the developmental achievements of their countries from their own point of view and trying to articulate a new and more realistic strategy for bringing about broader based improvements in the life of their people. It is the hope that this present volume will be a valuable addition to that literature and would contribute in a modest way to the growing realization that the development effort must be evaluated only in so far as it enhances the capacity of individuals and societies to cope effectively with the changing circumstances of their lives.

15 Postscript

Since the first edition of this book was published in 1980, many significant changes have taken place in the circumstances of underdeveloped countries. These have given greater salience to some of the concepts and ideas expounded in the book. In my own country – Nigeria – these changes have, in fact, induced a re-appraisal of the development process and directed efforts along some of the novel lines indicated in this book. I have been privileged to be directly involved in some of these efforts and believe it worthwhile in this postscript to dwell at some length on this Nigerian experience.

Perhaps the most important change that has occurred in the circumstances of underdeveloped countries since 1980 has been the worsening of their state of international indebtedness. Between 1970 and 1984, the outstanding medium- and long-term debt of underdeveloped countries grew almost tenfold from about $70 billion to over $686 billion despite the decline in capital flows since 1981.[1] The most striking feature of this growth was the surge in lending by foreign commercial banks whose share of total new capital flows increased from 15 per cent in 1970 to 36 per cent in 1983. Nonetheless, because of the vagaries of the market for export commodities from underdeveloped countries, their debt-servicing ability deteriorated drastically, particularly after 1974, as the magnitude of their debt increased. For many of the countries, the ratio of debt to GNP more than doubled, from 14 per cent in 1970 to almost 34 per cent in 1984. The situation was more gruelling for the low-income African countries where by 1980 the ratio of debt to GNP was already 40 per cent and rose to 55 per cent by 1984. In terms of the ratio of debt to exports, the position for low-income African countries went from 75 per cent in 1970 to nearly 280 per cent in 1984.

All of this was to underscore the fact that most underdeveloped countries were by the 1980s living clearly beyond their means. The consequence, not unexpectedly, was that creditor-nations no longer showed any willingness to continue to sustain their apparent profligacy. In spite of the various debt-rescheduling exercises for many of these countries, the situation of economic recession and individual deprivation became overwhelming. Various reasons have been given for this turn of events. These need not concern us here. What is important is that for nations, particularly those in Africa whose development had been outward-oriented, the situation meant a sharp drop in the range of imports they could now pay for and therefore a basic shift inwards to look for substitutes. In short, in spite of themselves, these countries have been constrained to adopt policies and programmes tending towards what has been referred to in the book as 'selective closure'.

'Selective closure' and structural adjustment programmes

In the conventional literature, the more commonly accepted terminology for the ensuing situation is 'structural adjustment'. The International Monetary Fund (IMF) has been mainly responsible for many developing countries adopting programmes of structural adjustment as a means of correcting the distortions in their economic situation that have weakened their ability to meet their trade obligations. The role of the IMF in this respect is strongly reinforced by Western creditor nations which repose considerable confidence in ministration to troubled economies. This is underscored by the fact that they will not resume normal economic relations with debtor countries without some re-assurance of a 'clean bill of health' from the IMF.

The standard prescriptions of the IMF under a structural adjustment programme are a set of reforms of a wide-ranging nature whose objective, it is claimed, is to remove operational bottlenecks and institutional rigidities which militate against the efficient and competitive performance of an economy. However, whilst there is no conflict as to objective, considerable controversy surrounds the general orientation and instruments of policy designed by the IMF to bring this about. From the point of view of international financial institutions and creditor countries, this general orientation should aim at two outcomes: first, the promotion of strong international interdependence through further opening up of the economy of the structurally adjusting country to the free flow of goods and capital investment on the basis of the principle of comparative advantage; second, the inculcation of a free market approach to production organization and economic transactions.

To these ends, certain policy instruments are preferred. These include the use of indirect monetary and fiscal policies rather than direct administrative controls; the maintenance of realistic, flexible and responsive foreign exchange rates, especially at levels which facilitate export promotion; the erection of low and uniform rates of effective protection to domestic producers in any given sectors or in relation to any given product; the establishment of factor prices such that real wage increases are justified only by rising productivity while interest rates are kept positive in real terms; the minimization of taxation, subsidies and general governmental expenditure through the application of cost-recovery principles in pricing infrastructural services; the eschewing of inflation-inducing budget deficits at all times, and the avoidance of the underpricing of agricultural produce.

In specific terms, structural adjustment is thus seen as involving the movement of real resources to agriculture in order to correct the prevailing urban bias of development policies. It also entails redressing price distortions through eliminating the administration of quotas and import licences, reducing tariffs considerably, pursuing tight money, credit and fiscal policies, containing capital flight, dampening inflation, rationalizing parastatal enterprises through closing down or privatizing inefficient ones, de-regulating the marketing of agricultural commodities, re-aligning the exchange rate to international equilibrium level through devaluing the domestic currency and liberalizing trade generally.

These prescriptions, if implemented, would mean a drastic re-orientation of national economies along strictly capitalist lines. Given the nascent nature of the economies of most underdeveloped countries, particularly those in Africa, and their inherent vulnerability, it is understandable that many governments have had real difficulties accepting the new orthodoxy of the IMF. The feeling that, under the guise of structural adjustment, underdeveloped countries are being made to sacrifice preferred social values, cultural self-esteem and authentic humanist vision that their leaders have striven to develop and propagate has led to a certain (reluctance and) resistance to accept all of what is now commonly regarded as 'the shock treatment' of the IMF.

President Julius Nyerere of Tanzania and President Kenneth Kaunda of Zambia were both very vocal in protesting at the social and political insensitivity of the IMF structural adjustment programme. The latter President, in fact, in response to the build-up of national disaffection with the programme in Zambia, has had to terminate its implementation at least for the moment. But Nigeria has been unique not only in rejecting the IMF loan, particularly in view of its implication that that agency will manage or directly oversee the structural adjustment of the country's economy but also in providing an alternative programme package for achieving that goal and further securing the long-term transformation of socio-economic conditions. From the Nigerian point of view, the basic difference between the goal of the IMF prescription and that of its alternative programme is thus that between enhanced dependency and emergent self-reliance. To clarify the distinction requires a brief history of the background to the present structural adjustment programme of the Nigerian economy.

Adjustment conditionalities and self-reliant development in Nigeria

Up to the 1960s, Nigeria was pre-eminently an agricultural country.[2] Agriculture was the mainstay of the economy, providing the bulk of both government revenue and the nation's foreign exchange earnings. Yet, since the 1970s, agriculture has suffered severe neglect because of a petroleum oil-boom. The country developed an almost insatiable appetite for imported goods and commodities. Even food items such as maize, rice and vegetable oil, which the country used to produce in abundance, came to be imported. The import bill for food items rose from under 2 million *naira* in 1962 to over 2 billion *naira* by 1984, more than a 1000 per cent increase. In 1984 wheat alone accounted for over 20 per cent of total food imports.

Industrial production also rose dramatically between 1960 and 1980, although its contribution to gross domestic production remained relatively low, rising from under 5 to 9 per cent during the period. Capital investment in manufacturing grew astronomically, going from 64 million *naira* in 1975 to nearly 900 million *naira* in 1980. Much of this was accounted for by such traditional sub-sectors as breweries, textiles, vehicle assembly, cement, sugar refining and flour-milling, as well as by state-directed efforts in iron and steel, pulp and paper, petrochemical and liquefied natural gas. The main component of industrial costs, however, was that of raw material, most of which was imported. In most years, this accounted for more than 75 per cent of total cost of production.

All of these heavy investments and importations were, of course, made possible by the phenomenal growth in petroleum exports, especially from 1973. Up to that year, government revenue from oil was never more than 1 billion *naira*. It rose to 4.2 billion *naira* in 1974 and was over 13 billion *naira* in 1980. However, by 1981, the drop to 9.6 billion *naira*, due to the glut in the international oil market, was the first sign of the fragility and nose-diving decline of the economy. Total foreign exchange earnings of the country dropped from 14.2 billion *naira* in 1980 to 7.6 billion *naira* in 1983. The balance of payments position swung from a surplus of about 2 billion *naira* in 1980 to a deficit of nearly 3 billion *naira* in 1984. The problems of the external sector were further compounded by high debt service burden and accumulation of short-term trade arrears on documentary credit and bills on open account. The total external debt of the country, which was just over 1 billion *naira* in 1979 escalated to 12.8 billion *naira* by 1984, of which 62.5 per cent belonged to the federal government, 25.1 per cent to state governments collectively and 12.4 per cent to the private sector.

By that time, the situation had become so critical that the country was having real difficulties in raising the necessary credit to cover her foreign trade transactions. These difficulties had dramatic impact on the manufacturing sector. Industrial output suffered a sharp decline of 21 per cent between 1983 and 1984 due to drastic shortage of industrial raw material imports. This resulted from the tight administrative controls on the spending of the greatly reduced foreign exchange earnings. Other measures, including direct levies, were imposed to curb the Nigerian penchant for imported goods and services. But by then the economy was in no position to respond positively to these measures.

It was in these circumstances that on 18 April 1983, Nigeria approached the IMF for a three-year extended loan facility of between $1.9 and $2.4 billion. This was essentially to satisfy a condition set by some of its creditors for refinancing its trade arrears. Because of the brazen economic mismanagement which had characterized the oil-boom years in Nigeria, the IMF, quite naturally, insisted on wide-ranging reforms of the conventional structural adjustment type.

These reforms, which formed the 'conditionalities' attaching to the IMF intervention, were by and large accepted by the Nigerian government. On two issues, however, the government was unyielding. The first was the adjustment of the rate of exchange to correct for the over-valuation of the local currency; the second was the liberalization of trade policy through abolishing the tight administrative controls in the area of international trade. The IMF considered the local currency as over-valued by about 60 per cent as at May 1984 and had recommended a 25–30 per cent initial devaluation so as to bring it to parity with the US dollar. This would be followed by quarterly reviews and possible further devaluation depending on the performance of the economy after the adoption of the structural adjustment programme and until the element of over-valuation was removed. The Nigerian government had doubts about the expected benefits from devaluation. It noted that Nigeria's export prices, notably those of petroleum, were usually quoted in foreign currencies, that locally manufactured goods had high import content and that there was generally a high propensity of demand for imported goods. With regard to trade liberalization, the government maintained that in a situation of mounting trade imbalance it would be economically suicidal to open the floodgates to the importation of all sorts of goods.

This was where matters stood when on 27 August 1985 the country experienced a change of leadership in its administration. The new military administration saw the issue of whether Nigeria should or should not accept the IMF loan on the basis of final agreement to the remaining conditionalities as a matter of grave national significance. It therefore decided to throw it open to public debate. All segments of the population – bankers, businessmen, academics, trade unionists, politicians, journalists, small-scale industrialists, professionals, market women, farmers and students – participated in the debate. By the end of it, there was no doubt that the overwhelming view in the country was against taking the loan even when it was agreed that this was bound to result in serious hardship and to call for great sacrifice on the part of the populace. According to Oyejide and others, the issues that had swung the majority against the loan included those of upholding the sovereignty of the country, of commitment to self-reliant development, fear of additional loan repayment burden, fear of further economic mismanagement and the lessons of international experience, especially of African countries which had accepted IMF loans.[3] Consequently, the Nigerian government decided to reject the loan, and, in the words of the nation's President, to seek solutions to the social and economic problems of the country 'through our own efforts, at our own pace and on our own volition, consistent with our long-term national interest'.[4]

Following on this rejection, the government proceeded to ban immediately further importation of food items such as rice, maize and vegetable oil and served notice to ban wheat importation a year later. It imposed a 30 per cent levy on all imports and reduced the level of subsidy on petroleum products by 80 per cent. This last decision was expected to generate a revenue of over 900 million *naira*. Half of this amount was to be made available to a newly created Directorate of Food, Roads and Rural Infrastructure based in the Office of the President and headed by a member of the Armed Forces Ruling Council. The Directorate was expected principally to promote a framework for the grassroot social mobilization of the rural population and mount a vigorous programme of engaging them actively in the expansion of food and agricultural raw material production as well as in the construction of roads and other rural infrastructure.

But perhaps the most crucial decision of the new economic package was to reject a simple devaluation of the Nigerian currency and opt instead for the introduction of a foreign exchange market (FEM) through which the currency was expected to find its own parity level. What this decision has done is to link the import propensities of Nigerians with their capacity to produce specifically for export and particularly in the non-petroleum sectors of the economy. Furthermore, apart from easing government control of the economy and reducing bureaucratic hassle with its attendant encouragement of corruption, the new foreign exchange market reinforced the trend towards curbing the import orientation of consumption patterns among Nigerians and induced industrialists to look more seriously for local sources of raw materials. Although it is perhaps too early to regard 1986 as a watershed in Nigeria's development process, it is already possible to identify certain impacts the new policies are having on social and economic activities in both the urban and the rural areas of the country.

The urban impact of structural adjustment

In looking at the urban impact of the structural adjustment programme, it is useful to consider both the consumption and production dimensions. With regard to the former, the most immediate impact had been on food preferences. In a sense, changes away from imported items have been greatly facilitated by the fortunate succession of good harvests in the preceding three years. This has meant that local staples were abundant in the market and relatively cheap. By contrast, manufactured food items based on imported raw materials found consumer price resistance a major constraint to their viability especially as higher prices simply provoked shifts to local substitutes. This development has been particularly noticeable in the case of wheat bread whose consumption has dropped significantly.

At the same time, government has also been engaged in encouraging product substitution of various types. The activities of industrial research institutes have achieved greater national salience and many transnational corporations have themselves started to pay more attention to local research and development. As the 4 January 1988 edition of a local weekly news magazine noted:

> The structural adjustment programme (SAP) has given a fillip to indigenous technology as manufacturers in all areas seek local replacements for parts that used to be imported. Breweries with giant manufacturing machines are courting local foundries to cast replacements of damaged implements. Transnational food processing companies have resorted to tapping locally produced grains and crops for brand products. Industries are gravitating towards research. Businessmen who used to have the propensity to import now buy up locally fabricated machinery at the Project Development Agency (PRODA), Enugu, faster than the agency produces. The Nigerian Stored Products Research Institute (NISPRI), the Federal Institute of Industrial Research (FIIRO), Oshodi, the Forest Research Institute of Nigeria (FRIN), and some other institutes all disclose that they have been engaged by more customers in the past one year than ever before.[5]

The transitional situation has also meant staff retrenchment in both the public and private sector enterprises and a rapidly rising level of unemployment, especially among young school leavers and fresh graduates. The government has established a National Directorate of Employment to articulate and implement various schemes designed to facilitate self-employment generation among different categories of unemployed. These schemes include the National Open Apprenticeship Programme, which is designed to provide vocational training to secondary school leavers and other unemployed youths. On this scheme participants are attached to government institutions and private enterprises recognized by the government for on-the-job training for periods long enough for them to acquire enough skills to become craftsmen and artisans of various trades, including building, mechanical, electrical/electronics trades, computer technology, draughtsmanship, woodwork, music, and so on.

Another of the schemes is the small-scale industries and graduate entrepreneurship programme. This is designed to provide unemployed graduates and others with bright and viable business ideas with financial assistance and other support that will enable them to establish small-scale industries that are employment generating. The financial assistance is in the form of a loan for which beneficiaries do not have to provide any security beyond their certificate. Prior to providing such assistance, however, those selected for this scheme undergo a two-week intensive programme in entrepreneurship development. The objective is to inculcate in them the spirit of self-enterprise and creativity and teach them the basics of how to manage their own small enterprise and make it grow.

A third scheme of the National Directorate is the Graduates and non-Graduates Agricultural Self-Employment Programme. This seeks to provide self-employment in farming for those who have sufficient interest and experience in agriculture. Farmland is allocated to beneficiaries of this scheme who are also provided with agricultural loans in cash and kind usually through banking institutions which are meant to ensure the usual discipline associated with bank loans.

Already, the effects of these various schemes are starting to be felt in the nation's economy. The garment industry has blossomed as never before since the importation of ready-made clothes from Europe and elsewhere is now no longer a viable proposition. Food processing activities are also growing in variety and sophistication. Imported wine and sparkling wines like champagne are being substituted for by local raw materials such as kolanuts. Printing and book publishing, which tended to be dominated from abroad, now show a high proportion of products of local publishing houses. Metal fabricating industries are starting to grow in number and capability all over the country. In short, Nigerians are starting to accept the inevitability of adjusting their consumption pattern to what they can produce.

One other area in which the impact of the new dispensation is being seriously felt is in urban transportation. In the heyday of the oil boom, the volume of vehicles on Nigerian roads rose phenomenally. With the introduction of the FEM prices for vehicles have more than quadrupled, putting them well beyond the purchasing power of the vast majority of the populace. High prices also restricted the easy availability of spare parts. On top of this has come the government's withdrawal of the petroleum subsidy. All this has curbed the excessive mobility of the Nigerian population, eased urban traffic congestion and greatly facilitated the maintenance of urban environmental standards. More importantly, it has helped to reinforce the emergence of a 'maintenance culture' which is itself a critical factor for a society seriously committed to enhancing its technological capabilities for development. Furthermore, the collapse of private transportation has sharply emphasized the need to develop public transport facilities both for intra- and inter-urban movement. Various state governments are now establishing appropriate institutions to ensure efficient management of mass transit systems in their major urban centres.

The rural impact of structural adjustment

The rural impact of the structural adjustment programme has been diverse and compelling. As life became more difficult in the cities, many urban families turned to part-time farming to supplement their income. Around every major city, there is now noticeable an expanding zone of cultivation, largely by urban residents. Some of these cultivators are organized groups of graduates and non-graduates whom various state governments, under the Agricultural Self-Employment Scheme of the National Directorate of Employment, are assisting to settle down to farming.

In general, government has also encouraged major Nigerian and multinational companies as well as wealthy Nigerian businessmen to go into large-scale agricultural production. With the banning of the importation of various agricultural produce, such ventures are proving less financially risky than they used to be. Even wheat is now being produced internally and on a rising scale. The expanding urban market for poultry and other animal products has meant in particular an insatiable demand for such crops as maize, rice and cassava for which farmers can now get very good prices.

Of considerable importance in this respect has been the dissolution of the various commodity marketing boards. These boards were set up in the colonial days to protect farmers against the wild fluctuations of prices on the international market for export crops such as cocoa, palm produce, cotton, groundnuts and rubber. These boards have gone through a chequered history of increasing unresponsiveness to the real needs of farmers and had reached a point of actually depressing farmers' productivity when they were taken over by the federal government in 1974. Since then, similar boards had been set up for domestic crops especially grains. But by 1984 it was already clear that the boards were not only being managed incompetently but were of little value to the farmers.

At any rate, the dissolution of the boards has meant that for export crops farmers are now getting close to the full international market price whilst for domestic crops, especially those like maize and rice, which are becoming industrial raw materials, they are not doing badly. There is, however, no doubt that some form of institutionalized marketing arrangement will be required. The need is to create an organization which would be close enough to the farmers and very responsive to their needs. There is, besides, the question of how intimately farmers themselves should participate in the running of such an organization. These matters are currently being examined at various forums and there can be no doubt that some new level of resolution will be achieved.

But perhaps the most important impact of the new dispensation in the rural area has come from the establishment of the Directorate of Food, Roads and Rural Infrastructure based in the Office of the President and with enormous powers to co-ordinate all activities connected with rural and agricultural development in the country. Four elements characterize the operations of the Directorate to date. First, there is the concern with involving grassroot participation in its activities through community mobilization. Although community organization and community governance are real facts of life in Nigeria, present administrative arrangements accord them hardly any recognition. The formal administrative structure goes down to the level of local government, of which there are only 304 in the country. In terms of area, each of these local governments is very extensive and embraces numerous communities whose leadership and territoriality are only vaguely appreciated. The Directorate has been changing this situation, especially as it tries to ensure that all its activities are seen as assistance to individual communities, which are expected to make their own contribution to the effort. For this reason, the Directorate is engaged in identifying the number of communities in each local government area as well as the list of settlements belonging to each community. Already, preliminary returns give the number of settlements as just over 97,000. This information, when finally determined, is meant to facilitate the commitment of the Directorate to monitoring effectively all activities in the rural areas for which it is responsible.

Apart from organizing and mobilizing rural communities, especially through community development associations which are now being instituted, the Directorate has been engaged in two major activities, notably rural infrastructural development and farm and non-farm rural production. Infrastructural development has so far concentrated on feeder roads, water and electricity supply and rural housing improvement. Forty per cent of the annual budget of 500 million *naira* of the Directorate has been devoted to ensuring the construction of 30,000 kilometres of rural feeder roads in each of the last two years. Every local government area has been involved in this activity on the basis of identified priorities and the establishment of clear construction specifications. Similarly, 20 per cent of the budget was devoted to providing some 5000 rural communities with potable water in each of the last two years. The programme for rural electrification has just started whilst that for rural housing is dedicated largely to enhanced training in the production and use of improved local material for rural housing construction.

With regard to productive activities, the emphasis of the Directorate has been in disseminating rapidly through community organizations not only improved seeds of major national staples, but also fruit seedlings and oil seeds to form a solid basis for rural industrial development. A vigorous programme of aquaculture for local fish production in each local government area is also being pursued to improve on the level of protein consumption in the country. With animal husbandry, emphasis is being placed on widespread development of farms for ruminants and small animals such as rabbits.

A third element in the operation of the Directorate is to involve all existing institutions and organizations in the country. This is particularly noticeable with respect to its programme for expanding food producing capacity in the country. It is clear that for this to be feasible, tremendous effort must be directed at large-scale multiplication of seeds, fingerlings and breeder stock as well as to the production of other inputs needed by farmers. This has necessitated the mobilization of all federal and state research institutes, state ministries and local government departments of agriculture as well as large-scale private sector enterprises.

Finally, there is the element concerned with the promotion of appropriate technology. It was found, for instance, that little attention was paid in previous programmes to disseminating in the rural areas technology of a type that rural residents can relate to. A case in point was the rural water supply scheme. The earlier programme of the Ministry of Water Resources was based on a capital intensive and technologically sophisticated system involving the use of electrical generating plants and submersible pumps for transferring water into large 'Braithwaite' overhead tanks from which reticulated pipes distributed it to a number of villages and homes. However, once the generating plant broke down, which was not infrequent, the whole scheme collapsed. For the Directorate, therefore, the better option was to promote the provision of hand-operated pumps which can be made by many fabricating companies in the country and maintained by village craftsmen offered a short course as to what to do. The same strategy works for agricultural processing machinery and equipment, especially as it is increasingly being realized that processing is as crucial as storage for stimulating higher levels of agricultural productivity.

The whole of this rural development strategy is thus based on a mobilization approach rather than simply on an investment programme. The funds available to the Directorate in any one year are not large relative to the total budget and are treated as distinct from the funds of the traditional sectoral Ministries of Agriculture, Industries, Works, Education, Health, and so on. Their role is to be used as catalyst to provoke and stimulate wide-ranging activities by rural communities to engage in their own development. This is why the mobilization and the enlightenment of all rural communities is a central feature of the present rural development strategy in the country. Enlightenment in this connection involves the vigorous prosecution of a functional adult education programme as well as the active promotion of consciousness-raising activities through the community development associations.

The mobilization strategy, however, does not end with the rural communities. It is extended to embrace all agencies involved in the development of the rural areas. Thus, considerable effort has gone into mobilizing all agricultural research institutes and agencies in the country to participate in the production of biological inputs required by the farming communities and to help resolve problems arising from their vastly expanded productive activities. Other institutes, notably those in roads and building construction research as well as in industrial research, have been equally mobilized. Commercial and development banks hold periodic meetings with the Directorate to evaluate progress in the expansion of rural productive activities. Small-scale industrialists as well as large-scale multinational organizations have been mobilized to ensure greater integration between the activities of the myriads of small farmers and the marketing of their produce for industrial processing and transformation.

The first two years have witnessed tremendous activities in the Nigerian rural areas of a type that has never been seen before. Heightened social identification both of 'sons abroad' in the cities and of local leaders with the imperatives of developing community facilities for the welfare and productive activities of the rural populace has been a novel experience everywhere. The Directorate of Food, Roads and Rural Infrastructure has been under continuing pressure to formulate, articulate and implement programmes to extend its activities to all communities in the country. The results achieved to date have been truly phenomenal, especially in terms of national re-orientation as to what development entails. It is certainly a far cry from the days when development was equated with the acquisition of foreign factories and structures that have no immediate relevance to the exploitation and utilization of local resources.

It is consequently this promise that, through a vigorous mobilization strategy, much effective transformation in the socio-economic circumstances of a people can be accomplished that makes the present Nigerian experience so challenging and illustrative of an approach for inducing wide-ranging changes in underdeveloped countries.

Conclusion: can the Nigerian experience be generalized?

It is, of course, too early to judge the Nigerian experience. For one thing, the changes in development strategies need to work themselves out and become integral to day-to-day economic life. For another, other consequential structural adjustments still need to be undertaken to ensure that present efforts at correcting the short- to medium-term imbalances in the country's external accounts and domestic budgets do not compromise the nation's long-term development potential and performance. Such a compromise may arise if adequate attention is not paid to issues of the social relations of production, especially as between the emerging class of local capitalists and the rest of the population, as well as of socio-political equity in the process of achieving fundamental re-structuring of the nation's economy.

Of equal interest is the question whether the Nigerian experience can be generalized to embrace all underdeveloped countries. Here it is of interest to record briefly the very similar experience of the Republic of Korea (South Korea) in Far East Asia, which occurred just over a decade ago. In the period up to 1971, the Republic had been facing serious economic problems resulting from a national development strategy which over-emphasized outward-oriented industrialization and resulted in deepening rural–urban disparities. Whilst the deliberate allocation of resources in favour of the industrial sector had led to the achievement of remarkable economic growth, it had, at the same time, precipitated widespread agricultural stagnation, with ever-increasing inequalities between the urban and the rural areas of the country.

It was in these circumstances that, under the direct motivation and dynamic leadership of its President, Park Chung Hee, the country introduced what has now come to be known as the 'Saemaul Undong' movement.[6] The movement sought to rectify the undesirable trend of urban–rural imbalances through channelling greater development effort into the rural areas and the agricultural sector of the economy. It began basically as a rural asset-formation strategy directed at utilizing marginally idle resources in the rural areas, especially those of land and labour. Within a short time, this original concept gave way to one that saw the movement as an all-out development programme for fostering a fundamental adjustment in the environment of rural existence, in the methods of production and in the rural way of life. Its essence was the mobilization of the entire rural population through co-operative endeavours of their community members based on two overriding strategies: local initiative and the use of locally supplied labour, materials and skills. Government investment in the programme, although substantial, was provided only to supplement locally organized and mobilized resources and was therefore considered secondary in importance. Thus, in 1971 when the movement began, direct government investment was 4.1 billion *won*, while locally induced contribution was estimated at 12.2 billion won. By 1978, the former had risen to 338 billion *won* whilst the latter had also increased impressively to 634 billion *won*. Over the period, some 80,000 kilometres of rural and farm roads were constructed, over 31,000 ponds and some 2 million village wells were developed, village warehouses were built, many rural communities were electrified whilst the number and resources of co-operatives increased phenomenally.

The scope of the movement was later widened to cover the urban sector of the country and is now considered as the critical strategy for changing the entire nation into a 'new society' based on the ideals of diligence, co-operation and self-reliance. As Fu-chen and Byung-Nak put it:

> What may be very important about the Saemaul Undong is that it is turning out to be [for the nation] a training ground for practical democracy. In the Saemaul Undong, villages learn through free discussion and voting procedures democratic ways of electing their leaders, selecting village projects and carrying out those projects through voluntary participation. . . . They learn and become accustomed to democratic ways not through books, but through hard work and productive co-operation.[7]

If the Nigerian experience had thus been already foreshadowed by the Saemaul Undong of the Republic of Korea does this mean it is applicable to or replicable in all underdeveloped countries, particularly those in sub-Saharan Africa? This latter group of countries, for instance, is characterized by population and markets which are small in size and by economies which remain very vulnerable and exposed. Compared to Nigera with, in 1984, a population of some 97 million and a gross domestic product of over $70 billion, 38 of the other 47 countries of sub-Saharan Africa have population of less than 10 million each and a gross domestic product individually of less than $5 billion. The remaining 9 have population of less than 45 million each and, with the exception of South Africa, a gross domestic product individually of less than $8 billion. Industrially, these countries remain some of the most lagging in the world with the average share of manufacturing in their gross domestic product being below 5 per cent. Indeed, in spite of their possessing an abundance of potential natural resources, many of them constitute some of the poorest of underdeveloped countries with some of the lowest per capita incomes and living standards.

Yet, when all this has been noted, the fact remains that the path out of their present predicament requires serious re-structuring of both their economies and their societies through a mobilization approach not too different from that charted from Nigeria and the Republic of Korea and more broadly adumbrated in this book. It is true that many of these countries have so far survived largely on foreign aid. Moreover, the approach of the international aid community to their restructuring effort has been generally shaped by the view that what is needed is to restore their import capacity which had been badly damaged by foreign indebtedness and years of drought and failed harvests. There are various aggregate estimates of this capacity. But it is widely accepted that to restore, for instance, the import capacity of sub-Saharan Africa in the second half of the 1980s to its level in the early 1980s would require a minimum gross external resource flow into the region of about $20 billion a year. When this gross flow is netted off after taking account of the reverse flow due to debt servicing and payment, the demand–supply gap is, if anything, likely to further widen in the foreseeable future. Already, the net resource transfer to sub-Saharan Africa has fallen dramatically from $12.6 billion in 1982 to only $1.7 billion in 1985.[8] Consequently, grave doubts must be expressed as to whether even the restoration of an historically determined import capacity can obviate the need for the inevitable structural adjustment of the economies of these countries.

What is indisputable, therefore, is that whilst the Nigerian experience may not be directly replicable in many of these countries because of the less robust nature of their economies, their development process can hardly be expected to get under way until they have attempted to do two things: first, re-structure the consumption profile of their population (particularly of their elite) to bring it more in accord with the resource situation of the country; and second, mobilize the rural resources of land and labour for increased productivity and transforming the income-earning capacity and life style of the rural majority. There can be no single or simple way to achieving such mobilization. Every country has to look into itself, into its own history and the cultural traditions of its people, to draw inspiration for a viable and widely accepted mobilization strategy for inducing its development process. The need is to inculcate a new ethos and tradition of hard work and productive co-operation so that these countries may start first by engendering the capacity to be relatively self-sufficient in food production and, through more diversified effort, accumulate agricultural and other surpluses necessary for laying the foundation for a more secure and self-sustaining development process.

Clearly, what is called for is appreciating the need for a new strategy of development of the type explored in the present book. This insists on a mobilization strategy which concentrates developmental attention on the exploitation and utilization of local resources and on the moderation of consumption patterns to make them more compatible with the resource endowment of the individual countries concerned. Improved agricultural production organization and increased rural productivity are central to such a strategy. Equally, important is a more imaginative restructuring of the manufacturing sector based not only on the idea of greater integration with agricultural production but also with the strong possibility of export. In this regard, greater attention must be paid to creating opportunities for upgrading local technological skills through fostering a more diversified industrial structure while providing ample room for small- and medium-scale local industrialists in addition to the large-scale multinational enterprises. Above all, greater emphasis must be placed on the mobilization of the population, in both their rural and urban setting, on the basis of a new spatial organization which facilitates their close and effective involvement in the multi-faceted and enduring process of self-reliant development.

Notes and references

1 Introduction

1 See, for example, J. Friedmann and W. Alonso, *Regional Development and Planning: A Reader* (Cambridge, Mass., 1964), and various other publications on regional planning.

2 For a summary of the growing literature in this area, see Richard Peet (ed.), *Radical Geography: Alternative Viewpoints on Contemporary Social Issues* (Chicago, 1977).

3 Ismael-Sabri Abdalla, 'Heterogeneity and differentiation – the end for the Third World?', *Development Dialogue,* 1978/ 2, pp.4–5. The figures by Abdalla were derived from the *World Bank Atlas* (Washington, D.C., 1977) but they do not agree with the latter as to definition.

4 Wallerstein dates the origin of the 'modern world system' to the sixteenth century. See Immanuel Wallerstein, *The Modern World System: Capitalist Agriculture and the Origins of the European World-Economy in the Sixteenth Century* (New York, 1974).

5 See, for instance, Douglas H. K. Lee, *Climate and Economic Development in the Tropics* (New York, 1957).

6 Karl Marx, *Das Kapital* (Chicago, 1905), vol. 1, p.13.

7 Much of this summary of the neo-classical economic development paradigm is from Janet Abu-Lughod, 'Development and urbanization', *Habitat International,* vol. 2, no. 5/6 (1977), pp.417–26.

8 This section derives largely from David Slater, 'Geography and underdevelopment', *Antipode: A Radical Journal of Geography,* vol. 9, no. 3 (December 1977), pp.1–31.

9 Gustav Ranis, 'Economic development: A suggested approach, *Kyklos,* vol. 1, fasc. 3 (1959), p.440.

10 See particularly chs. 6 and 7 of Ian Roxborough, *Theories of Underdevelopment* (London, 1979).

11 See, for instance, Hugo Radice (ed.), *International Firms and Modern Imperialism* (Penguin Books, London, 1975).

12 Dudley Seers, 'The new meaning of development', *International Development Review,* no. 3 (1977), p.3.

13 Gary and Marilyn Gates, 'Uncertainty and developmental risk in Pequena irrigation decisions for peasants in Campeche, Mexico', *Economic Geography,* vol. 48 (April 1972), pp.140–57.

2 Defining development

1 Dudley Seers, 'The new meaning of development', *International Development Review,* no. 3 (1977), p.3.
2 Dudley Seers, 'The meaning of development', *International Development Review,* no. 4 (December 1969), p.3.
3 Seers, 'The new meaning of development', (1977), p.5.
4 Richard A. Easterlin, 'Overview on economic growth' in D. L. Sills (ed.), *International Encyclopedia of the Social Sciences* (New York, 1968), vol. 4, p.395.
5 W. Arthur Lewis, *The Theory of Economic Growth* (London, 1955), pp.225–6.
6 Daniel Lerner, 'Social aspects of modernization', in D. L. Sills (ed.), *International Encyclopedia of the Social Sciences* (New York, 1968), vol. 10, p.387.
7 J. M. Buchanan, *The Demand and Supply of Public Goods* (New York, 1968).
8 E. J. Mishan, *Welfare Economics: Ten Introductory Essays* (New York, 1969), p.164.
9 M. S. Ahluwalia and H. Chenery, 'The Economic Framework' in H. Chenery, M. S. Ahluwalia, C. L. G. Bell, J. H. Duloy and R. Jolly, *Redistribution with Growth* (London, 1974), p.47.
10 David Harvey, *Social Justice and the City* (London, 1973), p.110.
11 Harvey, p.199.
12 G. Lukács, *Lenin* (London, 1971), p.45.
13 See, for example, contributions by B. Hindess and P. Q. Hirst, *Pre-Capitalist Modes of Production* (London, 1975); Aidan Foster-Carter, 'The modes of production controversy', *New Left Review,* no. 107 (1978) pp.47–77; and J. G. Taylor, *From Modernization of Modes of Production* (London, 1979).
14 See J. A. Hobson, *Imperialism: A Study* (London, 1902); and V. I. Lenin, *Imperialism, the Highest Stage of Capitalism* (1917; Moscow, 1964).
15 See Ian Roxborough, *Theories of Underdevelopment* (London, 1979).
16 See A. G. Frank, *Capitalism and Underdevelopment in Latin America* (New York, 1967). Note also that it has been pointed out that participation in the world economic system is not the same thing as production under the capitalist mode of production, and that 'dependency theory' confuses the two: see E. Laclau, 'Imperialism in Latin America', *New Left Review,* no. 67 (May/June 1971).
17 Frank, p.166.
18 A. K. Cairncross, 'Interrnational trade and economic development', *Economica,* vol. 28, no. 109 (February 1961), p.250.
19 Gunnar Myrdal, *Asian Drama: an Inquiry into the Poverty of Nations* (New York, 1968), vol. 3, p.1868.

20 See Immanuel Wallerstein, *The Modern World System: Capitalist Agriculture and the Origins of the European World Economy in the Sixteenth Century* (New York, 1974).
21 Roxborough, pp.x–xii.
22 H. Brunschwig, *L'Avènement de l'Afrique Noire* (Paris, 1963), p.212. See also K. O. Dike, *Trade and Politics in the Niger Delta, 1830–1885* (Oxford, 1956).
23 See, for instance, J. L. and Barbara Hammond, *The Town Labourer, 1760–1832* (Anchor Books edition, 1968), pp.84–97. Also E. J. Hobsbawm, *Labouring Men: Studies in the History of Labour* (London, 1971), pp.64–125.
24 Karl Polanyi, *The Great Transformation: the Political and Economic Origins of Our Time* (New York, 1944), p.73.
25 Polanyi, p.157.

3 Geographic space and development

1 P. Vidal de la Blache, *The Personality of France* (London, 1928), p.14.
2 For the various types of space see David Harvey, *Social Justice and the City* (London, 1973), pp.13–14.
3 Milton Santos, 'Society and space: Social formation as theory and method', *Antipode: A Radical Journal of Geography,* vol. 9, no. 1 (February 1977), p.5.
4 Santos, p.5.
5 Harvey, p.11.
6 Paul Vieille, 'L'espace globale du capitalisme d'organisation', *Espaces et Sociétés,* no. 12 (mai 1974), p.32.
7 John D. Nystuen, 'Identification of some fundamental spatial concepts' in Brian J. L. Berry and Duane F. Marble (eds.), *Spatial Analysis: A Reader in Statistical Geography* (Englewood Cliffs, 1968), pp.37–8.
8 Richard Hartshorne, *Perspective on the Nature of Geography* (New York, 1959), p.12.
9 Gunnar Myrdal, *Economic Theory and Underdeveloped regions* (New York, 1957), pp.11–18.
10 Myrdal, p.17.
11 Myrdal, p.19.
12 See J. G. Williamson, 'Regional inequality and the process of national development: a description of the patterns', *Economic Development and Cultural Change,* vol. 13, (1965), pp.3–43; and the critique of his position in Alan G. Gilbert and David E. Goodman, 'Regional income disparities and economic development: A critique', in Alan Gilbert (ed.), *Development Planning and Spatial Structure* (London, 1976), pp.113–41.
13 Torsten Hägerstrand, *The Propagation of Innovation Waves,* Lund

Studies in Geography, Series B, Human Geography (Lund, 1962), IV.

14 Paul Ove Pedersen, 'Innovation diffusion in urban systems' in Torsten Hägerstrand and A. Kuklinski (eds.), *Information Systems for Regional Development,* Lund Studies in Geography, Series B, Human Geography (Lund, 1971), no. 37, pp.137–47.

15 Edward, W. Soja, *The Geography of Modernization in Kenya* (Syracuse, 1968), p.3.

16 Wilbert E. Moore, *Man, Time and Society* (New York, 1963), p.7.

4 Collapse of traditional rural structures

1 George Rosen, *Peasant Society in a Changing Economy* (Chicago, 1975), p.173. See also J. W. Mellor *et al., Developing Rural India: Plan and Practice* (Ithaca, New York, 1968), pp.49–55.

2 Rosen, p.85.

3 R. J. Alexander, *Agrarian Reform in Latin America* (New York, 1974), p.16.

4 See, for instance, Ronald Cohen, 'From empire to colony: Bornu in the nineteenth and twentieth centuries' in V. Turner (ed.), *Colonialism in Africa 1870–1960* (Cambridge, 1971), 3rd ed., p.100; and Olga Linares de Sapir, 'Agriculture and the Diola society', in Peter F. M. McLoughlin (ed.), *African Food Production Systems: Cases and Theory* (Baltimore, 1970), pp.207–8.

5 Youssouf Gueye, 'Essai sur les causes et les conséquences de la micropropriété au Fouta-Toro', *Bulletin de l'IFAN,* B Series, vol. 19 (1957), pp.28–42.

6 A. G. Hopkins, *An Economic History of West Africa* (London, 1973), p.24.

7 M. Fried, *The Evolution of Political Society* (New York, 1967), p.109.

8 F. M. Stenton, *Anglo-Saxon England* (Oxford, 1943), pp.313–14.

9 Hopkins, p.21.

10 Karl A. Wittfogel, *Oriental Despotism: a Comparative Study of Total Power* (New Haven, 1957), 556pp.

11 Jack Goody, *Technology, Tradition and the State in Africa* (London, 1971), p.27.

12 Allan G. B. Fisher and H. J. Fisher, *Slavery and Muslim Society in Africa* (London, 1970), pp.111–12.

13 See Patrick Manning, 'An economic history of southern Dahomey 1880–1914' (Unpublished PhD thesis, University of Wisconsin, 1969). p.54.

14 Hopkins, p.71.

15 See ch. 9 of J. R. Harlan, *Crops and Man* (Madison, Wisconsin, 1975), for a useful discussion of the evidence relating to the domestication of crops in Africa.

16 Peter M. Weil, 'The introduction of the ox-plow in Central Gambia', in

Peter F. M. McLoughlin (ed.), *African Food Production Systems: Cases and Theory* (Baltimore, 1970), p.251.

17 See, for instance, Conrad C. Reining, 'Zande subsistence and food production', in Peter F. M. McLoughlin (ed.), *African Food Production Systems: Cases and Theory* (Baltimore, 1970), pp.140–9.

18 Polanyi, p.178.

19 B. Ohlin, *Inter-regional and International Trade* (Stockholm, 1935), p.42.

20 Polanyi, p.180.

21 Hopkins, pp.238–9.

22 J. R. V. Prescott, *The Geography of Frontiers and Boundaries* (London, 1965), p.159.

23 Lord Hailey, *An African Survey* (revised 1956), (London, 1957), pp.803–4.

24 Anthony N. Allott, 'Legal development and economic growth in Africa', in J. N. D. Anderson (ed.), *Changing Law in Developing Countries* (London, 1963), pp.194–209.

25 John C. de Wilde, *Experiences with Agricultural Development in Tropical Africa,* vol. 1: *The Synthesis* (Baltimore, 1967), pp.143–4.

26 de Wilde, vol. 2, p.60.

27 R. K. Udo, *Migrant Tenant Farmers of Nigeria* (Lagos, 1975), p.32.

28 Polly Hill, 'The Myth of the Amorphous Peasantry: a Northern Nigerian Case Study', *Nigerian Journal of Economic and Social Studies,* vol. 10, no. 2 (July 1968), p.251.

29 District Commissioners' unpublished annual report for Fort Hall, 1948. See also D. R. F. Taylor, 'Agricultural change in Kikuyuland' in M. F. Thomas and G. W. Whittington (eds.), *Environment and Land Use in Africa* (London, 1969), pp.463–94.

30 T. C. Mbagwu, 'The oil palm economy in Ngwaland (eastern Nigeria)' (Unpublished PhD thesis, University of Ibadan, 1970), pp.184–8.

31 Margaret Haswell, *The Nature of Poverty: A Case History of the First Quarter Century after World War II* (London, 1975), 234pp.

32 W. H. Beckett, *Akokoaso: A Survey of a Gold Coast Village,* Monograph on Social Anthropology, no. 16 (London School of Economics and Political Science, London, 1956), p.40.

33 R. K. Udo, 'Disintegration of the nucleated settlement in eastern Nigeria', *Geographical Review,* vol. 55 (1965), pp.53–67.

34 W. B. Morgan and J. C. Pugh, *West Africa* (London, 1969), p.427.

35 *Colonial Annual Reports* (Lagos, 1900–1). p.6.

36 J. F. M. Middleton and D. J. Greenland, 'Land and population in West Nile District, Uganda', in R. M. Prothero (ed.), *People and Land in Africa South of the Sahara* (London, 1972), pp.167–9.

37 Polly Hill, *Rural Hausa: A Village and a Setting* (Cambridge, 1972), p.184.

38 J. K. Olayemi, 'Some economic characteristics of peasant agriculture in the cocoa belt of western Nigeria', *Bulletin of Rural Economics and Sociology,* vol. 7, no. 2 (1973), p.199.
39 See J. Rweyemamu, *Underdevelopment and Industrialization in Tanzania* (Nairobi, 1973), pp.153–4.
40 Michael Todaro, 'Income expectations, rural–urban migration and employment in Africa', *International Labour Review,* vol. 104, no. 5 (November 1971).
41 W. A. Lewis, *Development Planning: The Essentials of Economic Policy* (London, 1966), p.46.

5 The nature of rural development

1 Uma Lele, *The Design of Rural Development* (London, 1975), pp.12–19.
2 Compare figures cited in Paul Richards, 'Farming systems, settlement and state formation: the Nigerian evidence', in David Green, Colin Haselgrove and Matthew Spriggs (eds.), *Social Organisation and Settlement,* British Archaeological Report International Series (Supplementary) 47 (Oxford, 1978), Table 3, p.488, with estimates for North American agriculture in Michael Perelman, 'Efficiency in agriculture: the economics of energy', in R. Merrill (ed.), *Radical Agriculture* (New York, 1976), p.79.
3 Lele, p.12.
4 H. A. Oluwasanmi, *Agriculture and Nigerian Economic Development* (Ibadan, 1966), pp.118–19.
5 John C. de Wilde, *Experiences with Agricultural Development in Tropical Africa,* vol. 1, *The Synthesis* (Baltimore, 1967), p.20.
6 Lele, p.17.
7 J. M. Cohen, 'Rural change in Ethiopia: The Chilalo agricultural development unit', *Economic Development and Cultural Change,* vol. 22, no. 4 (July 1974), p.600.
8 See, for example, K. Griffin, 'Policy options for rural development', *Oxford Bulletin of Economics and Statistics,* vol. 35, no. 4 (November 1973), pp.245–6 and R. P. Shaw, 'Land tenure and the rural exodus in Latin America', *Economic Development and Cultural Change,* vol. 23, no. 1 (October 1974), pp.123–30.
9 C. Eicher, T. Zalla, J. Kocher and F. Winch, *Employment Generation in African Agriculture* (Institute of International Agriculture, Michigan State University, East Lansing, 1970), Research report no. 9, pp.33–4.
10 Peter Dorner, *Land Reform and Economic Development* (Penguin Books, New York, 1972), p.26. Also see Keith Griffin, *The Green Revolution: An Economic Analysis* (UNRISD, Geneva, 1972), pp.50–6.
11 Lele, p.176.
12 Marilyn Gates and Gary Gates, 'Proyestismo: the ethics of organised

change', *Antipode: A Radical Journal of Geography,* vol. 8, no. 3 (September 1976), pp.72–82.

13 Ismail Ajami, 'Land reform and modernisation of farming structure in Iran', *Oxford Agrarian Studies,* vol. 2, no. 2, n.s. (1973), p.126.

14 See J. D. Montgomery, 'Allocation of authority in land reform programs: A comparative study of administration processes and output' (Agricultural Research Council, Research and Training Network, New York, March 1974); also Solon Barraclough, *Agrarian Structure in Latin America* (Lexington, Mass., 1973).

15 Griffin, pp.254–5.

16 See, for instance, B. W. Blouet, 'Factors influencing the evolution of settlement patterns' in P. J. Vcko, R. Tringham and G. W. Dimbleby (eds.), *Man, Settlement and Urbanism* (London, 1972), pp.3–5; also W. B. Morgan and J. C. Pugh, *West Africa* (London, 1969), pp.32–46.

17 See, for example, D. Grossman, 'The emergence of settlement duality: The case of the tenant camp of Nikeland', *Nigerian Geographical Journal,* vol. 17, no. 2 (1974), pp.93–110.

18 K. M. Barbour, *The Republic of the Sudan* (London, 1961), p.19.

19 F. M. Stenton, *Anglo-Saxon England* (Oxford 1943), p.289.

20 See, for instance, M. G. Smith, *The Economy of Hausa Communities of Zaria* (London: HMSO, 1955), Colonial Research Series, no. 16, p.13; also Polly Hill, *Rural Hausa: A Village and a Setting* (Cambridge, 1972), p.331.

21 Frank H. H. King, *A Concise Economic History of Modern China, 1840–1961* (New York, 1968), p.192.

22 Arie S. Shachar, 'Israel's development towns: Evaluation of a national urbanization policy', *Journal of the American Institute of Planners,* vol. 37 (1971), pp.362–72.

23 Doreen Warriner, *Land Reform in Principle and Practice* (Oxford, 1969), p.225.

24 Some of these problems have been studied and reported in various publications. See, for example, Inayatullah (ed.), *Co-operatives and Planned Change in Asian Rural Communities: Case Studies and Diaries* (UNRISD, Geneva, 1970).

25 R. Apthorpe, *Rural Co-operatives and Planned Change in Africa: An Analytical Overview* (UNRISD, Geneva, 1972).

26 See Klaus-Joachim Michalski, *Landwirtschafthche Genossenschaften in afro-asiatischen entwicklungslandern* (Berlin, 1973), English summary, pp. 418-21.

27 Gunnar Myrdal, *Asian Drama: An Inquiry into the Poverty of Nations* (New York, 1968), vol. 3, p.1899.

28 Uma Lele, *The Design of Rural Development: Lessons from Africa* (Washington, 1975), p. 189.

6 Strategy of rural development

1 See David Sills (ed.), *International Encyclopedia of the Social Sciences* (New York, 1968), vol. 11, p.568.
2 K. D. S. Baldwin, *The Niger Agricultural Project: An Experiment in African Development* (Oxford, 1957), pp.68–70.
3 Benjamin L. Whort, *Language, Thought and Reality* (Cambridge, Mass., 1956).
4 Muzfer Chefit, 'A study of some social factors in perception', *Archives of Psychology*, vol. 27, no. 187 (1935).
5 Kurt Lewin, *Field Theory in Social Sciences: Selected Theoretical Essays,* Dorwin Cartwright (ed.), (New York, 1951), pp.228–9.
6 G. W. Allport, 'Catharsis and the reduction of prejudice', *Journal of Social Issues* (New York), vol. 1, no. 3 (1945), pp.3–10.
7 Other thresholds have been characterized as the industrial and the transportation revolutions. That these three revolutions took place over more or less the same period emphasizes the close inter-relationships of these three sectors and the reinforcing effects of positive changes in one on the others.
8 Arthur Birnie, *An Economic History of Europe 1760–1939* (London, 1930), p.19.
9 E. C. K. Gonner, *Common Land and Inclosure* (1912, reissued, London, 1966), p.118.
10 W. G. Hoskins, *The Making of the English Landscape* (London, 1957), p.143.
11 Hoskins, p.139.
12 Paul Bairoch, *The Economic Development of the Third World Since 1900,* Cynthia Postan (trans.) (London, 1975), p.41.
13 Walter W. Jennings, *A History of Economic Progress in the United States* (London, 1925), p.242.
14 Hildegard Binder Johnson, *Order Upon the Land: the US Rectangular Land Survey and the Upper Mississippi Country* (New York, 1976), p.39.
15 J. Wreford Watson, *North America: Its Countries and Regions* (London, 1963), p.650.
16 Johnson, p.44.
17 Jennings, p.133.
18 Johnson, p.78.
19 Jennings, p.308.
20 *Documentary History of American Industrial Society,* vol. 5, pp.46–7.
21 *Documentary History of American Industrial Society,* vol. 7, pp.305–7.
22 H. F. Williamson (ed.), *The Growth of the American Economy* (New York, 1944), p.373.
23 F. J. Turner, *The Frontier in American History* (New York, 1920).

24 H. U. Faulkner, *American Political and Social History* (New York, 1946), p.431.
25 H. C. Allen, 'F. J. Turner and the Frontier in American History' in H. C. Allen and C.P. Hill (eds.), *British Essays in American History* (London, 1957), p.166.
26 W. E. Mosse, 'Stolypin's Villages', *Slavonic and East European Review*, vol. 43 (1965), pp.257–74.
27 Mosse, pp.257–74.
28 Maurice Dobb, *Soviet Economic Development since 1917* (London. 1948), pp.223–4.
29 Dobb, p.208.
30 Dobb, p.209.
31 Dobb, p.249.
32 Dobb, p.253.
33 Roger A. Clarke, *Soviet Economic Facts 1917–1970* (London, 1972), pp.10–11.
34 T. R. Tregear, A Geography of China (London, 1965), pp.162–3.
35 Kenneth R. Walker, *Planning in Chinese Agriculture Socialization and the Private Sector, 1956–62* (London, 1965), p.5.
36 Frank H. H. King, *A Concise Economic History of Modern China, 1840–1961* (New York, 1968), pp.184–6.
37 Walker, p.12.
38 Walker, p.13.
39 Derek T. Healey, 'Chinese real output 1950–1970', *Bulletin* (Institute of Development Studies, University of Sussex), vol. 4, no. 2/3 (June 1972), pp.49–59; see also K. Sarwar Lateef, *China and India: Economic Performance and Prospects* (Institute of Development Studies, University of Sussex), Communication 118 (December 1975), pp.14–23.
40 See Ramon H. Myers, review of Alexander Eckstein, *China's Economic Revolution* (Cambridge, 1977), in *Economic Development and Cultural Change*, vol. 27, no. 3 (April 1979), pp.558–9.
41 John Gittings, *How to Study China's Socialist Development* (Institute of Development Studies, University of Sussex), Communication 117 (August 1975).
42 Gittings, p.16.
43 J. Gray, 'The Chinese Model', in A. Nove and B. Nuti (eds.), *Socialist Economics* (Penguin Books, 1972), p.501.
44 J. Gray, 'The Two Roads: Alternative Strategies of Social Change and Economic Growth in China', in S. R. Schram (ed.), *Authority, Participation and Cultural Change in China* (Cambridge, 1973), p.114.
45 R. M. van Hekken and H. U. E. Thoden van Velzen, *Land Scarcity and Rural Inequality in Tanzania: Some Case Studies from Rungwe District* (The Hague, (Mouton), 1972), Communications no. 3.

46 Julius K. Nyerere, *Socialism and Rural Development* (Dar-es-Salaam, 1967).
47 Julius K. Nyerere, *Freedom and Socialism* (Dar-es-Salaam, 1968), p.422.
48 Overseas Development Group, University of East Anglia, UK, *Iringa Region, Tanzania – Integrated Rural Development Proposals for the Third Five-Year Plan 1976–81* (United Nations Food and Agriculture Organization, Rome, 1976), vol. 1, p.59.
49 Claes-Fredrik Claeson and J. E. Moore (eds.), *Mwanza Integrated Regional Planning Project* (Stockholm, 1976), vol. 3, ch. 5, pp.6–9.
50 R. N. Blue and J. H. Weaver, 'A Critical Assessment of the Tanzanian Model of Development', *Agricultural Development Council Inc.*, Reprint Series, no. 30 (July).
51 Blue and Weaver, p.6.

7 Emergent urbanization

1 See David Harvey, *Social Justice and the City* (London, 1973), p.231, for a fuller discussion of the relation between urbanization and the accumulation of social surplus value.
2 In the English colonial empire, this superimposition is most clearly indicated by the 'indirect rule' system best described with special reference to Africa in F. J. D. Lugard, *The Dual Mandate in British Tropical Africa* (London, 1922).
3 Paul Wheatley, *The City as Symbol* (London, 1969), Inaugural lecture, University College, London, p.31.
4 For a more detailed characterization of the colonial city in Africa, see A. L. Mabogunje, 'Urbanization patterns in Africa' in John Paden and Ed Soja (eds.), *The African Experience*, vol. 1 (Evanston, 1969), pp.383–420.
5 See H. C. Brookfield, 'Some geographical implications of the apartheid and partnership policies in southern Africa', *Transactions*, Institute of British Geographers, no. 23 (1957), pp.243–5.
6 This condition is not rigorously enforced in the third type of colonial city as in the other two. Indeed, in the case of South Africa, although there was no overt policy discouraging industrial development it was not until after the Union in 1910 that the cities became major centres of industrial activities. See M. H. de Koch, *The Economic Development of South Africa* (London, 1935), pp.70–99.
7 Leo Africanus, *The History and Description of Africa*, J. Pory (trans.), 1600, (London, 1896), vol. 3, p.828.
8 D. Hinderer, *Journals* (September 1851), CMS Archives, CA2/049.
9 S. F. Nadel, *A Black Byzantium: The Kingdom of Nupe in Nigeria*, (London, 1942), pp.259–69.

10 See for instance, D. K. Rangnekar, *Poverty and Capital Development in India* (London, 1958), pp.92–6.

11 J. H. Clapham, *An Economic History of Modern Britain,* Book 3 (Cambridge, 1932), pp.458–9.

12 H. Hazama, 'Formation of the Management System in Meiji Japan: Personnel Management in Large Corporations', *The Developing Economies,* vol. 15, no. 4 (December 1977), p.405.

13 Anthony D. King, *Colonial Urban Development: Culture, Social Power and Environment* (London, 1976), p.59.

14 King, p.65.

15 The figures are from UNESCO, *Statistical Yearbook* (Paris, 1963 and 1975). For Asia, they exclude USSR, mainland China, People's Republics of Korea and North Vietnam.

16 See A. O. Hirschman, *The Strategy of Economic Development* (New Haven, 1957).

17 George Beier, A. Churchill, M. Cohen and B. Renand, 'The task ahead for the cities of the developing countries', *World Development,* vol. 4, no. 5 (May 1976), p.388.

18 See D. Morawetz, 'Employment implications of industrialization in developing countries', *Economic Journal,* vol. 84, no. 335 (September 1974), p.501 footnote.

19 Ian Little, T. Scitovsky and M. Scott, *Industry and Trade in some Developing Countries* (OECD, Paris, 1970), p.42.

20 United States Senate (Committee on Finance), *Implications of Multinational Firms for World Trade and Investment and for US Trade and Labor* (Washington, 1973).

21 US Senate p.29. See also C. V. Vaitsos, 'Power, knowledge and development policy: relations between transnational enterprises and developing countries', in G. K. Helleiner (ed.), *A World Divided* (London, 1976), pp.113–46.

22 Helen Hughes, 'Debt and development: The role of foreign capital in economic growth', *World Development,* vol. 7. no. 2 (February 1979), p.107. See also World Bank, *World Development Report, 1985* (Washington D.C., 1985), pp.202–5.

23 See World Bank, *Prospects for Developing Countries, 1978–1985* (Washington, 1977), Appendix 1, p.78.

24 Paul Streeten, 'Technology gaps between rich and poor countries', *Scottish Journal of Political Economy,* vol. 19, no. 3 (November 1972), pp.213–30.

25 Hans Singer, 'The development outlook for poor countries: Technology is the key', *Challenge,* vol. 16, no. 2 (May–June 1973), pp.42–8.

26 Andre Gunder Frank, *Capitalism and Underdevelopment in Latin America* (New York, 1967), pp.207, 211.

27 See S. K. Subramanian, 'An approach to the science and technology

plan in India', *World Development,* vol. 1, no. 7 (July 1973), pp.23–9.
28 Examples of such countries are most common in East and South Africa.
29 Salah El-Shakhs, 'Development, primacy and systems of cities', *Journal of Developing Areas,* vol. 7, no. 1 (October 1972), pp.11–36.
30 Norton Ginsburg, *Atlas of Economic Development* (Chicago, 1961).
31 N. V. Sovani, 'The analysis of overurbanization', *Economic Development and Cultural Change,* vol. 12, no. 2 (January 1964), pp.113–22.
32 Little *et al.,* Table 2.12, p.73.
33 E. A. J. Johnson, *The Organization of Space in Developing Countries* (Cambridge, Mass., 1970), pp.154–6.
34 David Harvey, *Social Justice and the City* (London, 1973), p.232.
35 Much of this section is from Beier *et al.,* pp.363–410.
36 With respect to the implication of such high standards for the provision of housing for the urban populations, see A. L. Mabogunje, J. Hardoy and R. P. Misra, *Shelter Provision in Developing Countries* (New York, 1977).
37 Leo A. Orleans, 'China's experience in population control: The elusive model', *World Development,* vol. 3, no. 7/8 (July–August 1975), pp.497–526.
38 See Andrzej Krassowski, *Development and the Debt Trap: Economic Planning and External Borrowing in Ghana* (London, 1974).
39 A. O. Hirschman, 'The political economy of import substituting industrialization in Latin America', in C. T. Nisbet (ed.), *Latin America: Problems in Economic Development* (New York, 1969), pp.255–6.

8 Urban crisis of underdevelopment

1 United States Government, *The Ribicoff Hearings: Report of the National Commission on Urban Problems, 1966* (Washington, 1966), p.25.
2 T. G. McGee, *The Southeast Asian City* (London, 1967), p.15.
3 E. F. Arriaga, 'Components of city growth in selected Latin American countries', *Milbank Memorial Fund Quarterly,* vol. 46, no. 2, pt 1 (April 1968), p.241.
4 Kingsley Davis, 'The urbanization of the human population' in G. Breese, *The City in Newly Developing Countries* (New York, 1969), p.12.
5 World Bank, *Urbanization: Sector Working Paper* (Washington, June 1972), p.80.
6 See Sally Findley, *Planning for Internal Migration: A Review of Issues and Policies in Developing Countries* (US Bureau of Census, Washington, 1977), pp.34–6.
7 See, for instance, McGee, p.117.

8 Paul Bairoch, *The Economic Development of the Third World since 1900,* Cynthia Postan (trans.) (London, 1975), p.150.
9 Ray Bromley, 'The urban informal sector: Why is it worth discussing', *World Development,* vol. 6, no. 9/10 (October 1978), pp.1033–40.
10 See Dipak Mazumdar, 'The urban informal sector', *World Development,* vol. 4, no. 8 (August 1976), pp.655–80.
11 Inaia M. M. Carvallo, 'Urban employment: A case study of Bahia', *Antipode: A Radical Journal of Geography,* vol. 9, no. 3 (December 1977).
12 T. W. Merrick, *The Informal Sector in Belo Horizonte: A Case Study* (ILO, Geneva, 1973).
13 See, for example, Richard Webb, *The Urban Transitional Sector in Peru* (IBRD mimeograph, 1975).
14 Archie Callaway, 'Nigeria's indigenous education: The apprentice system', *ODU Journal of the University of Ife, Institute of African Studies,* vol. 1, no. 1 (July 1964), p.63.
15 Victor E. Tokman, 'An exploration into the nature of informal–formal sector relationships', *World Development,* vol. 6, no. 9/10 (September–October 1978), pp.1065–76.
16 T. G. McGee, *The Urbanization Process in the Third World* (London, 1971), p.74.
17 See Mazumdar, p.675; and Webb. Also J. Weeks, 'Policies for expanding employment in the informal urban sector of developing countries', *International Labour Review,* vol. 3 (1975), pp.1–15.
18 For a good discussion of the two positions see C. O. N. Moser, 'Informal sector or petty commodity production: Dualism or dependence in urban development', *World Development,* vol. 6, no. 9/10 (September–October 1978), pp.1041–64.
19 Andre G. Frank, *Capitalism and Underdevelopment in Latin America* (London, 1967), p.110.
20 Keith Buchanan, 'Profiles of the Third World', *Pacific Viewpoint,* vol. 15, no. 2 (1974), p.108.
21 See Milton Santos, *L'Espace Partagé* (Paris 1975).
22 Sir Arthur Lewis, 'Unemployment in developing countries', *The World Today,* no. 1 (January, 1967), pp.21–2.
23 McGee, *The Urbanization Process in the Third World,* p.84.
24 A. L. Mabogunje, J. Hardoy and R. P. Misra, *Shelter Provision in Developing Countries* (New York, 1977), p.10.
25 Otto Koenigsberger, 'Housing in the national development plan: an example from Nigeria', *Ekistics,* vol. 30 (1970), pp.393–7.
26 Kenneth Hubbell, 'The provision and pricing of public utilities for the urban poor in less developed countries', *ILO Urban Poverty Task Force* (November 1974).
27 John F. C. Turner, 'Uncontrolled urban settlement: Problems and

policies' in Gerald Breese (ed.), *The City in Newly Developing Countries* (Englewood Cliffs, 1969), pp.507–34. See also J. F. C. Turner and R. Fichter, *Freedom to Build* (New York, 1972).

28 F. Engels, *The Housing Question* (New York, 1935 ed.). pp.74–7.

29 It is possible to detail other environment problems such as the long and protracted journey to and from work of the low income group due to inadequate provision of public transport facilities, the problem of water and solid waste pollution, and in some cases, air and noise pollution, the absence of green spaces and the overall urban landscape. In a sense these problems are at best of a second order of importance and can more easily be resolved once the basic one of employment, housing and the general alienation from the urban society of which they are a part has been resolved.

30 Charles Abrams, *Squatter Settlements: The Problem and the Opportunity* (Office of International Affairs, Department of Housing and Urban Development, Washington, 1966).

31 Kenneth Little, *West African Urbanization* (Cambridge, 1965), pp.96–7.

32 Manuel Castells, *The Urban Question: A Marxist Approach,* Allan Sheridan (trans.), (London, 1976), pp.365–6.

33 World Bank, *Urbanization: Sector Working Paper* (Washington, 1972), p.48.

34 W. A. Cornelius and R. V. Kemper (eds.), *Latin American Urban Research,* vol. 6 (London, 1978), p.20. For a good example from Africa, see Otto Koenigsberger *et al., Metropolitan Lagos* (United Nations Publication, New York, 1964).

35 Alan Gilbert, 'Bogota: Politics, planning and the crisis of lost opportunities' in W. A. Cornelius and R. V. Kemper (eds.), *Latin American Urban Research,* vol. 5 (London, 1978), p.109.

9 Urban system and national development

1 E. A. Johnson, *The Organization of Space in Developing Countries* (Cambridge, Mass., 1970), p.171.

2 Thomas C. Smith, *Agrarian Origins of Modern Japan* (Stanford, 1959), p.212.

3 Walter Christaller, *Central Places in Southern Germany,* Carlisle W. Baskin (trans.), (Englewood Cliffs, N.J., 1966). Also August Lösch, *The Economics of Location,* 2nd rev. ed. W. Stolper (trans.), (New Haven, 1954).

4 See Brian J. L. Berry, 'City size and economic development', in L. Jakobson and V. Prakash, *Urbanization and National Development* (New York, 1971). The modification relates to a situation especially in developed countries of relatively small size where the largest city has

some of the characteristics of primacy but cities below them show a clear rank-size pattern.

5 R. J. Johnston, 'Regarding urban origins, urbanization and urban patterns', *Geography*, vol. 62, pt 1, no. 274 (January 1977), pp.1–8.

6 Eliezer Brutzkus, 'The scheme for spatial distribution of 5-million population in Israel', *The Developing Economies*, vol. 13, no. 3 (September 1975), pp.302–17.

7 B. J. L. Berry, *The Human Consequences of Urbanization* (London, 1973), p.107.

8 Efraim Orni and Elisha Efrat, *Geography of Israel* (Jerusalem, 1964), pp.206–11.

9 Berry, p.107; see also A. S. Schachar, 'Israel's development towns: Evaluation of a national urbanization policy, *Journal of the American Institute of Planners*, vol. 37 (1971), pp.362–72.

10 Robert S. Merrill, 'The study of technology' in David L. Sills (ed.), *International Encyclopedia of the Social Sciences*, vol. 15 (New York, 1968), pp.576–7.

11 Peter Kilby, *Industrialization in an Open Economy: Nigeria 1954–1966* (Cambridge, 1969), p.75.

12 Stanley A. Hetzler, *Applied Measures for Promoting Technological Growth* (London, 1973), pp.92–3.

13 Johannes Hirschmeier, *The Origins of Entrepreneurship in Meiji Japan* (Cambridge, Mass., 1964), pp.125–7.

14 Hirschmeier, p.96.

15 Gustav Ranis, 'Planning for resources and planning for strategic change', *Weltwirtschaftliches Archiv*, vol. 95, pt 1 (1965), p.30.

16 William W. Lockwood, *Economic Development of Japan: Growth and Structural Change, 1868–1938* (Princeton, NJ, 1954), p.561.

17 Gustav Ranis, p.30.

18 See, for instance, Albert Waterson, 'Viable model for rural development', *Finance and Development*, vol. 2, no. 4 (December 1974), and D. B. W. M. van Dusseldorp, *Planning of Rural Service Centres in Rural Areas of Developing Countries* (International Institute for Land Reclamation and Improvement, Wageningen, Netherlands, 1971), publication no. 15.

19 Marc Penouil, 'Growth poles in underdeveloped regions and countries', in A. Kuklinski and R. Petrella (eds.), *Growth Poles and Regional Policies* (The Hague, 1972), p.138.

20 Alan Gilbert, *Latin American Development: A Geographical Perspective* (Penguin Books, 1974), p.262.

21 Allan Pred, 'Diffusion, organizational spatial structure and city system development', *Economic Geography*, vol. 51, no. 3 (July 1975), pp.252–68.

22 G. J. Afolabi Ojo, *Yoruba Culture: A Geographical Analysis* (London, 1966), pp.119–20.

23 Prue Dempster, *Japan Advances: A Geographical Study* (London, 1967), p.262.

24 D. A. Davis and A. Whinston, 'The economics of complex systems: The case of municipal zoning', *Kyklos,* vol. 27 (1964), pp.419–46.

25 S. J. Makielski, *The Politics of Zoning* (New York, 1966).

26 I. D. Sherrard (ed.), *Social Welfare and Urban Problems* (New York, 1968).

27 M. Olson, *The Logic of Collective Action* (Cambridge, Mass., 1965), p.128.

28 M. Kotler, *Neighbourhood Government: the Local Foundations of Political Life* (Indianapolis, 1969), p.71. See also David Harvey, *Social Justice and the City* (London, 1973), pp.73–95.

29 J. F. C. Turner, *Housing by People* (London, 1976).

30 Rod Burgess, 'Petty commodity housing or dweller control? A critique of John Turner's views on housing policy', *World Development,* vol. 6. no. 9/10 (September–October 1978), pp.1105–33.

31 Burgess, p.1129.

32 E. Brutzkus, statement made at the 1971 Rehovot Conference on Urbanization in Developing Countries, quoted in B. J. L. Berry, *Human Consequences of Urbanization* (London, 1973), p.109.

10 Integrating the national population

1 Goran Chlin, *Population Control and Economic Development* (Paris, 1967), p.14. See also Peter Newman, *Malaria Eradication and Population Growth* (London, 1965), p.14.

2 R. M. Titmuss and B. Abel-Smith, *Social Policies and Population Growth in Mauritius* (London, 1961).

3 Food and Agriculture Organization, *The State of Food and Agriculture 1973* (Rome, 1973); also *The State of Food and Agriculture 1985* (Rome, 1986), p.19.

4 Winifred Weekes-Vagliani *et al., Family Life and Structure in Southern Cameroon* (Development Centre, OECD, Paris, 1976), pp.32–3.

5 Harley L. Browning, 'Some sociological considerations of population pressure on resources', in W. Zelinsky, L. A. Kosinski and R. M. Prothero (eds.), *Geography and a Crowding World* (London, 1970).

6 C. Geertz, *Agriculture Involution: the Process of Ecological Change in Indonesia* (Berkeley and Los Angeles, 1963), pp.99–100.

7 H. C. Brookfield, 'Population, society, and the allocation of resources', in W. Zelinski, L. A. Kosinski and R. M. Prothero (eds.), p.142.

8 Brookfield, p.143.

9 A. L. Mabogunje, 'A typology of population pressure on resources in West Africa', in W. Zelinsky, L. A. Kosinski and R. M. Prothero (eds.), p.123.

10 Ester Boserup, 'Environment, population and technology in primitive societies', *Population and Development Review*, vol. 2, no. 1 (March 1976), pp.21–36.
11 Ester Boserup, *The Conditions of Agricultural Growth* (London, 1965), p.118.
12 Willam Allan, *The African Husbandman* (London, 1965), p.89.
13 Pierre Gourou, *The Tropical World: Its Social and Economic Conditions and its Future Status*, S. H. Bearer and E. D. Laborde (trans.) (London, 1953), pp.16–30.
14 Thomas T. Poleman, 'World Food: Myth and Reality', *World Development*, vol. 5, nos. 5–7 (May–July 1977), pp.383–94.
15 C. A. de Vries, J. D. Ferwerda and M. Flach, 'Choice of food crops in relation to actual and potential production in the tropics', *Netherlands Journal of Agricultural Science* (November 1967), p.246.
16 Paul Bairoch, *The Economic Development of the Third World since 1900*, Cynthia Postan (trans.) (London, 1975), pp.138–9.
17 Bairoch, p.138.
18 Bairoch, p.137.
19 W. Petersen, 'A general typology of migration', in Clifford J. Jansen (ed.), *Readings in the Sociology of Migration* (London, 1970), pp.49–68.
20 Stanley Barrett, *Two Villages on Stilts* (London, 1974).
21 A. Akinbode, 'The location of kola production in southwestern Nigeria: A study of the spatial diffusion of an agricultural innovation'. (Unpublished PhD thesis, University of Ibadan 1973), pp.27–58.
22 Myron Weiner, 'Socio-political consequences of interstate migration in India' in W. H. Wriggins and J. F. Guyot (eds.), *Population, Politics and the Future of Southern Asia* (New York, 1973), pp.190–228.
23 W. Petersen, p.63.
24 A. L. Mabogunje, 'Migration policy and regional development in Nigeria', *Nigerian Journal of Economics and Social Studies*.
25 Karl J. Pelzer, 'The agricultural foundation' in B. Glassburner (ed.), *The Economy of Indonesia* (Ithaca, 1971), p.138.
26 A. L. Mabogunje, 'Systems approach to a theory of rural-urban migration', *Geographical Analysis*, vol. 2 (1970), pp.1–18.
27 R. J. Pryor, 'Migration and the process of modernization', in L. A. Kosinski and R. M. Prothero (eds.), *People on the Move: Studies of Internal Migration* (London, 1975), pp.35–6.
28 See, for instance, A. L. Mabogunje, *Regional Mobility and Resource Development in West Africa* (Montreal, 1972).
29 For the importance attached to this problem as socialist development, see R. J. Fuchs and G. J. Demko, 'Spatial population policies in the socialist countries of Eastern Europe', *Social Science Quarterly*, vol. 58, no. 1 (June 1977), pp.60–73.

30 The demographic transition theory derives largely from the work of Frank Notestein in 1945. In recent years various modifications especially as to the factors underlying the transition have been made to aspects of this theory on the basis of increasing data from underdeveloped countries. One of the most recent is that by J. C. Caldwell, 'Toward a restatement of demographic transition theory', *Population and Development Review,* vol. 2, nos. 3 and 4 (September–December 1976), pp.321–66. Caldwell argues that an important mechanism in the transition is change in the direction and magnitude of inter-generational wealth flows as between parents and children.

31 See, for instance, Ansley J. Coale and Edgar M. Hoover, *Population Growth and Economic Development in Low-Wealth Countries: A Case Study of India's Prospects* (Princeton, 1958).

32 George Zeidenstein, *Report of the Population Council for 1978* (New York, 1979).

33 For projected effects of a basic needs policy on population growth, see David Morawetz, 'Basic needs policies and population growth', *World Development,* vol. 6, no. 11/12, (November–December 1978), pp.1251–60. Morawetz, however, argues that these policies only make family planning programmes more effective. They do not obviate the need for them.

11 Information flows

1 See Clifford Geertz, 'The integrative revolution, primordial sentiments and civil politics in the new states', in Clifford Geertz (ed.), *Old Societies and New States* (New York, 1963), pp.109ff.

2 Myron Weiner, 'Political Integration and Political Development', *Annals of the American Academy of Political and Social Science,* vol. 358 (March 1965), pp.52–64.

3 See for example, Daniel Lerner, *The Passing of Traditional Society* (Glencoe, Ill., 1958), pp.55–6.

4 Nora C. Quebral, 'What do we mean by "development communications"?' *International Development Review,* vol. 15, no. 2 (1973/2), p.25.

5 Richard L. Meier, *Communications Theory of Urban Growth* (Cambridge, Mass., 1962), pp.150–2.

6 Torsten Hägerstrand, 'Introduction' in Torsten Hägerstrand and A. Kuklinski (eds.), *Information Systems for Regional Development,* Lund Studies in Geography, Series 8, Human Geography, no. 37 (Lund, 1971), p.1.

7 Tormod Hermansen, 'Information systems for regional development planning: Issues and problems', in T. Hägerstrand and A. Kuklinski (eds.), p.8.

8 N. C. Quebral, p.25.

9 Karl W. Deutsch, *Nationalism and Social Communiction* (Cambridge, Mass., 1953), p.101.

10 See T. Griffith Jones, 'Promoting agricultural change,' *Span,* vol. 18, no. 2 (1975), pp.54–6, for similar successful private sector effort at agricultural extension in Thailand, Portugal and Nigeria.

11 Food and Agriculture Organization, *Agricultural Development in Nigeria: 1965–1980* (Rome, 1966), p.300.

12 Montagu Yudelman, *Africans on the Land* (London, 1964), pp.145–6.

13 R. P. Misra, 'Monte Carlo simulation of spatial diffusion: Rationale and application to Indian conditions', in R. P. Misra (ed.), *Regional Planning* (Mysore, 1969), pp.251–76.

14 Pierre D. Sam, 'Le groupe de travail en tant qu'unite de base de la formation au village', *International Development Review, Focus,* vol. 18, no. 2 (1976/2), pp.17–20.

15 Akhter H. Khan, 'The Comilla projects – A personal account', *International Development Review,* vol. 16, no. 3 (1974/3), p.6.

16 Lawrence Bass, 'The role of technologic institutes in industrial development', *World Development,* vol. 1, no. 10 (October 1973), pp.27–32.

17 Lawrence W. Bass, 'Technical and managerial help for small enterprises', *World Development,* vol. 4, no. 4 (April 1976), pp.339–48.

18 See also UNIDO, *Small-Scale Industry: Industrialization of Developing Countries: Problems and Prospects* (New York, 1969).

19 Jon Sigurdson, 'Rural industrialization: A comparison of development planning in China and India', *World Development,* vol. 6, no. 5 (May 1978), pp.667–80.

20 Gunnar Tornqvist, *Flows of Information and the Location of Economic Activities,* Lund Studies in Geography, Series B, Human Geography, no. 30 (Lund, 1968), p.101.

21 Paul Streeten, 'Technology gaps between rich and poor countries', *Scottish Journal of Political Economy,* vol. 19, no. 3 (November 1972).

22 V. V. Bhatt, 'On technology policy and its institutional frame', *World Development,* vol. 3, no. 9 (September 1975), pp.654–5.

23 James M. Utterback, 'The role of applied research institutes in the transfer of technology in Latin America', *World Development,* vol. 3, no. 9 (September 1975), pp.665–73.

24 Edward W. Soja, 'Communications and territorial integration in East Africa: An introduction to transaction flow analysis', in R. E. Kasperson and J. V. Minghi (eds.), *The Structure of Political Geography* (London, 1970), p.231.

25 Karl W. Deutsch, *Nationalism and Social Communication* (Cambridge, Mass., 1953), pp.77–8.

26 S. M. Kimani and D. R. F. Taylor, *Growth Centres and Rural Development in Kenya* (Thika, Kenya, 1973), p.18; see also Jan Lundqvist,

Local and Central Impulses for Change and Development (Göteborg, 1975), pp.27–8.

27 Merle Fainsod, 'The structure of development administration', Irving Swerdlow (ed.), *Development Administration: Concepts and Problems* (Syracuse, 1963), p.2. See also Fred W. Riggs, 'The context of development administration', in F. W. Riggs (ed.), *Frontiers of Development Administration* (Durham, 1971), p.73.

28 Irving Swedlow, *The Public Administration of Economic Devlopment* (Praeger, New York, 1975), p.347.

29 B. B. Schaffer, 'The deadlock in development administration', in Colin Leys (ed.), *Politics and Change in Developing Countries* (Cambridge, 1969), pp.190–2.

30 Schaffer, p.184.

31 Milton J. Esman, 'Development administration and constituency organization', *Public Administration Review,* vol. 38, no. 2 (March–April 1978), pp.166–72.

32 The countries involved are Bangladesh, China, Egypt, India, Indonesia, Japan, Malaysia, Pakistan, the Philippines, South Korea, Sri Lanka, Taiwan, Thailand, Turkey and Yugoslavia. Nineteen monographs were produced as well as a summary volume by Norman Uphoff and Milton J. Esman, *Local Organization for Rural Development: Analysis of Asian Experience* (Ithaca, New York, 1978).

33 Tormod Hermansen, p.13.

12 Movements of goods and services

1 See Bertil Ohlin, *Inter-regional and International Trade* (Cambridge, Mass., 1933), pp.3–64.

2 Karl Polanyi, *Primitive, Archaic and Modern Economies: Essays of Karl Polanyi,* G. Dalton (ed.), (Boston, 1968), pp.148–9.

3 See, for instance, A. Emmanuel, *Unequal Exchange: A Study of the Imperialism of Trade,* B. Pearce (trans.) (London, 1972), especially ch. 4.

4 Lord Lugard, *The Dual Mandate in British Tropical Africa* (London, 1922), p.5.

5 For examples of the restrictive legislation on road transport development, see J. L. Lougbottom, *An Analysis of the Motor Traffic Legislation of the Colonies* (Accra, 1933).

6 R. C. Harkema, 'The ports and access routes of land-locked Zambia', *Geografisch Tijdschrift,* vol. 6, no. 3 (1972), pp.223–31.

7 UN Economic Commission for Africa, *A survey of economic conditions in Africa,* E/CN. 14/397 (Addis Ababa, 1967).

8 R. J. Harrison-Church, 'The evolution of railways in French and British West Africa', *Congrès International de Géographie* (Lisbon, 1949), tome 4, p.113.

9 B. W. Hodder, 'Tin Mining on the Jos Plateau', *Economic Geography,* vol. 35, no. 2 (April 1959), p.110.

10 A. G. Frank, *Capitalism and Underdevelopment in Latin America* (London, 1967), p.290.

11 Tarlok Singh, *India's Development Experience* (London, 1974), p.275.

12 A. M. Hay and R. H. T. Smith, *Inter-regional Trade and Money Flows in Nigeria, 1964* (Ibadan, 1970), pp.51–3.

13 P. T. Bauer, *West African Trade: A Study of Competition, Oligopoly and Monopoly in a Changing Economy* (Cambridge, 1954), p.22.

14 Guy Hunter, *Modernizing Peasant Societies: A Comparative Study in Asia and Africa* (London, 1969), pp.168–9.

15 W. D. Jones, *Marketing Staple Food Crops in Tropical Africa* (Ithaca, New York, 1972), p.261.

16 Abner Cohen, *Custom and Politics in Urban Africa: a Study of Hausa Migrants in Yoruba Towns* (London, 1969).

17 Polly Hill, 'Landlords and brokers: a West African trading system', *Cahiers Études Africaines,* vol. 23, no. 3 (1966), pp.349–66.

18 P. T. Bauer, pp.389–90.

19 C. Guthrie, 'Voluntary chains of retail food stores in Latin America', *Development Digest,* vol. 12, no. 4 (October 1974), pp.64–73.

20 *Statement on the Future Marketing of West African Cocoa,* Cmd. 6950, (HMSO, London, 1946).

21 J. C. Abbott and H. Creupelandt, 'Agricultural marketing boards in the developing countries: Problems of efficiency appraisal', *FAO Monthly Bulletin of Agricultural Economics and Statistics,* vol. 16, no. 9 (September 1967), pp.1–9.

22 P. T. Bauer, p.65.

23 E. J. Taaffe, R. L. Morrill and P. R. Gould, 'Transport expansion in underdeveloped countries: a comparative analysis', *Geographical Review,* vol. 53, no. 4 (1963), pp.503–29.

24 M. K. McCall, 'Political economy and rural transport: An appraisal of Western misconceptions', *Antipode: A Radical Journal of Geography,* vol. 9, no. 3 (December 1977), p.102.

25 J. Muller, 'Labour-intensive methods in low-cost road construction: a case study', *International Labour Review,* vol. 101, no. 4 (1970), pp.359–75. See also his *Choice of Technology in Underdeveloped Countries* (Copenhagen, Technical University of Denmark, 1973).

26 R. D. Wolff, *Britain and Kenya, 1870–1930: The Economics of Capitalism* (Nairobi, 1974).

27 See A. L. Mabogunje, 'Changing pattern of rural settlement and rural economy in Egba Division, southwestern Nigeria', (Unpublished MA thesis, University of London, 1958), p.140.

28 M. K. McCall, 'Political economy and rural transport: A reappraisal of transportation impacts', *Antipode: A Radical Journal of Geography,* vol. 9, no. 1 (February 1977), p.59.

29 International Bank for Reconstruction and Development, *Roads and Road Transport in Tanzania: Agriculture and Rural Development Sector Study,* IBRD Report 541a – TA vol. 2, Annex 8, (Washington, DC, 1974).

30 H. G. Van der Tak and J. DeWeille, *Reappraisal of a Road Project in Iran,* World Bank Staff Occasional Paper no. 7 (Washington, DC, 1969).

31 D. Dunant, J. Escher, B. Gruber and K. Lieberherr, *Yemen Arab Republic Feeder Road Study, Report Phase I (Tgez-Torbah Road),* (Zurich, for IBRD, Washington, DC, 1973), p.136.

32 M. K. McCall, p.56.

33 E. A. J. Johnson, *The Organization of Space in Developing Countries* (Cambridge, Mass., 1970), pp.235–41.

34 J. M. Healey, 'Economic overheads: Co-ordination and pricing', in Paul Streeten and Michael Lipton (eds.), *The Crisis of Indian Planning* (London, 1968), p.164.

35 K. M. Buchanan, *The Transformation of the Chinese Earth* (London, 1970), p.257.

36 John P. Emerson, *Non-agricultural Employment in Mainland China, 1949–1958,* International Population Statistics Reports, Series p.90. no. 21 (Washington, DC, US, Department of Commerce, Bureau of Census). See also 'Manpower absorption in the non-agricultural branches of the economy of communist China', *The China Quarterly* (July–September 1961).

37 L. J. Zimmerman, 'Non-monetary capital formation and rural development', *World Devlopment,* vol. 3, no. 6 (June 1975), pp.411–20.

38 See, for instance, Lalit Sen (ed.), *Readings on Micro-Level Planning and Rural Growth Centres* (National Institute of Community Development, Hyderabad, India, 1972); and E. M. Kulp, *Rural Development Planning: Systems Analysis and Working Method* (Praeger, New York, 1970).

39 Tarlok Singh, *India's Development Experience* (London, 1974), pp.279–81.

40 W. B. Morgan and J. C. Pugh, *West Africa* (London, 1969), pp.98–100. See also W. B. Morgan, 'The distribution of food crop storage methods in Nigeria', *Journal of Tropical Geography,* vol. 13 (1959), pp.58–64.

41 Olga Linares de Sapir, 'Agriculture and Diola society' in Peter F. M. McLoughlin (ed.), *African Food Production System, Cases and Theory,* (Baltimore, 1970), pp.217–18.

42 See, M. Lipton, I. Cook and N. Nair, 'Cost-benefit analysis of crop storage improvements: a south Indian pilot study', *Eppo Bulletin,* vol. 4, no. 4 (1974), pp.447–53; also 'Food shortage and nutrition in the ECAFE region', *Economic Bulletin for Asia and the Far East,* vol. 25, no. 1 (1974), pp.21–32.

43 See, E. Reusse, 'Economic and marketing aspects of post-harvest systems in small farmer economies', *FAO Monthly Bulletin of Agricultural Economies and Statistics*, vol. 25, no. 9 (September 1976), pp.1–7.

44 See H. L. Cook and T. Cook, 'Organization of trade in one tropical municipality of Vera Cruz, Mexico', *Land Tenure Centre Report*, Michigan State University, no. 48 (September 1972); also Development Alternatives Inc., 'Fourteen case study summaries of small farmer and rural development projects in Latin America' (unpublished manuscript, 1975); and Uma Lele, *The Design of Rural Development: Lessons from Africa* (Baltimore, 1975).

45 A. O. Hirschman, *The Strategy of Economic Development* (New Haven, 1958), pp.198–9.

46 P. E. Jacob and J. V. Toscano (eds.), *The Integration of Political Communities* (New York, 1964), pp.16–45.

47 Karl Deutsch, *Nationalism and Social Communication* (Boston, 1962), pp.90–6.

48 Irma Adelman and George Dalton, 'A factor analysis of modernization in village India', *Economic Journal*, vol. 81 (1971), pp.563–79.

13 External relations

1 See Karl W. Deutsch and R. I. Savage, 'A statistical model of gross analysis of transaction shows', *Econometrica*, vol. 28 (July 1960), pp.551–72.

2 F. S. Northedge, *The Foreign Policies of the Powers* (London, 1968), p.15.

3 Samir Amin, *Neo-Colonialism in West Africa* (Penguin Books, London, 1973).

4 R. J. Harrison-Church, *West Africa* 5th ed. (London, 1966), p.197.

5 André Van Haeverkeke, *Renumération du travail et commerce extérieur* (Leuven, 1970), p.15.

6 BCEAD, *Bulletin*, May 1969 and May 1970.

7 Samir Amin, pp.9–19.

8 IBRD, *Situation et perspective économique du Sénégal*, mimeograph, (Washington, DC, 1970), vol. 1, Tables 5, 6 and 7.

9 See, for example, A. G. Frank, *Capitalism and Underdevelopment in Latin America* (New York, 1967); also *Latin America: Underdevelopment or Revolution* (New York, 1969); C. Furtado, *Subdesarrollo y Estancamiento en America Latina* (Buenos Aires, 1966); and F. H. Cardoso and E. Faletto, *Dependency and Underdevelopment in Latin America*, US ed. (New York, 1977).

10 Gabriel Palma, 'Dependency: A formal theory of underdevelopment or a methodology for the analysis of concrete situations of underdevelopment', *World Development*, vol. 6, nos. 7–8 (July/August 1978), p.908.

11 For the critical role of Central Banks in promoting development in underdeveloped countries, see V. V. Bhatt, 'Some aspects of financial policies and central banking in developing countries', *World Development,* vol. 2, nos. 10–12 (October/December 1974). pp.59–68.
12 For a good review of the problem see G. Maynard and G. Bird, 'International monetary issues and the developing countries: A survey', *World Development,* vol. 3, no. 9 (September 1972), pp.609–31.
13 For ways of achieving a degree of regional independence in monetary and financial matters, see Clark W. Reynolds, 'Achieving greater financial independence for Latin America: A proposal', *World Development,* vol. 3, nos. 11 and 12, (November–December 1975), pp.839–44.
14 United Nations, *Review of International Trade and Development 1973* (New York, 1973), sales no. E-74-II-D-14.
15 ibid.
16 L. K. Mytelka, 'Licensing and technology dependence in the Andean group', *World Development,* vol. 6, no. 4 (April, 1978), p. 456.
17 James Moxton, *Volta: Man's Greatest Lake* (London, 1969).
18 See Sanjaya Lall, 'Transfer pricing and developing countries: some problems of investigation', *World Development,* vol. 7, no. 1 (January 1979), pp.59–72.
19 See Jack N. Behrmann. 'The multinational enterprises and economic internationalism', *World Development.* vol. 3, nos. 11–12 (November–December 1975), pp.845–56.
20 For examples of countries in Africa caught in this type of situation, see Bela Balassa's contribution to the discussion of A. L. Mabogunje, 'International circumstances affecting the development and trade of developing countries', in Bertil Ohlin, Per-Ove Hesselborn and Per Magnus Wijkman (eds.), *The International Allocation of Economic Activity* (London, 1977), pp.499–500.
21 Harry G. Johnson, 'The theory of international trade', paper presented at the International Congress on the Future of International Economic Relations (Montreal, Canada, 2–7 September 1968).
22 See M. Marois (ed.), *Towards a Plan of Actions for Mankind* (Amsterdam, 1974), p.135.
23 M. O. Oyawoye, 'Development and management of mineral resources', Presidential address, Geological Society of Africa (Khartoum, 1975).
24 Dudley Seers, 'The new meaning of development', *International Development Review,* no. 3 (1977), pp.2–7.
25 *The Haslemere Declaration. A radical analysis of the relationships between the rich world and the poor world* (the Haslemere Declaration Group, April 1968). See also Judith Hart, *Aid and Liberation* (London, 1973), pp.238–9.
26 A. Bendavid and L. Bendavid, 'Developed and underdeveloped: A radical view of constructive relationships', *International Development Review,* vol. 16, no. 1 (1974), pp.13–14.

27 Peter Bauer, 'Foreign aid: Necessary? useful? damaging?', *New Scientist,* vol. 52 (30 December 1971), p.252. The reference to 'partners in development' was to the Pearson Commission Report, *Partners in Development,* report of the Commission on International Development, made to the president of the International Bank for Reconstruction and Development (Pall Mall Press, London, 1969).

28 B. W. T. Mutharika, *Toward Multinational Economic Co-operation in Africa* (Praeger, New York, 1972), pp.6–7.

29 For a detailed study of some of these difficulties, see C. V. Vaitsos, 'Crisis in regional economic co-operation (integration) among developing countries: A survey', *World Development,* vol. 6, no. 6 (June 1978), pp.719–69.

30 Mutharika, pp.45–51.

31 United Nations, 'Policies relating to technology of the countries of the Andean Pact: Their foundations', *Proceedings of the United Nations Conference on Trade and Development: Third Session, Santiago de Chile, 13th April to 21st May, 1972,* vol. 3, *Financing and Invisibles* (New York, 1973). TD/180, pp.122–36.

14 Conclusion

1 See K. Boulding, 'Toward a general theory of growth', *General Systems Yearbook,* vol. 1 (1956), pp.66–75, for a discussion of different types of growth.

2 William W. Lockwood, *The Economic Development of Japan: Growth and Structural Change 1868–1938* (London, 1955), p.499.

3 T. S. Eliot, 'Tradition and the individual talent', *Points of View* (London, 1942), pp.23–8.

4 Oscar Gish, 'Health planning in developing countries', *Development Digest,* vol. 9, no. 3 (July 1971), pp.67–76.

5 It has also been argued that even if they do not return, such highly skilled persons may contribute in many other diverse ways to the development of their home region. See, for instance, H. B. Grubel and A. D. Scott, 'The international flow of human capital', *American Economic Review,* vol. 56, no. 2 (May 1966), pp.268–74.

6 See Gunnar Myrdal, *Asian Drama: an Inquiry into the Poverty of Nations,* appendix (New York, 1968).

7 Barbara Bradby, 'The destruction of natural economy', *Economy and Society,* vol. 4, no. 2 (May 1975), p.148.

8 K. Buchanan, 'The Third World – its emergence and contours', *New Left Review,* no. 18 (January–February 1963), p.7.

9 David Slater, 'Geography and underdevelopment – Part II', *Antipode: A Radical Journal of Geography,* vol. 9, no. 3 (December 1977), pp.19–20.

10 See, for instance, Harry W. Blair, 'Rural development, class structure and bureaucracy in Bangladesh', *World Development,* vol. 6, no. 1 (January 1978), pp.65–82.

11 Suzanne Keller, 'Elites', *International Encyclopedia of the Social Sciences,* David L. Sills (ed.), vol. 5 (New York, 1968), p.26.

12 United Nations Economic Commission for Africa, *Africa's Strategy for Development in the 1970s,* E/CN.14/RES/218X (Addis Ababa, 1971), p.3.

15 Postscript

1 The World Bank, *World Development Report, 1985* (Washington, D. C.), p.2.

2 Most of the information on the Nigerian economy is from L. A. Alli, *Nigerian Economic Review*, Lagos, 1986.

3 T. A. Oyejide, A. Soyode and M. O. Kayode, *Nigeria and the IMF* (Ibadan, 1985).

4 *Address to the Nation on the 1986 Budget* by Major-General Ibrahim Babangida, President, Commander-in-Chief of the Nigerian Armed Forces, Lagos, 31 December 1985.

5 *This Week*, vol. 7, no. 3 (4 January 1988), p.15.

6 Soo Young Park, 'Approaches to small area development: the case of the Republic of Korea', in R. P. Misra (ed.), *Regional Development: Essays in Honour of Masahiko Honjo* (Maruzen Asia, 1982), pp. 217–29.

7 Fu-Chen Ho and Byung-Nak Song, *The Saemaul Undoing: The Korean Way of Rural Transformation*, (WP 79-09) United Nations Centre for Regional Development, Nagoya, Japan, 1979.

8 Carol Lancaster and John Williamson (eds.), *African Debt and Financing* (Washington, D.C), 1986, p.202.

Index